Urban Sustainability

The Urban Sustainability Book Series is a valuable resource for sustainability and urban-related education and research. It offers an inter-disciplinary platform covering all four areas of practice, policy, education, research, and their nexus. The publications in this series are related to critical areas of sustainability, urban studies, planning, and urban geography.

This book series aims to put together cutting-edge research findings linked to the overarching field of urban sustainability. The scope and nature of the topic are broad and interdisciplinary and bring together various associated disciplines from sustainable development, environmental sciences, urbanism, etc. With many advanced research findings in the field, there is a need to put together various discussions and contributions on specific sustainability fields, covering a good range of topics on sustainable development, sustainable urbanism, and urban sustainability. Despite the broad range of issues, we note the importance of practical and policy-oriented directions, extending the literature and directions and pathways towards achieving urban sustainability.

The series will appeal to urbanists, geographers, planners, engineers, architects, governmental authorities, policymakers, researchers of all levels, and to all of those interested in a wide-ranging overview of urban sustainability and its associated fields. The series includes monographs and edited volumes, covering a range of topics under the urban sustainability topic, which can also be used for teaching materials.

Shovan Ghosh · Sushobhan Majumdar ·
Ali Cheshmehzangi
Editors

Cities of Tomorrow: Urban Resilience and Climate Change Preparedness

Springer

Editors
Shovan Ghosh
Department of Geography
Diamond Harbour Women's University
Kolkata, West Bengal, India

Sushobhan Majumdar
Department of Geography
Pandit Raghunath Murmu Smriti
Mahavidyalaya, Bankura University
Bankura, India

Ali Cheshmehzangi
School of Architecture, Design
and Planning
The University of Queensland
Brisbane, QLD, Australia

ISSN 2731-6483 ISSN 2731-6491 (electronic)
Urban Sustainability
ISBN 978-981-97-9657-1 ISBN 978-981-97-9658-8 (eBook)
https://doi.org/10.1007/978-981-97-9658-8

This Springer imprint is published by the registered company Springer Nature Singapore Pte Ltd.
The registered company address is: 152 Beach Road, #21-01/04 Gateway East, Singapore 189721, Singapore

Contents

Part I
Understanding Urban Resilience and Climate Change

Chapter 1
Integrating Urban Master Planning with Environmental Impact Assessment for Sustainable and Resilient Cities

Khushi Yadav and Harshita Jain

Abstract This chapter highlights the need of incorporating Environmental Impact Assessments (EIAs) into urban master planning processes, with a particular focus on sustainable urban growth. With the increasing understanding of the need for sustainable practices in urban development, incorporating environmental impact assessments (EIAs) into master planning is an essential strategy to mitigate adverse environmental consequences and boost the resilience of urban ecosystems. To elucidate the fundamental concepts, challenges, best practices, and benefits of merging environmental impact assessment (EIA) with urban master planning, this chapter offers a thorough review of the literature. It also delves into global case studies to draw attention to efficient integration strategies and glean valuable insights. In order to accomplish the objectives of sustainable urban development, this chapter emphasises the significance of cross-disciplinary collaboration, stakeholder involvement, data-driven decision-making, and adaptive management.

Keywords Environmental impact assessment · City master planning · Sustainable development · Urban resilience

1.1 Introduction

The necessity of addressing environmental issues, fostering resilience, and ensuring the welfare of urban people has made sustainable urban development more and more clear in recent years. Integrating EIAs into urban master planning procedures is essential to reaching this objective. As global urbanisation picks up speed, cities are under increasing pressure to manage expanding populations while also lowering environmental pollution and promoting sustainability [23]. Controlling the social, physical, and economic growth of metropolitan regions requires the use of metropolitan

K. Yadav · H. Jain (✉)
Amity Institute of Environmental Sciences, Amity University Noida, Noida 201301, Uttar Pradesh, India
e-mail: hjain@amity.edu

S. Ghosh et al. (eds.), *Cities of Tomorrow: Urban Resilience and Climate Change Preparedness*, Urban Sustainability, https://doi.org/10.1007/978-981-97-9658-8_1

planning. However, environmental concerns are often overlooked or undervalued in conventional master planning, which has detrimental effects on human health, biodiversity, air and water quality, ecosystems, and so on. In response to these concerns, the EIAs as a holistic approach to urban development has increased in popularity within urban planning processes. It has become clear that including EIA into municipal master planning may support sustainable urban expansion. Rapid urbanisation and population expansion have heightened the need for robust urban planning frameworks that carefully consider environmental problems [4]. The environmental impacts of conventional master planning strategies are frequently ignored, exacerbating their detrimental effects on both human health and natural ecosystems [18].

Integrating the EIA into urban planning procedures is essential to guaranteeing the resilient and sustainable growth of cities (Table 1.1). Through the EIA process, proposed plans, programmes, or projects are rigorously examined for potential environmental implications before to implementation. By identifying and addressing any negative environmental consequences, this systematic evaluation helps decision-makers reduce environmental dangers and improve the overall sustainability of urban development initiatives. Decision-makers may proactively address environmental issues and support the long-term sustainability of urban areas by including EIA into the city's master plan. Cities are better equipped to identify and address environmental pressures, such climate change, by integrating adaptation measures into their development goals thanks to this integration [19]. Cities can maximise the positive effects of urban development projects, minimise negative effects, and optimise resource utilisation by taking environmental considerations into account early in the planning process. Given the complex interactions that exist between urban systems and the environment, the significance of incorporatingEIAs into urban planning becomes even more obvious. Urban regions are linked ecosystems where natural processes and human activity combine to produce complex dynamics that may have long-term effects on the environment. Urban development projects may be better understood and managed by cities by integrating EIA into the design stage, minimising adverse effects and optimising good results. Furthermore, decision-makers can avoid costly and ecologically damaging errors by including EIA into the city's master plan [8]. Cities may create more resilient and sustainable urban environments by identifying possible risks and vulnerabilities early in the planning phase through comprehensive environmental evaluations. By taking a proactive stance, communities may prevent future costly restorations or cleanup projects, sparing them money and environmental harm [38].

Cities are also able to support fair development and proactively address environmental challenges when they integrate EIA into the larger framework of urban planning. Decision-makers may guarantee that the benefits of development are dispersed equitably across all communities, regardless of socioeconomic level or geographic location, by taking the environmental effects of proposed projects and policies into account. In the process, more livable and sustainable urban settings are created for all citizens, fostering social fairness and inclusion. To encourage resilient and sustainable urban growth, environmental impact assessments must be incorporated into urban planning procedures [4, 5]. Resilient urban landscapes may endure future

Table 11.1 EIA Integration into City Master Planning Processes [21]

Aspect	Description	Importance	Key Steps in EIA process	Examples/Cases
Objective	To assess potential environmental impacts of proposed projects, policies, or plans before implementation, ensuring informed decision-making and sustainable development	– Identify and mitigate environmental risks – Enhance sustainability and resilience	1. Screening: Determine if EIA is required 2. Scoping: Define the scope and boundaries of assessment 3. Impact Assessment: Identify and evaluate potential impacts 4. Mitigation: Propose measures to mitigate adverse impacts	– EIA for urban infrastructure projects – EIA for land use planning initiatives
Stakeholder Engagement	Engage a diverse range of stakeholders throughout the EIA process, including local communities, NGOs, government agencies, and businesses, to ensure transparency, inclusivity, and accountability	– Promote transparency and inclusiveness – Gather diverse perspectives and expertise	– Public Consultation: Seek input from stakeholders – Collaboration: Work with relevant agencies and organizations – Communication: Ensure clear and accessible information sharing	– Stakeholder engagement in EIA for city infrastructure projects – Involvement of localcommunities in land use planning assessments

(continued)

Table 11.1 (continued)

Aspect	Description	Importance	Key Steps in EIA process	Examples/Cases
Data Collection and Analysis	Gather comprehensive data on environmental factors, including air and water quality, biodiversity, land use, and socio-economic conditions, to assess potential impacts accurately and inform decision-making	– Ensure accuracy and reliability of information. –Identify potential impacts effectively	– Environmental Baseline Studies: Assess current environmental conditions – Impact Analysis: Analyze potential effects on various aspects – Socio-Economic Assessment: Evaluate social and economic implications	– Data collection for EIA of city infrastructure projects – Analysis of environmental factors in land use planning assessments

(continued)

Table 11.1 (continued)

Aspect	Description	Importance	Key Steps in EIA process	Examples/Cases
Impact Assessment and Mitigation	Identify, assess, and mitigate potential environmental impacts, proposing measures to minimize adverse effects and enhance sustainability	– Minimize adverse effects on the environment and human health – Promote sustainability and resilience	– Identification: Identify potential impacts on the environment and communities – Assessment: Evaluate significance and severity of impacts – Mitigation: Propose measures to avoid, minimize, or compensate for adverse effects	– Mitigation measures for EIA of urban infrastructure projects – Strategies for minimizing environmental impacts in land use planning assessments
Monitoring and Adaptive Management	Implement monitoring programs to track environmental changes and project performance over time, enabling adaptive management and continual improvement	– Ensure effectiveness of mitigation measures – Adapt to changing environmental conditions	– Monitoring: Implement monitoring programs to track environmental indicators – Evaluation: Assess effectiveness of mitigation measures – Adaptation: Adjust strategies based on monitoring results and changing conditions	– Monitoring and adaptive management in EIA for city infrastructure projects – Continual evaluation and adjustment in land use planning assessments

environmental uncertainties when cities identify and minimise environmental threats, optimise resource utilisation, and foster equitable growth. Cities are able to address environmental challenges in a proactive manner and build more sustainable and equitable urban settings for future generations by including EIA into the larger framework of urban planning. Incorporating EIAs into municipal master plans is an innovative approach to promote sustainable urban expansion, given the growing urbanisation and environmental concerns. By systematically evaluating and resolving possible environmental risks and consequences at the commencement of the planning process, this synergistic approach has the potential to revolutionise conventional paradigms in urban planning. As a result, cities will be better able to withstand and adapt to shifting environmental challenges and grow to be more resilient, egalitarian, and environmentally sustainable [16]. The important function of EIAs in urban master planning procedures is thoroughly explored in this chapter. The chapter aims to provide a nuanced understanding of the opportunities and complexities involved in creating resilient and sustainable cities for the future by examining the pressing need for sustainable urban development, highlighting the significance of integrating EIA, and outlining the goals that guide this integration. Cities may use the potential of EIA to negotiate the intricacies of urban growth and steer towards a more resilient and sustainable urban future by collaborating across disciplinary boundaries, engaging stakeholders, and making decisions based on data.

1.2 EIA in Urban Planning

EIA is a vital process for assessing the environmental implications of policies, plans, or projects before they are enacted, helping decision-makers minimize adverse effects. EIAs aim to identify potential environmental hazards and provide information about the environmental impacts of proposed measures, crucial for sustainable urban development [32]. Integrating EIA into urban master planning ensures sustainable growth by recognizing and mitigating environmental risks, thus preserving ecosystems, air and water quality, and human health (Fig. 1.1). This proactive approach allows for comprehensive evaluations during the planning stage, preventing costly errors and enhancing urban development sustainability by optimizing resource use. Incorporating EIA strengthens urban resilience against environmental pressures like climate change, ensuring cities can withstand and recover from environmental shocks [28]. Early incorporation of EIA principles fosters interdisciplinary collaboration, stakeholder involvement, and data-driven decision-making, promoting transparency and inclusivity. This comprehensive approach, combined with adaptive management, enhances urban planning's effectiveness in addressing dynamic environmental and societal challenges, leading to resilient and sustainable urban ecosystems. Furthermore, cities can respond to stakeholder feedback promptly, integrate new data, and adapt to these changes by implementing flexible and iterative processes [20].

Fig. 1.1 Stakeholders engagement in city projects

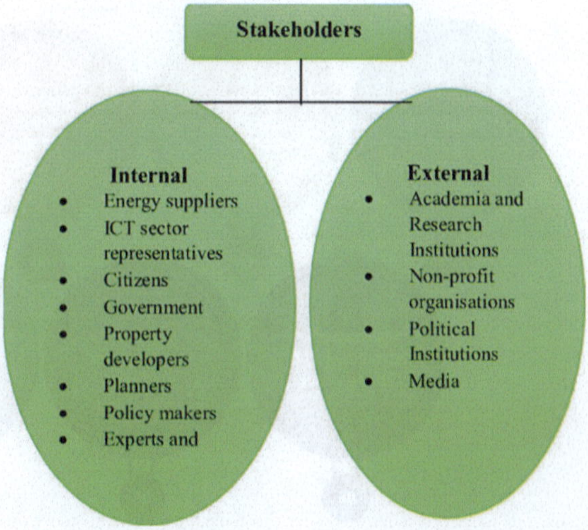

1.3 Proactive Planning and Consideration of Long-Term Effects

Environmental assessments (EIA and SEA) are critical tools in preventing irreversible environmental damage by allowing decision-makers to identify potential risks early in the planning phase and implement policies to mitigate negative impacts, preserving environmental integrity for future generations [1]. Proactive planning, which incorporates EIA and SEA concepts, helps cities recognize vulnerable ecosystems and natural resources at risk, enabling early intervention to protect these assets and promote sustainable development. Preventive actions, as emphasized are crucial in avoiding permanent environmental harm, such as biodiversity loss and habitat destruction, by prioritizing environmental considerations during the design stage, thus saving time and costs associated with later repairs and mitigation measures [7]. Furthermore, proactive planning enhances urban resilience against environmental threats, including resource depletion and climate change, by integrating strategies to reduce risks and adapt to changing conditions. This approach supports long-term sustainability by fostering economic equilibrium and balancing sustainability with economic growth, social fairness, and ecological preservation. Policymakers, by considering long-term environmental impacts, can promote resource efficiency, minimize waste, and encourage sustainable economic activities, leading to healthier and more inclusive urban environments [30]. Addressing long-term environmental impacts also improves public health and well-being by reducing pollution exposure and increasing access to green spaces (Fig. 1.2). Therefore, adhering to environmental regulations and best practicesreduces the risk of legal disputes and environmental liability, ensuring cities remain committed to sustainability and environmental stewardship. Proactive planning strategies like EIA and SEA are thus essential for sustainable development, protecting the environment, enhancing resilience, and fostering

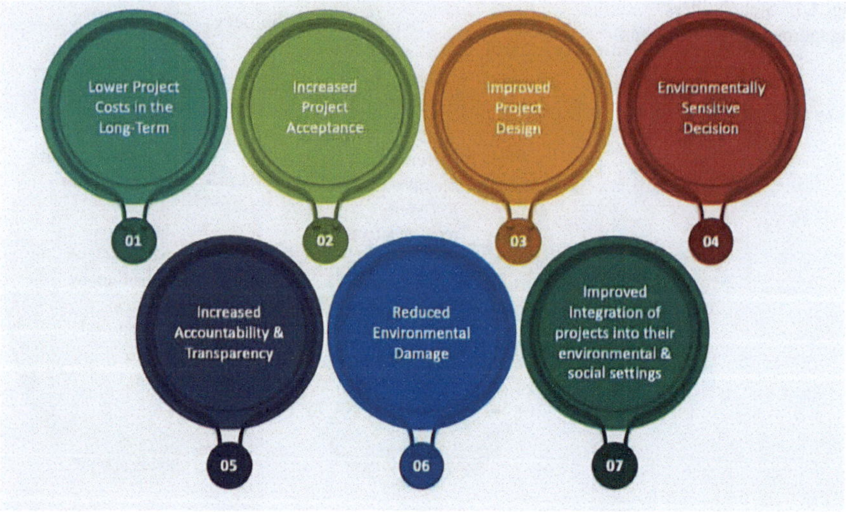

Fig. 1.2 Benefits of environmental impact assessment (EAI) and strategic environmental assessment (SEA)

inclusive urban settings that balance economic growth, social equity, and public health [36].

1.4 Data-Informed Decision-Making and Flexible Administration

Data-driven decision-making and adaptive management are essential in supporting EIAs and urban planning by providing crucial insights into environmental conditions and future impacts. Data-driven decision-making involves collecting, analysing, and interpreting various types of data, such as environmental, socioeconomic, demographic, and spatial data, to inform and support decisions. This approach helps identify sensitive areas, potential risks, and environmental hotspots, while socioeconomic data provides insights into the needs of local communities. Spatial data, like GIS, helps visualize relationships and assess spatial trends. Additionally, data-driven approaches enable tracking and evaluating the outcomes of EIA and urban planning interventions over time. Key indicators such as biodiversity, land use changes, air and water quality, and greenhouse gas emissions can be monitored to assess the effectiveness of decisions and adjust strategies accordingly [27]. Adaptive management, on the other hand, emphasizes flexibility, adaptability, and learning in response to uncertainties and changing conditions. It involves cycles of planning, implementation, monitoring, and evaluation, allowing decision-makers to continuously refine their strategies based on new information and feedback. This approach is crucial for addressing the complexities and uncertainties inherent in environmental systems and urban planning. It also enhances resilience by improving the ability to respond to unforeseen events and disruptions, such as climate variability and natural disasters. It also promotes collaboration and stakeholder engagement by providing opportunities for dialogue, learning, and shared decision-making. This inclusive approach builds trust, transparency, and legitimacy in the planning process. Therefore, integrating data-driven decision-making and adaptive management into EIAs and urban planning improves the effectiveness, resilience, and sustainability of urban development projects. These strategies help cities better anticipate and address environmental challenges, enhance stakeholder participation, and achieve more sustainable and equitable outcomes for the future [6].

1.5 Challenges and Solutions in Integration

EIAs must be included in municipal master planning to promote sustainable and inclusive urban growth. However, a lack of knowledge, competing interests, resources, and unwillingness to change may prevent its integration. This article clarifies these issues and suggests solutions by citing relevant literature and essential ideas. First, a city's master plan may not include an EIA due to a lack of data on biodiversity, ecosystem health, and air and water quality. When knowledge is scarce, environmental effects and mitigation techniques are difficult to predict [24]. This data shortage makes it difficult for decision-makers to accurately assess development projects' environmental impacts. Without sufficient information, balancing environmental protection and economic growth is difficult. Due to competing interests and objectives among stakeholders, government agencies, developers, community organisations, environmental organisations, and industry representatives, incorporating EIAs into city master plans may be difficult. Development ambitions, environmental protection policies, and land use can conflict, causing project cancellations or delays [9]. Since social fairness, environmental preservation, and economic advancement sometimes conflict, reaching agreement is difficult and time-consuming. EIA integration with city master plans may be difficult due to inconsistent local, regional, national, and international legislative frameworks. Uncertain regulations, permissions, and approvals can make decision-making harder and increase compliance costs for developers and designers. Harmonising regulatory frameworks and optimising approval processes might help city planners include Environmental Impact Assessments (EIAs). Finances, human resources, and technology can hinder EIA incorporation into city master planning. Lack of institutional support, finance, and experience might hinder mitigation plans, stakeholder engagement, and environmental assessments. Resource constraints impair environmental evaluations and insufficient consideration of environmental impacts, reducing planning effectiveness. Government agencies, planning departments, and other stakeholders may resist EIA in city master plans. Bureaucratic inertia, entrenched interests, and resistance to new methods and technology can stifle sustainable urban development innovation [2].

Overcoming change resistance requires strong leadership, active stakeholder participation, and capacity-building to develop a collaborative and sustainable culture. EIA may also be challenging to include into municipal master planning due to data accessibility, quality, and trustworthiness. Limited data sources hamper planning transparency, cooperation, and public participation, while incomplete or obsolete data impairs environmental assessments. Improved data quality and accessibility necessitate infrastructure investments for data collection, analysis, and dissemination. Build capacity for technical skills and knowledge-sharing networks. Policymakers, planners, stakeholders, and society must work together to solve these problems. To tackle these issues, promote knowledge accessibility, cooperation, simplifying norms, wise resource distribution, transparency, and EIA integration into urban planning. Addressing these concerns can make cities more resilient, inclusive, and sustainable and improve future quality of life. Urban master planning is difficult. These issues demand a multi-stakeholder strategy that promotes cooperation, innovation, and strategic planning [29]. Best practices and solutions can help decision-makers overcome these issues and ensure EIA supports sustainable urban development. Lack of environmental data may prevent EIA inclusion in master planning. Decision-makers may sponsor data gathering and analysis programmes to increase environmental data quantity and quality. Machine learning, GIS, and remote sensing could speed up and improve environmental data collection and processing. Decision-makers can improve stakeholder collaboration and information exchange to make important data accessible and efficient. Environmentalists, local officials, and others can advise decision-makers. Decision-making becomes more involved and informed. Stakeholder interests and agendas may make EIA tougher to incorporate into master planning. Intergroup discussion and consensus can assist decision-makers address this issue during stakeholder meetings, community workshops, and public hearings. Decision-makers may resolve concerns, develop confidence, and encourage ownership of planning results by incorporating stakeholders early and regularly. Decision-makers can encourage accountability and openness by allowing stakeholders to learn, contribute, and engage. Openness and accountability can assist decision-makers create stakeholder trust and incorporate EIA into master planning [26]. Government regulatory framework overlap may also limit EIA inclusion in master planning. Decision-makers can simplify laws, permission needs, and regulatory approval processes to fix this. Decision-makers can cut bureaucracy and boost efficiency by streamlining regulations. Technical, financial, and human resource restrictions may also hinder master planning EIA integration. Leaders can overcome these limits by collaborating with outside groups, allocating resources wisely, and growing capability. Master planners can incorporate EIA by overcoming resource constraints, learning technical skills, and working with external partners. Government agencies, planning departments, and other stakeholders may resist change, making EIA incorporation into master planning challenging. To enhance EIA adoption, policymakers should educate stakeholders, train and equip them, and offer incentives and rewards. Innovation, learning, and adaptability can help decision-makers overcome change resistance and encourage EIA in master planning [24, 26, 29].

1.5.1 Policy and Governance for Integration

Legislation and effective regulations are essential for requiring EIAs to be included in urban planning procedures. The systematic integration of environmental factors into master planning is ensured by explicit standards and regulatory obligations. Coordination and cooperation can be enhanced by putting in place a strong institutional framework with clearly defined roles and responsibilities for various agencies [10]. Integrating environmental assessments and guaranteeing regulatory compliance need interagency cooperation. By increasing their knowledge of EIA procedures and the significance of environmental issues in urban development, planners, legislators, and other stakeholders can be better equipped. Sufficient skills and knowledge are ensured for stakeholders through ongoing professional development. Participating in the planning process with the general public and local communities guarantees that their opinions and concerns are taken into account. Participation from the public promotes accountability, openness, and inclusivity, which produces more acceptable and long-lasting results [17].

1.5.2 Innovative Approaches and Technologies

Digital tools and technologies for smart cities, like big data analytics and the Internet of Things, improve data monitoring, collection, and decision-making. With the application of these technologies, urban surroundings may be better understood in real time, which can maximise resource efficiency and enhance the results of urban planning. IoT devices may track environmental factors including noise levels, traffic flow, and the quality of the air and water, giving important information for environmental impact assessments. This information can be used to locate hotspots for environmental issues, monitor changes over time, and develop mitigation plans. Large amounts of environmental, social, and economic data can be processed by big data analytics, which can then be used to find patterns and trends that help guide decisions. Planners can better predict such effects and create mitigation strategies with the use of these insights [14]. Digital twins are computer-generated images of real-world settings that can be used to mimic various situations and evaluate the possible effects of urban development initiatives. They improve the precision and potency of EIAs by allowing planners to see and evaluate the results of different initiatives. Urban planning may improve resilience and lessen its negative effects on the environment by including green infrastructure and nature-based alternatives. Examples of ecosystem services that are offered by green roofs, urban forests, and wetlands include stormwater management, air purification, and biodiversity preservation. Urban planning that incorporates mitigation and adaptation measures for climate change guarantees that cities are ready to face its effects. This covers actions like expanding green areas, enhancing energy economy, and supporting environmentally friendly transportation [34].

1.5.3 Monitoring, Evaluation, and Continuous Planning

To assess the viability and success of urban planning projects, indicators are crucial. The quality of the air and water, biodiversity, the amount of green space, and greenhouse gas emissions are important indicators. Population density, social fairness, affordability of housing, and accessibility to public services (healthcare, education). rates of employment, earnings, diversification of the economy, and investment in infrastructure. Community resilience metrics, disaster preparedness, and climate adaption measures. These metrics aid in evaluating the long-term effects of urban development initiatives and guarantee that they support resilient and sustainable growth. M&E methods entail gathering and evaluating data in a methodical manner in order to monitor the development and results of urban planning projects. Important methods consist of monitoring biodiversity, changes in land use, and the quality of the air and water can be done with sensors, remote sensing, and GIS [3]. Performing Social Impact Assessments (SIA) and Environmental Impact Assessments (EIA) to estimate and assess possible project effects, monitoring municipal policies' and projects' performance on a regular basis in relation to predetermined goals and objectives. using citizen science and participatory monitoring initiatives to involve the public in monitoring efforts. In urban planning, adaptive management is a dynamic approach that promotes adaptability and learning. Developing and analysing various future scenarios to foresee potential changes and uncertainties, testing new strategies on a small scale before wider implementation, regularly engaging stakeholders to gather insights and adjust plans accordingly, and feedback loops are essential for ensuring continuous improvement in urban planning are all part of the process [11]. In order to maintain transparency and accountability, they entail collecting data on key indicators, analysing trends and patterns, sharing findings with stakeholders and the public, applying the knowledge gained to make informed adjustments to policies, plans, and projects, and promoting a culture of learning within planning institutions to continuously improve skills and knowledge. Cities can guarantee that their planning procedures continue to be responsive, efficient, and equipped to support resilient and sustainable urban development by putting these principles into practice.

1.6 Case Studies: Successful Integration of Environmental Impact Assessment (EIA) into Urban Master Planning

1.6.1 Case Study 1: Curitiba, Brazil

The Brazilian state of Paraná's capital, Curitiba, is a witness to the transforming potential of creative urban planning techniques, especially when it comes to incorporating environmental issues into the master plan (Fig. 1.3). Curitiba's master plan established the groundwork for sustainable urban development by placing a strong emphasis on compact urban form, mixed land use, effective public transportation, and the preservation of green spaces to improve the city's quality of life and reduce its negative environmental effects. Curitiba's urban planning triumph is mostly attributed to its groundbreaking Bus Rapid Transit (BRT) system, which is the world's first of its type [25]. When the BRT system was implemented in the city during the 1970s, it drastically reduced air pollution and traffic while enhancing citizen mobility, revolutionising public transportation. Curitiba prioritised the construction of an effective and easily accessible public transportation system, which not only addressed urgent environmental issues but also promoted social fairness by giving all citizens access to reasonably priced and dependable transportation alternatives [25].

Additionally, as essential elements of the urban landscape, parks, woods, and riverbanks were prioritised for preservation in Curitiba's master plan. In addition to boosting the visual appeal of the city, these green spaces fulfilled important ecological roles by reducing floods, enhancing air quality, and giving locals access to recreational activities [33]. Curitiba showed its dedication to sustainability and environmental care by preserving these natural resources, guaranteeing that future generations would be able to benefit from a resilient and green urban environment (Fig. 1.4). Most importantly, Curitiba adopted a participatory approach to urban planning, actively include locals in the process of making decisions via forums including citizen consultation commissions, community workshops, and public hearings. The adoption of an inclusive strategy empowered inhabitants to take part in creating the future of their city by instilling in them a feeling of ownership and responsibility. Curitiba's planning authorities were able to ensure that the master plan represented the needs and ambitions of the community it served by incorporating local knowledge and priorities into the process of gathering input from a varied range of stakeholders [15].

Curitiba's comprehensive approach to urban planning has produced several social, economic, and environmental advantages for the city and its citizens. Notably, the focus on a compact urban form and effective public transit resulted in a notable decrease in greenhouse gas emissions and traffic congestion, enhancing air quality and increasing public health outcomes. Furthermore, the maintenance of green areas

Fig. 1.3 Location of Curitiba in Brazil

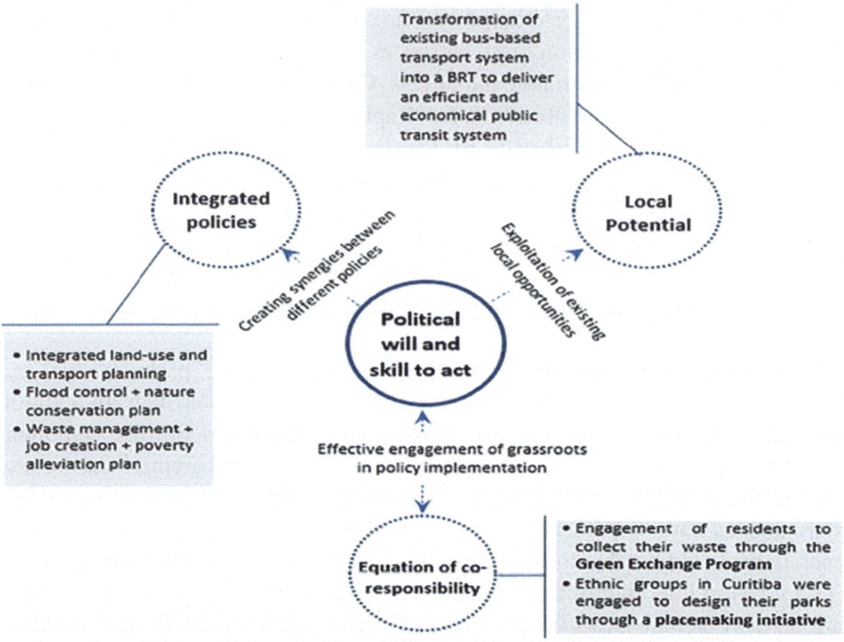

Fig. 1.4 Curitiba's sustainability transitions

improved biodiversity, gave animals a place to live, and provided recreational opportunities for locals, all of which improved their general well-being and quality of life. In addition, Curitiba's efforts in sustainable urban development have paid off financially, as the city's property prices have steadily increased over time [3]. Curitiba made a strong impression on businesses, citizens, and investors by prioritising social fairness and environmental sustainability in its infrastructure investments. The city's standing as a leader in sustainable urban planning brought affluence to its residents by promoting economic growth and job creation in addition to drawing tourists. To sum up, Curitiba's innovative attempts to incorporate environmental concerns into its master plan are a brilliant example of successful and forward-thinking urban planning. Prioritising fairness, sustainability, and community involvement has allowed Curitiba to evolve into a thriving, livable city that strikes a balance between environmental stewardship and economic progress. Curitiba's experience provides important lessons on the significance of proactive and coordinated urban planning in developing resilient and sustainable urban environments for future generations, as cities throughout the world struggle with the difficulties of growing urbanisation and climate change [35].

1.6.2 Case Study 2—Portland, Oregon

Portland, Oregon, is a model city for sustainable urban design because of its creative development techniques and strong focus on environmental care [12]. Portland's master planning process, which is renowned for its dedication to sustainability, incorporates environmental impact assessments to guide development choices and give priority to actions that reduce environmental consequences while improving the quality of life for locals. This article will examine Portland's approach to sustainable urban planning, with a particular emphasis on the South Waterfront District's growth as a noteworthy case study illustrating the city's integrated approach to master planning and environmental impact assessment [12]. The South Waterfront District, which is situated alongside the Willamette River, was formerly an industrial region distinguished by polluted terrain and abandoned warehouses (Fig. 1.5). Portland started planning to transform the region into a mixed-use, sustainable neighbourhood that would be a role model for urban development in the late 1990s. An extensive environmental impact study was carried out before to reconstruction in order to determine potential environmental problems and assess the ecological health of the land. City planners, environmental specialists, and community members worked together to evaluate important aspects including energy use, transportation infrastructure, habitat protection, and the quality of the air and water [22].

A master plan that prioritises sustainability and environmental stewardship was created to direct the rebuilding of the South Waterfront District based on the results of the environmental impact assessment. Brownfield rehabilitation was one of the main tactics included in the master plan to remediate contaminants and prepare the land for safe usage. This entailed restoring abandoned industrial property into a lively, livable neighbourhood by taking steps to reduce environmental concerns and maintain public safety. In order to control stormwater runoff, boost the quality of the water, and improve the urban ecology, the master plan also included green infrastructure components including bioswales, green roofs, and permeable pavement. In addition to reducing the impact of the urban heat island, these elements enhanced ecosystem health and biodiversity, resulting in a more resilient and sustainable urban environment [39]. In order to lessen reliance on automobiles, encourage walkability, and foster a lively, diversified neighbourhood, the South Waterfront District was

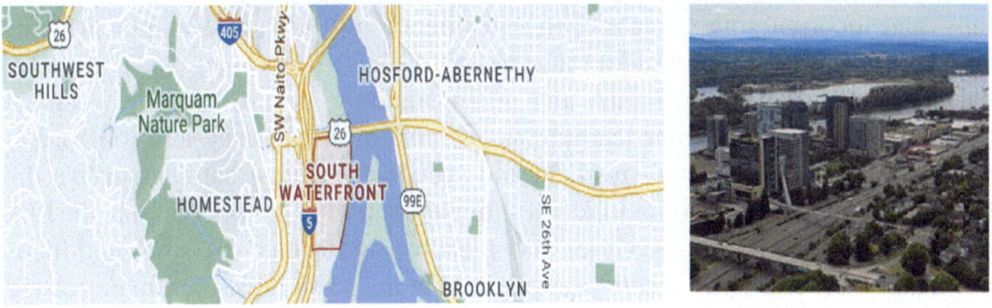

Fig. 1.5 Location of South Waterfront in Portland

planned to support mixed-use development, which includes residential, commercial, and recreational areas. The plan gave priority to funding light rail and streetcar lines as well as other public transit infrastructure in order to give locals easy access to a variety of transportation choices. Furthermore, in order to promote active transportation and lower greenhouse gas emissions, improvements were made to the infrastructure for bicycles and pedestrians [37].

In order to encourage the use of renewable energy sources and energy-efficient construction techniques, building regulations and design guidelines were created. Portland's dedication to environmental sustainability was further demonstrated by the provision of incentives for the installation of solar panels, energy-efficient appliances, and other sustainable technology in new building and renovation projects. An effective case study for sustainable urban development is the South Waterfront District revitalization in Portland, which highlights the significance of integrating environmental factors into the design process [31]. Portland has been able to reduce its negative environmental effects, increase its citizens' resistance to climate change, and improve their quality of life by including environmental impact assessments into master planning procedures and placing a high priority on sustainability. To sum up, Portland's South Waterfront District is a prime example of the city's dedication to environmentally conscious development and sustainable urban design. Portland, via cooperative and integrated approaches to master planning and environmental impact assessment, has turned abandoned industrial property into a thriving neighbourhood that strikes a balance between environmental preservation and economic growth. The South Waterfront redevelopment's success offers a crucial lesson for cities throughout the globe, emphasising how crucial it is to include environmental factors in urban planning procedures in order to build resilient, sustainable, and livable communities [13].

1.7 Future Directions and Opportunities

The necessity to include climate change concerns into environmental impact assessment (EIA) and urban planning procedures is becoming more pressing as cities throughout the globe struggle with the complex issues brought about by climate change. In the future, evaluating climate vulnerability, coming up with adaptation plans, and incorporating climate change resilience into urban development plans should be the key priorities. In addition, new approaches to improving data collecting, analysis, and visualisation in EIA and urban planning are provided by technical developments such as machine learning, geographic information systems (GIS), and remote sensing [16]. Creating resilient and sustainable cities that can survive the effects of climate change requires incorporating climate change concerns into the EIA and urban planning processes. Cities may make well-informed decisions and focus their adaptation efforts by identifying places that are vulnerable to climate-related hazards including floods, excessive heat, and sea-level rise through the assessment of climate vulnerability. In addition, incorporating climate change resilience into urban development planning guarantees that communities, buildings, and infrastructure are built to resist future climatic shocks, protecting people, property, and vital infrastructure [28].

Technological advancements like machine learning, GIS, and remote sensing have the ability to completely transform urban planning and EIA procedures while improving the accessibility, effectiveness, and precision of environmental assessments. High-resolution spatial data may be gathered thanks to remote sensing technology, making it easier to identify and track changing environmental variables over time. Planners are able to evaluate environmental consequences, pinpoint mitigation strategies, and visualise spatial linkages with the use of GIS systems, which offer strong analytical capabilities for spatial data processing [32]. Large datasets may be analysed using machine learning algorithms to find patterns, trends, and correlations that can help with decision-making and guide the creation of policies. Moreover, environmentally sound and sustainable methods of tackling urban environmental problems may be found in nature-based solutions (NBS), which include natural resource management, green infrastructure, and ecosystem restoration. Parks, green roofs, and urban forests are examples of green infrastructure that benefits communities recreationally and aesthetically while reducing stormwater runoff, mitigating urban heat islands, and enhancing biodiversity. Projects aimed at restoring ecosystems, such the rehabilitation of rivers and the restoration of wetlands, boost ecosystem services, restore habitat, and improve water quality. Cities may address the effects of climate change and improve urban sustainability, resilience, and livability by incorporating NBS into urban master planning procedures [1].

To guarantee that the advantages and costs of urban growth are distributed fairly among all socioeconomic strata, social equity and justice must be given top priority in EIA and urban planning procedures, in addition to environmental concerns. In addition to frequently bearing an unfair share of the weight from environmental threats, marginalised and vulnerable populations frequently lack access to green space and other vital resources. Subsequent investigations ought to delve into strategies for incorporating social equity into environmental evaluation and urban design, including matters like fair access to public facilities and green infrastructure, community relocation, and environmental justice [16]. The effectiveness of EIA and urban planning procedures depends heavily on stakeholder participation and engagement, especially in marginalised and disadvantaged groups. A meaningful engagement makes sure that all stakeholders' interests and priorities are taken into account, that a variety of viewpoints are taken into account, and that local knowledge is integrated into decisions. Novel strategies may empower communities, bolster different perspectives, and promote stakeholder collaboration. Examples of these strategies include digital engagement platforms, community-led projects, and participatory planning workshops. Cities may foster inclusive decision-making procedures to increase social cohesiveness, foster trust, and develop more equitable and resilient urban settings [7].

To sum up, the advancement of resilient and sustainable urban development depends on incorporating climate change concerns, utilising technology advancements, adopting nature-based solutions, giving social equality top priority, and enhancing stakeholder participation. Cities need to take a comprehensive and inclusive approach to urban planning and EIA going forward, making sure that social, cultural, and economic factors are taken into account when making decisions. Cities can create resilient, inclusive, and sustainable communities that can flourish in the face of urban issues such as climate change by embracing innovation, cooperation, and equity. Studies on long-term monitoring and evaluation of urban development projects can provide valuable information on the effectiveness of environmental impact assessment and urban planning strategies. in the achievement of the goals of sustainable development. This research can inform future policy development and decision-making processes. To address the health impacts of urban development decisions, health impact assessment must be integrated into environmental impact and urban planning processes [20, 27]. Research at the intersection of environmental, social and health impacts can inform policies and interventions designed to promote public health and well-being in urban areas. Investments in capacity development and training programs for urban planners and policy makers, and stakeholders to improve their skills and knowledge on environmental impact assessment and sustainable urban development. Research on effective capacity-building strategies and training interventions can support professional development and institution-strengthening efforts. Policy makers should explore innovative policy instruments, incentives and regulatory frameworks to encourage sustainable urban development practices and integrate environmental impacts. to urban planning processes. Research on policy innovation and institutional reform can identify opportunities to overcome barriers and promote more effective environmental management. Promoting cross-disciplinary collaboration and partnership between government agencies, academia, industry, civil society and communities for solving complex urban problems and promoting sustainable development. Exploring collaborative management models, multi-stakeholder partnerships and co-design processes can facilitate knowledge sharing, resource mobilisation and joint action towards common goals [1, 16, 21, 32, 36].

1.8 Conclusion

Combining the evaluation of the ecological impact with the general plan of the city from the point of view of sustainable urban development. Through the analysis of case studies, literature and best practices, some important conclusions and implications emerged. Early integration of EIA principles in urban planning processes is necessary to identify and mitigate potential environmental impacts and promote. sustainable development and improving the sustainability of urban development projects. Interdisciplinary cooperation, engagement of stakeholders, data-based decision-making and adaptive management are important guiding principles of the integration process. Challenges such as lack of information, conflicts between interest groups, regulatory complexity, and limitation of resources must be considered to effectively integrate EIA into the city's master plan. The successful integration of EIA into the city's master plan requires proactive planning, consideration of long-term impacts and a holistic approach that balances environmental, social, social and economic considerations. Capacity development, policy innovation and institutional reform are needed to overcome barriers and promote effective environmental governance in urban areas. Nature-based solutions, climate resilience, social equity and health aspects should be integrated into EIA and urban planning processes to respond. to emerging urban problems and opportunities. Promoting the integration of EIA into the master plan of sustainable urban development of cities requires the joint action of decision-makers, planners, interest groups and the wider community. Make sustainability a central guiding principle in the city's master planning processes, focusing on minimising environmental impacts, improving sustainability and promoting equitable and inclusive development. Allocate resources for capacity—construction. development, training programs and communication initiatives to strengthen the skills and knowledge of urban planners, policy makers and stakeholders involved in environmental assessment and urban planning. Promote interdisciplinary collaboration, stakeholder engagement and partnership between government, academia and industry. Empower communities, especially marginalised and vulnerable groups, to participate meaningfully in decision-making processes, amplify their voices and promote their interests to shape the future of the city. Finally adopting these principles and taking collective action. With Actions, we can promote the integration of EIA into the city's master plan and pave the way for a more sustainable, sustainable and liveable urban environment for current and future generations.

References

1. ABD Elrahman AS, Asaad M (2021) Urban design & urban planning: a critical analysis to the theoretical relationship gap. Ain Shams Eng J 12:1163–1173
2. Bahadorestani A, Naderpajouh N, Sadiq R (2020) Planning for sustainable stakeholder engagement based on the assessment of conflicting interests in projects. J Clean Prod 242:118402
3. Bibri SE, Krogstie J, Kärrholm M (2020) Compact city planning and development: emerging practices and strategies for achieving the goals of sustainability. Developments Built Environ 4:100021
4. Bush J, Doyon A (2019) Building urban resilience with nature-based solutions: How can urban planning contribute? Cities 95:102483
5. Choudhury U, Kanga S, Singh SK, Kumar A, Meraj G, Kumar P, Singh S (2024) Projecting urban expansion by analyzing growth patterns and sustainable planning strategies—a case study of Kamrup Metropolitan, Assam, North-East India. Earth [Online], 5
6. Cilliers DP, Retief FP, Bond AJ, Roos C, Alberts RC (2022) The validity of spatial data-based EIA screening decisions. Environ Impact Assess Rev 93:106729
7. Cortinovis C, Geneletti D (2019) A framework to explore the effects of urban planning decisions on regulating ecosystem services in cities. Ecosyst Serv 38:100946
8. Dhupper R, Jain H, Gupta AK, Acharya P, Kumar D (2023) Management of flood disasters in peri-urban ecosystems of noida using remote sensing and geographical information systems. Geoinformatics Sustain Urban Dev
9. Ganeshu P, Fernando T, Keraminiyage K (2023) Barriers to, and enablers for, stakeholder collaboration in risk-sensitive urban planning: a systematised literature review. Sustainability [Online], 15

10. George TE, Karatu K, Edward A (2020) An evaluation of the environmental impact assessment practice in Uganda: challenges and opportunities for achieving sustainable development. Heliyon 6:e04758
11. Giles-Corti B, Moudon AV, Lowe M, Cerin E, Boeing G, Frumkin H, Salvo D, Foster S, Kleeman A, Bekessy S, De Sá TH, Nieuwenhuijsen M, Higgs C, Hinckson E, Adlakha D, Arundel J, Liu S, Oyeyemi AL, Nitvimol K, Sallis JF (2022) What next? Expanding our view of city planning and global health, and implementing and monitoring evidence-informed policy. Lancet Glob Health 10:e919–e926
12. Goodling E, Green J, Mcclintock N (2015) Uneven development of the sustainable city: shifting capital in Portland, Oregon. Urban Geogr 36:504–527
13. Hagerman C (2007) Shaping neighborhoods and nature: urban political ecologies of urban waterfront transformations in Portland, Oregon. Cities 24:285–297
14. Hajjaji Y, Boulila W, Farah IR, Romdhani I, Hussain A (2021) Big data and IoT-based applications in smart environments: a systematic review. Comput Sci Rev 39:100318
15. Halecki W, Stachura T, Fudała W, Stec A, Kuboń S (2023) Assessment and planning of green spaces in urban parks: a review. Sustain Cities Soc 88:104280
16. Hariram NP, Mekha KB, Suganthan V, Sudhakar K (2023) Sustainalism: an integrated socio-economic-environmental model to address sustainable development and sustainability. Sustainability [Online] 15
17. Hegazy IR, Tohlob AA (2024) Strategic environmental assessment and SDG integration: enhancing sustainable development in Egypt. J Umm Al-Qura Univ Eng Archit
18. Jain H, Dhupper R, Kumar D (2023a) Advancing the interventions of nature- based solutions in cities for urban climate resilience. Geoinformatics for sustainable urban development
19. Jain H, Dhupper R, Shrivastava A, Kumar D, Kumari M (2023) AI-enabled strategies for climate change adaptation: protecting communities, infrastructure, and businesses from the impacts of climate change. Comput Urban Sci 3:25
20. Jones MG (1983) Environmental impact assessment at the planning level: two systems. Environmental impact assessment. Dordrecht: Springer Netherlands
21. Keysar E, Steinemann A (2002) Integrating environmental impact assessment with master planning: lessons from the US Army. Environ Impact Assess Rev 22:583–609
22. Kolmes SA (2022) Intended and unintended consequences of two paradigms of urban planning, and their social justice and human health impacts, in Portland, Oregon. Environments [Online], 9
23. Kong L, Mu X, Hu G, Zhang Z (2022) The application of resilience theory in urban development: a literature review. Environ Sci Pollut Res 29:49651–49671
24. Lawrence DP (1997) PROFILE: integrating sustainability and environmental impact assessment. Environ Manage 21:23–42
25. Macedo J (2013) Planning a sustainable city: the making of Curitiba, Brazil. J Plan Hist 12:334–353
26. Munaro MR, Tavares SF (2023) A review on barriers, drivers, and stakeholders towards the circular economy: the construction sector perspective. Cleaner Responsible Consum 8:100107
27. Nautiyal H, Goel V (2021) Chapter 3—sustainability assessment: metrics and methods. In: Ren J (ed) Methods in sustainability science. Elsevier
28. Oppio A, Dell'ovo M (2020) Strategic environmental assessment (SEA) and multi-criteria analysis: an integrated approach. In: Campeol G (ed) Strategic environmental assessment and urban planning: methodological reflections and case studies. Cham: Springer International Publishing
29. Saarikoski H, Primmer E, Saarela S-R, Antunes P, Aszalós R, Baró F, Berry P, Blanko GG, Goméz-Baggethun E, Carvalho L, Dick J, Dunford R, Hanzu M, Harrison PA, Izakovicova Z, Kertész M, Kopperoinen L, Köhler B, Langemeyer J, Lapola D, Liquete C, Luque S, Mederly P, Niemelä J, Palomo I, Pastur GM, Peri PL, Preda E, Priess JA, Santos R, Schleyer C, Turkelboom F, Vadineanu A, Verheyden W, Vikström S, Young J (2018) Institutional challenges in putting ecosystem service knowledge in practice. Ecosyst Serv 29:579–598

30. Sandu I, Ionescu C, Ursache MA (2005) Pilot system for environmental impact assessment of pollution caused by urban development and urban air pollution forecast. In: Faragó I, Georgiev K, Havasi Á (eds) Advances in air pollution modeling for environmental security. Dordrecht. Springer Netherlands, pp 317–326
31. Sarkheyli E, Rafieian M (2018) Megaprojects and community participation: South Waterfront project in Portland, Oregon. Housing and Society 45:104–117
32. Shepherd A, OrtolanO L (1996) Strategic environmental assessment for sustainable urban development. Environ Impact Assess Rev 16:321–335
33. Sikorska D, Łaszkiewicz E, Krauze K, Sikorski P (2020) The role of informal green spaces in reducing inequalities in urban green space availability to children and seniors. Environ Sci Policy 108:144–154
34. Talebkhah M, Sali A, Marjani M, Gordan M, Hashim SJ, Rokhani FZ (2021) IoT and big data applications in smart cities: recent advances, challenges, and critical issues. IEEE Access 9:55465–55484
35. Wantzen KM, Alves CB, Badiane SD, Bala R, Blettler M, Callisto M, Cao Y, Kolb M, Kondolf GM, Leite MF, Macedo DR, Mahdi O, Neves M, Peralta ME, Rotgé V, Rueda-Delgado G, Scharager A, Serra-Llobet A, Yengué J-L, Zingraff-Hamed A (2019) Urban stream and wetland restoration in the global south—a DPSIR analysis. Sustainability [Online], 11
36. Weiland U (1998) Environmental impact assessment (EIA) for types of land use. In: Breuste J, Feldmann H, Uhlmann O (eds) Urban ecology. Berlin, Heidelberg: Springer, Berlin, Heidelberg
37. Wu Y, Liu Y (2023) Transforming industrial waterfronts into inclusive landscapes: a project method and investigation of landscape as a medium for sustainable revitalization. Sustainability [Online], 15
38. Zeng X, Yu Y, Yang S, Lv Y, Sarker MN (2022) Urban resilience for urban sustainability: concepts, dimensions, and perspectives. Sustainability [Online], 14
39. Zheng B, Masrabaye F (2023) Sustainable brownfield redevelopment and planning: bibliometric and visual analysis. Heliyon 9:e13280

Chapter 2
City as Urban Ecology: Ecosystems as Climate Regulators and Sources of Resilience

Maud Cassaignau and Markus Jung

Abstract Due to its carbon impact, building has become increasingly questionable: famous architects and urban designers contend, the disciplines should either deploy mapping skills towards better decision-making, or focus on careful small-scale interventions. A combined approach to climate and resilience addresses contexts of population growth and housing scarcity: The method combines dense conversion of buildings to reduce carbon footprint, together with transformation and revegetation of urban spaces for climate mitigation. Two co-developed projects illustrate in combination a strategy mapping out such urban renewal potentials, testing design at precinct scale through precise interventions, and application through 1:1-prototypes. The design-research investigations propose to look beyond buildings into whole city approaches to urban resilience, which rehabilitate urban structures and spaces. Two complementary blueprint case-studies investigate how landscape can combine with urban renewal strategies to address challenges of climate (urban heat, flooding) and biodiversity (extinction) as well as the very human need for engaging urban spaces. The strategies combine architecture, urban and landscape design to re-envision city spaces as new urban ecologies. They test how to reappropriate underused urban spaces, creating blue-green urban networks of amenity as optimistic perspective for the future. The intertwining studies unpack how we may reconceptualise city spaces as resilient and adaptive urban ecosystems comprising flora, fauna and humans within urban environments. Such ecosystems are able to absorb heavy rainfalls and heat and thus reduce vulnerability to climate change and weather extremes. A first study focused on identifying underused spaces within Melbourne's existing urban fabric for ecosystems that could host new resilient infrastructures and programs. The redesigned spaces create a balance between urban and natural systems. A second study delved into how to design and implement such ecosystems within urban contexts though a precinct strategy and four 1:1 prototypes. Starting from specific

M. Cassaignau
RMIT, Melbourne, Australia
e-mail: maud.cassaignau@rmit.edu.au

M. Jung (✉)
Swinburne University, Melbourne, Australia
e-mail: majung@swin.edu.au

S. Ghosh et al. (eds.), *Cities of Tomorrow: Urban Resilience and Climate Change Preparedness*, Urban Sustainability, https://doi.org/10.1007/978-981-97-9658-8_2

urban spaces, it explored their capacity to host particular types of ecosystems framed around specific endemic animals. The article will develop how these explorations have culminated in an expansion of the one-health concept (WHO, 2024) applied currently in health sciences in the field of urban design and planning. A new expanded concept conceptualizes the interdependency between human, biological and infrastructural health in cities. Thereby, it highlights the importance of thinking the city as an interlinked ecosystem while redesigning urban spaces for resilient and optimistic futures.

Keywords Urban and landscape design · WSUD · Climate adaptation · More-than-human approaches · Wilding

2.1 Introduction

The question, how to adapt cities to the challenges of declining climate and biodiversity has increasingly dominated news cycles in recent years. On one hand intensifying climate events and their destructive repercussions on inhabitants and infrastructures have shown that innovative and comprehensive approaches are urgently needed. Old, engineered prevention methods have shown their vulnerability in increasingly harsh extreme weather events. Current conditions already surpass in magnitude the scope these built measures have been calibrated to. Levees prominently collapsed in New Orleans [47], become insufficient again after upgrades [19]. Or even exacerbate flooding dangers like in Melbourne [51]. Mechanical climate conditioning while cooling one space will exacerbate heat in adjacent spaces [58].

Therefore, nature-based solutions, as imagined by climate sensitive and water sensitive urban design (CSUD and WSUD) have gained prominence in the field to address the climate crisis. These approaches, that use natural means, such as the introduction of vegetation, natural water retention, natural shade, ventilation and evaporative cooling additionally address the biodiversity crisis on the other hand. But while such approaches have been well-studied, they often overlook two factors.

Firstly, water sensitive and climate sensitive designs are often seen from a human centered perspective and framed as infrastructures rather than as urban ecologies. This results in badly functioning ecosystems. This article aims to address this gap. It formulates an approach based on a novel concept of urban ecology with ecosystems that include the built environment as well as humans, plants and animals. This framework leads to a way to reconceptualise human, biotic and urban system health as intertwined.

Secondly, the article challenges the understanding of cities and ecosystems as divided and opposed entities. It aims to wholistically look at cities, built and open spaces, to reimagine how urbanity and ecology can coexist. Thereby it combines distinct approaches, in response to the call by some urban and architecture design practitioners to either focus on mapping skills towards better decision-making [26], or on careful small-scale interventions [57].

By using case study projects and a design research approach [20, 21] to discuss the above, this paper aims to delve into how we may retrofit city systems in such manner. These studies provide a deeper understanding of response to local nuances and implications. The design research method of using a blueprint test area to investigate new methods, is common in urban and landscape design. Blueprint test projects and case studies offer significant insights and methodological pathways for policy makers, planners, developers and urban design practice. A precedent here forehas been our Building Mixity! [2] blueprint project which influenced local [4] and state policy (Victoria State Government, 2018), or our Melbourne Cool Lines project which led to city initiatives like 'Cooling the City' and 'Cool Routes' [10].

Finally, the proposed case studies build on widely established models of Water Sensitive Urban Design [56] and Climate Sensitive Design [35]. The projects reflect on how such models can be implemented within existing contexts, cross-fertilizing them with methods from the field of Urban Regeneration and Adaptive Reuse. These urban and landscape methods are revisited with the lens of recent theoretical calls for addressing more-than-human perspectives, and health approaches that link human, animal and environmental health.

2.2 Method and Outcomes

The investigation used a design research methodology [20, 21] common in urban, landscape and architectural design. We had this approach in earlier transdisciplinary projects to test water sensitive urban design strategies [3]. This time we would use it to investigate how our novel take on water and climate sensitive design could be implemented in real sites in Melbourne. For each iteration a blueprint area was chosen to test the new approaches and reflect on the outcomes. Each iteration informed and complemented the next investigation scope. This successive testing enabled first the design of a general urban retrofitting approach across built up and open urban to spaces. It then led to the development of context driven ecosystem responses, creating a correlation of urban typology, ecology and key urban animals. Finally a last iteration tested the implementation and user response to four prototypical tests of such ecosystems framed around key indigenous animals.

2.2.1 Designing an Urban Retrofitting Approach

We collaborated with the Collaborative Research Centre for Water Sensitive Cities (CRCWSC) and the City of Melbourne to co-develop the first transdisciplinary research of Melbourne Cool Lines[1] (Fig. 2.1). The aim was to create an urban plan that

[1] The research was presented as a film at the 17th Venice Architecture Biennale in the virtual Pavilion, *CityX Venice* [3].

would tackle climate change, urban heat islands (UHI), and flooding in north-west Melbourne's low-lying post-industrial regions. Currently, these post-industrial areas lack vegetation, shade and open water elements, due to sealed surfaces. Thereby, it is well recognised that hard surfaces increase the urban heat island effect, as they generate more heat radiation than vegetated areas and water surfaces [15].

Fig. 2.1 The map displays three testing scales: Melbourne Cool Lines (MCL) in blue for metropolitan strategy, Wilding the Campus (WTC) in aqua for precinct design area, and Wild Streets prototypical testing in white. *Source* Melbourne Cool Lines Mapping, illustration by Maud Cas-saignau, Markus Jung and Duy Phan 2024, 2024

Under the guidance of the CRCWSC, our project aimed at increasing vegetated areas and (re)introducing open water as these improve evaporative cooling. It is a mechanism that mitigates heat events and thermal discomfort. Simultaneously, the introduction of vegetation, wetlands and water retention enables the absorption, delay, filtration, and storage of rainwater [26] These design elements and principles prevent negative impacts of heavy rainfalls and floods, while retained water and humidity help mitigate later droughts. These water sensitive design and sponge city approaches would form the underlying rationale for the two projects described, while we would underpin them with more-than-human and wilding ideas.

The primary urban study devised a strategy to detect neglected areas and convert them into blue-green spaces, forming a resilient network throughout the city. By collecting rainwater in creeks, wetlands, and raingardens, these blue-green oases could create continuous refreshing and water capturing recreational spaces for Melbourne, connecting with existing pedestrian areas and parks.

Connecting communities, the pedestrian network of climate-mitigating urban oases would offer shaded amenity corridors, cooling water landscapes, and vegetated breezeways throughout the city. The concept of networking was developed to empower neighborhoods to take ownership of their spaces and amplify the impact of interventions through broader ecological connections. Existing pedestrian routes were combined with local interventions to create uninterrupted paths throughout the study area. With time, the expansion could widen to include additional metropolitan areas, linking residents and walking and biking paths. These paths would further twenty-minute neighborhoods and connect to public transport to increase resilience and sustainable modes of living. The use of an incremental strategy with precise interventions would further enable local testing, solutions, and funding, while facilitating progressive and flexible metropolitan adaptation strategies.

2.2.2 *Discovering Potentials for Urban Oases*

The location and nature of proposed interventions were determined by the circumstances of each neighbourhood. In our proposal for the Melbourne Cool Lines, we actively sought input from communities and conducted a thorough analysis of the surrounding contexts. This led to the discovery of neglected areas and perfect spots for blue-green spaces and pedestrian linkages. The study highlighted the specific needs for urban infrastructures and sheltered open spaces in north-west Melbourne. Aspects of landscape, infrastructure, and architectural, climatic and water sensitive urban design were combined in various ways, as shown in the following examples. CRCWSC's involvement in the project helped shape the design of hybrid landscapes, incorporating ecosystems, wetlands, swales, water gardens, urban forests, and gardening to improve climate resilience. These strategies' focus on maximising water retention and reuse, evaporative cooling, and shading [7] would benefit the entire neighbourhood, by reducing the impact of heatwaves, rain events, and flooding while improving indoor and outdoor comfort.

2.2.3 *Design Responses Tailored to Specific Found Spaces*

Indigenous Estuary Park

An initial intervention (Fig. 2.2) exemplified the renaturation of a recurrently flooded industrial site, presenting a sole link between the recently built docklands and West Melbourne. Located near a canalised creek and beneath a highway, tidal saltwater inflows were causing damage to the ecosystems. Reintroducing long-removed indigenous dam practices would enable the separation of salt and sweet water ecologies in the estuary. Reintroducing a chain of ponds would further enhance the ecological diversity of plants and animals by separating salt and freshwater environments and renaturing the canalized waterway. The reclaimed ecosystems containing a chain of ponds would serve as recreational areas for park-deprived urban neighbourhoods nearby. The enlarged estuary would absorb flood danger and heat waves, as well as improve the ecological value of the area. Moreover, the new water park would foster a reconnection to our indigenous history. The parkland would incorporate an indigenous center and boardwalk system, providing opportunities for cultural practices, education, ecological workshops, and an open-air museum.

Oasis Interchange

Another proposition (Fig. 2.3) offered to address the repetitive flash-flooding of a historic Melbournian road built over a former creek [11]. A large reservoir park would be integrated into an overpass that bridges a crowded city roundabout near the

Fig. 2.2 Indigenous Estuary Park containing chain of ponds. *Source* Melbourne Cool Lines Interventions, illustrations by Maud Cassaignau, Markus Jung and Duy Phan, 2020. Interactive VR versions: "Cool Lines Exhibition", XPACE, accessed April 24, 2024. https://www.xpace.cc/projects/urban-design/cool-lines-exhibtion/

hospital precinct. Not only would it absorb excess rainwater from Elizabeth Street, but it would also create a shaded and sheltered oasis for wildlife in the otherwise barren and hot urban area. The vine-covered overpass, which links to the hospital precinct, suggests the creation of an open-air shopping area that revolves around the reservoir ecosystem for staff and visitors. It would function as a tranquil, umbrageous and airy oasis, naturally ventilating adjacent spaces.

Water Housing Factory

Water Housing Factory (Fig. 2.4) put forward an alternative solution for housing expansion in flood-prone areas, a big issue in Melbourne ([9], 4–5, 53). Conventionally, these expansions have, until recently, mostly been realized without flood responsive building strategies and typologies. Often, such constructions also overlook the attraction of wetland ecologies for nearby inhabitants, and the potential for water and ecology related activities. The demonstration showcased the development and construction of modular elevated and floating housing near existing road and rail infrastructures, with testing conducted alongside Moonee Ponds Creek. These typologies extended on flood-resistant built examples from the Netherlands (Schuetze 2015, 1). These would coexist with indigenous riverine ecologies and vegetation, most adapted to resist local weather extremes. The changing ecosystems

Fig. 2.3 Interchange Oasis overpass integrating a wetland reservoir. *Source* Melbourne Cool Lines Interventions, illustrations by Maud Cassaignau, Markus Jung and Duy Phan, 2020. Interactive VR versions: "Cool Lines Exhibition", XPACE, accessed April 24, 2024. https://www.xpace.cc/projects/urban-design/cool-lines-exhibtion/

Fig. 2.4 The Water Housing Factory for flood-adapted housing integrating wetlands. *Source* Melbourne Cool Lines Interventions, illustrations by Maud Cassaignau, Markus Jung and Duy Phan, 2020. Interactive VR versions: "Cool Lines Exhibition", XPACE, accessed April 24, 2024. https://www.xpace.cc/projects/urban-design/cool-lines-exhibtion/

would offer a fascinating backdrop and spaces for recreation and discovery for new inhabitants, who would profit from direct access to water and the relative coolness of the area.

2.3 Analysis

2.3.1 Ecosystems as Regulators of Climate

The ecosystems of Melbourne Cool Lines had been designed in consultation with ecologists, climatologists, and water engineers of the CRC for Water Sensitive Cities. A maximisation of vegetation and reintroduction of open water aimed at reducing urban heat islands in the currently rather dry and sealed landscape of north and west Melbourne. While hot in summer and lacking recreational amenity, the low-lying neighborhoods have also been prone to recurrent flooding, as many sit on reclaimed swamp land. With climate change, filled-in waterways and flood plains are coming back. Therefore, finding strategies to reintroduce water ecologies in ways that work with the existing built fabric is crucial for resilience, flood prevention and ecological restoration. It is also the economic way to go, as climate damage cost often exceeds those of transformations (Serra-Llobet et al. 2022). The propositions advanced in this paper aimed at addressing the two opposed but interlinked sides of the climatic pendulum, between dry El Niño weather patterns of heat and drought and wet la Niña extremes of flood and heavy weather events. Replacing asphalt with vegetation and water retention next to shaded amenities would help absorption, filtering, and retention of water in one period and reusing it consecutively for cooling and watering surrounding parklands in drought and heat waves. The gaussian urban heat island curve (Fig. 2.5) would, instead of showing increased temperature bumps over sealed surfaces, reveal cool temperature dents in the same blue greened locations.

The suggested ecosystems would help create a healthier and more robust environment and temperate climate that benefits all generations, especially vulnerable people. Its reduced heat impact would mitigate health issues, hospitalisations and fatalities. The ecosystems' capacity to absorb water though wetlands and reservoirs

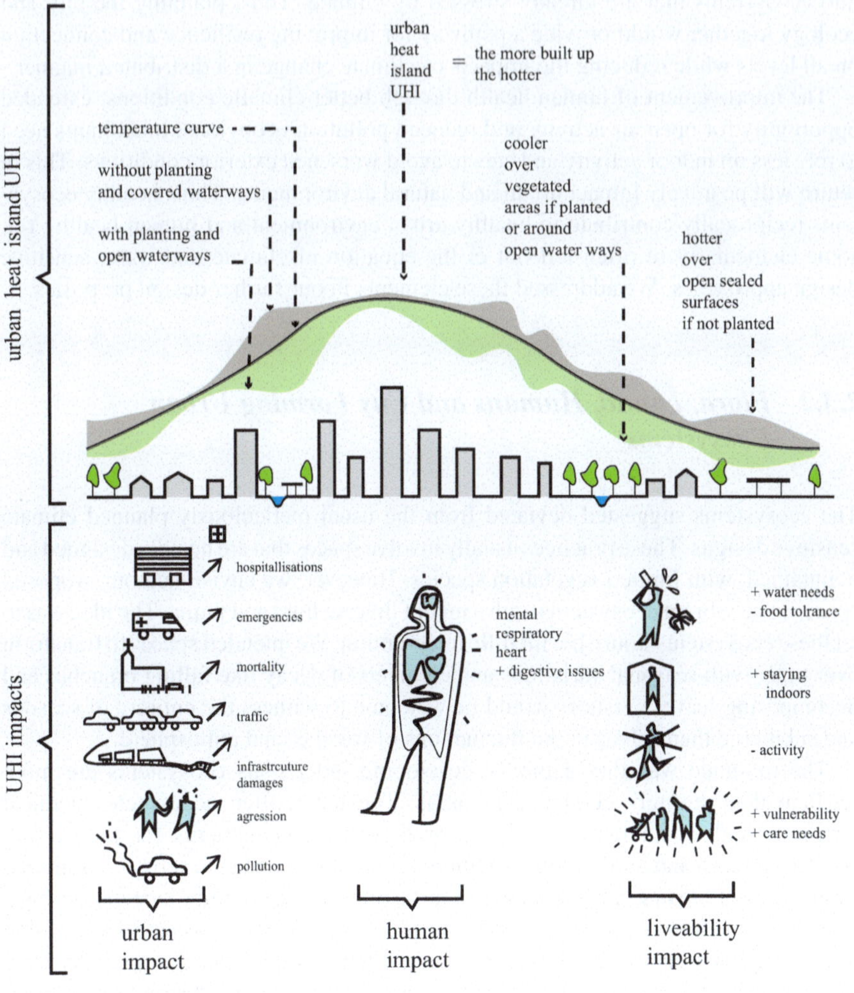

Fig. 2.5 Urban Heat Island curve with and without blue-green interventions and health and urban consequences of UHI. *Source* The Urban Heat Island and its Consequences, illustration by Maud Cassaignau and Markus Jung, 2024

and heat through shade and cooling would further prevent an overload of urban infrastructures such as wastewater pipes, air conditioners, electric systems, and transport thus improving urban system sustainability. By maintaining pedestrian comfort, such environments would encourage alternative circulation and reduce traffic stress and emissions. In return, this would prevent pollution of atmosphere and groundwater, as we can already witness in Melbourne after extreme weather events when there are discharges into the bay ([18], 3, 6). Distributed ecosystems would thus improve the health and wellbeing of humans, plants, and animals as well as infrastructures. In return, infrastructural integrity would prevent the aftereffects of system overflows into ecosystems that are already stressed by climate. Thus, planning the city and ecology together would provide a pathway for improving resilience and connection on all levels while reducing the impacts of climate change in a distributed manner.

The improvement of human health through better climatic conditions, extended opportunity for open air activity and reduced pollution is obvious. Inhabitants need to rely less on indoor activity and cars to avoid worsened exterior conditions. This in return will positively impact urban and natural environments. Thus, healthy ecosystems reciprocally contribute to healthy urban environment and human health. Yet, some elements were often left out of the equation of climate and water sensitive design approaches. We addressed these elements in our further design proposals.

2.3.2 Flora, Fauna, Humans and City Forming Urban Ecosystems

The ecosystems suggested deviated from the usual meticulously planned climate sensitive designs. These practices usually involve spaces that are heavily designed and maintained, with limited vegetation species. However, we envisioned our proposed spaces to be vibrant ecosystems with a mix of diverse flora and fauna. The idea was to let these ecosystems flourish with little intervention. We intended spaces to transform over time, with seasonal variations and elements of decay like falling branches and decomposing leaves. Visitors would be welcome to witness the entirety of seasons and submerge themselves in the fluctuations of weather and atmosphere.

The rationale was that naturally developing indigenous ecosystems are more resilient than strongly curated park areas. The latter often necessitate chemical fertilisers and weedkillers, which are problematic contaminants for fauna, soil, aquatic systems, and food chains. Additionally, carefully managed ecosystems necessitate intervention that can disrupt vital natural processes and symbiotic relationships. The Emerald Necklace Park in Boston is an example where non-native vegetation growth regularly obstructed the creek. As a result, the water ecologies were deprived of oxygen and flooded. It required heavy maintenance to remove obstructions, replace exotic with local species and restore ecological balance [45]. This example demonstrates that an approach building on native and fully functioning ecosystems best enables natural processes, adaptation and resilience. Our climatic and water sensitive approaches were envisioned with this kind of approach in mind.

Humans and urban contexts were further crucial in shaping these ecological designs in the city, where the ecosystems included plants, animals, people, water, soil, weather, and seasonality. The Emerald Necklace park network by Olmstead, which employed the concept of 'hubs and links', was successful in fighting ecological fragmentation. In an urban setting, it provides today a home for various native animals, from deer to snapping turtles, along with many replaced indigenous tree species. Therefore, it serves as a fascinating model for new urban designs that promote adaptation [14]. The surrounding urban context, its morphology, ground conditions, and overshadowing buildings played vital roles in the makeup of our proposed ecosystems. They addressed two gaps found in complementary disciplines that discuss the interaction between humans and ecosystems.

2.4 Discussion

2.4.1 Overcoming Fragmentary Ecosystems, Plant and Animal blindness

On one hand, in urban and landscape design, animals are rarely considered in the greening of urban spaces and the discourse on urban climate adaptation. Urban wildlife also remains unseen, sparsely documented, oftentimes beyond considerations of urban planning. Rather than being perceived as an asset to sustainable design, urban animals are seen as an impediment to be contended with or otherwise ignored. However, they form an integral part of native ecosystems.

While the use of indigenous plant species has become part of the accepted canon [16], most greenspaces are still designed around limited species of vegetation, even though they do not thrive in isolation. Yet, seminal forestry books [53] and talks [39] have popularised ideas that trees and fungi communicate and exchange nutrients in 'wood wide webs',[2] in a symbiotic relationship between flora and fauna, but also between larger species and microorganisms [41, 133–164]. These ideas are yet to be explored in their full complexity and importance. Despite the recognition of complex interactions between species of vegetation, fungi, microorganisms and the animal world, urban and landscape designers often overlook non-plant protagonists in the design of ecosystems in cities. Thereby they do not use the full potential and reduce by design the resilience of these ecosystems.

This results in a myopic view and only partially functioning ecosystems, which need heavy maintenance, fertilising and heavy-handed intervention to compensate

[2] Term coined by the cover of the August 1997 issue of *Nature* in reference to the paper: "Net transfer of carbon between tree species with shared ectomycorrhizal fungi" [40].

for the lack of natural nutrient exchange. Working with more diverse and self-reliant ecosystems would allow these to develop better interspecies collaboration and evolve into stable, self-regulating, and symbiotic biomes. They would have stronger resilience and better resist changing weather and climate conditions. Multispecies approaches would draw on nature's capacity to self-regulate, evolve and grow, while reducing the ongoing need for human intervention beyond initiating these ecosystems. Such ecosystems comprising flora and fauna would provide animal habitats where humans could encounter wildlife, as imagined in the imagery of Melbourne Cool Lines, teeming with birds and butterflies.

This, in return, would overcome a typical problem with revegetation strategies. The public is often blind to plant life, leading to disengagement with the relevance of plants [28, 3–51] and consequently, with revegetation strategies. Many members of the public will easily recognise and connect with different native species of birds, while unable to name the same number of vegetative species. This 'plant blindness' is a reason why physical encounters with wildlife will be much more memorable and meaningful to the public than the mere immersion in plant life. Therefore, the design of vegetation strategies as animal habitats promises to create deeper engagement through unforgettable experiences of more complete ecosystems. Ironically, to overcome our anthropocentrism, animals could form an entry point to talk about the importance of plants and forward whole ecosystem narratives, climate and revegetation strategies in cities.

On the other hand, health domains have been developing concepts that discuss the mutually beneficial interaction between humans, animals and plant environments and the interrelationship with their health. But these concepts often exclude the consideration of urban contexts where most populations reside. The 'one health' concept[3] contends that human physical and mental health is dependent on the health of flora and fauna. It discusses the interrelation between animals, ecosystems that somewhat irrationally separate out animals and environments, from medical and agricultural perspectives. How to transfer this understanding of mutually dependent and potentially symbiotic relationships to a civic realm and urban studies, adding a built environment dimension, became the subject of further exploration. We must go beyond simply retrofitting cities with vegetative ecosystems to achieve true climate resilience. In order to achieve long-term robustness of these systems, urban ecologies must encompass all domains of life.

Consideration of humans and urban contexts has been crucial in shaping the proposed ecological designs in the city. Proposed ecosystems include plants, animals and people with consideration of water, soil, weather, and seasonality. The surrounding urban context, its morphology, ground conditions, and overshadowing buildings, played vital roles in the proposed approaches. This perspective challenges the notion that ecology is incompatible with urbanity. Cities contain people, built

[3] 'ONE health' is a concept that links human, animal and environmental health. It artificially separates animals from environments they are part of, while also overlooking the role of urban environments. [53], 'One-health', accessed April 25, 2024,https://www.who.int/news-room/questions-and-answers/item/one-health.

objects and materials, but also many life forms ([30], 1–14) that are strongly interconnected. Therefore, we need to enhance our comprehension of how to coexist in mutually beneficial ways [37], while developing concepts of how these interdependences work to care for non-human species [24]. The redesigned One Health Model below (Fig. 2.6 right) aimed at integrating the animal realm into ecological city design while encompassing flora and fauna within resilient native ecosystems. It further aimed to extend the original one health (Fig. 2.6 left) concept from medical and agricultural domains to the realm of urban design and planning, with the objective of moving beyond unhelpful anthropocentric and binary paradigms of modern city planning. Developing a sense of interconnectedness and empathy towards other life forms [38] would support us in thinking and planning in more lateral ways. It would help us understand how our human actions affect other species' health as well as our own. Anthropogenic change demonstrably drives diseases though biodiversity loss, climate change, pollution, and landscape transformations [33]. Positive change, improving biodiversity and climate though naturally developing ecosystems, in return would improve health and resilience. Further, developing such an interconnected understanding would foster the potential for transdisciplinary collaboration between design and health disciplines to address these crises of our time, crises that demand new, complex, and open-minded approaches [32].

2.4.2 More-Than-Human Model and Experiences

Understanding all species, including humans, as interconnected and co-dependent helps us conceptualize ways of designing robust cities with all their interests in mind. How we may create spaces in cities where other protagonists can thrive, develop, evolve, and carve out ecological niches and habitats, became the driver for further urban landscape design studies and prototypical testing. These follow the paradigm of producing greater climate resilience for all. These are described in the second part of this article. Their approaches contrast with ideas that nature should be curated and tamed to serve human interests [55], as often implied in climate and water sensitive practice [5].

Incorporating ecosystems of plants and animals into city planning creates unexpected urban experiences of animal encounters, of natural processes and changing atmospheres that illustrate intertwined manmade and natural systems [24]. These shape unexpected perspectives and synesthetic experiences of ecosystem resilience in times of climatic and biodiversity crises. For instance, listening to aquatic wildlife from elevated board walks, seeing water birds land in wetland reservoirs, discovering multicolored parrots within a cool shaded rainforest, or witnessing the dance of bees and butterflies over wildflowers can nurture optimistic outlooks of resilience and ecologic niches.

Each urban heaven, meanwhile, would contribute to absorbing downpours, mitigating heat and sun with shading vegetation, evaporative cooling, and improving air, water, and urban noise. Visitors would witness changing atmospheres showcasing

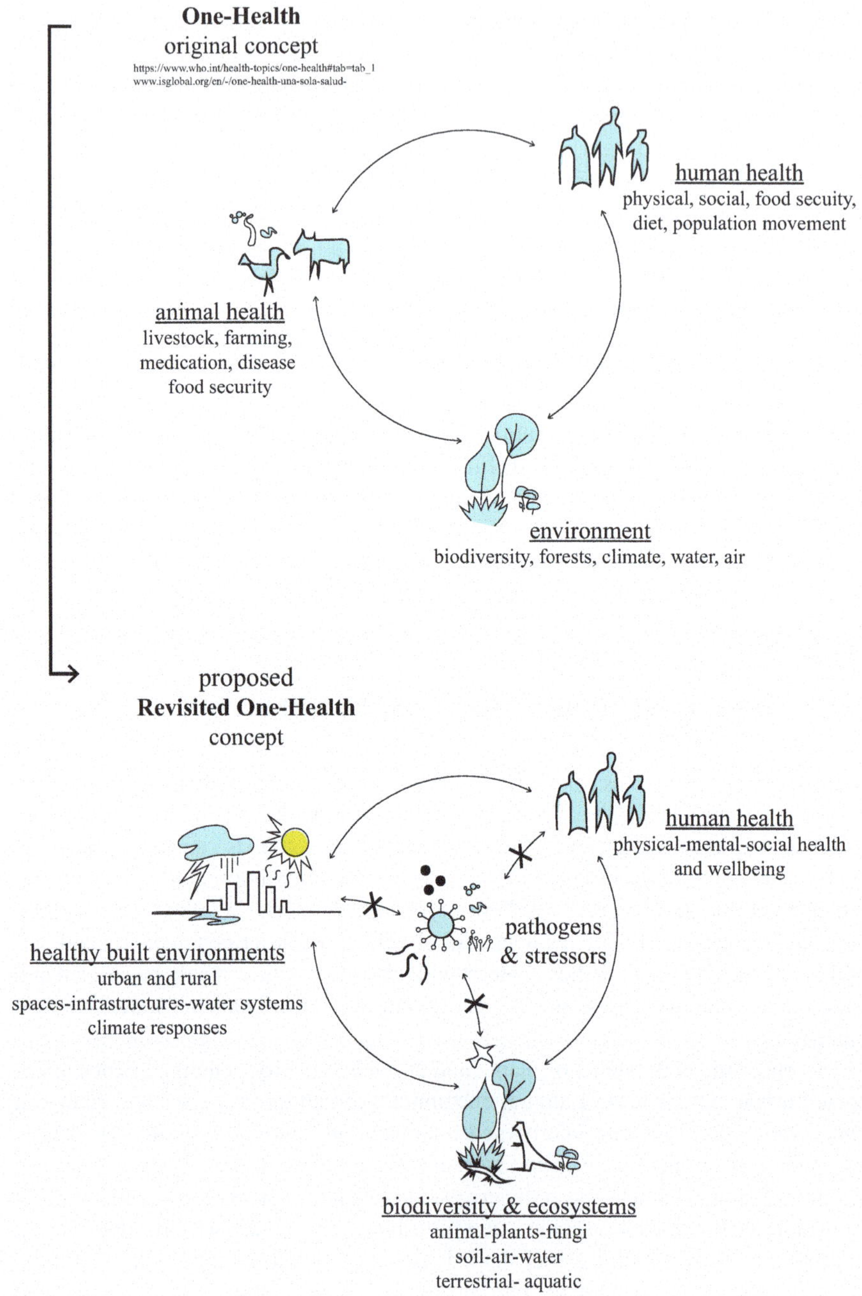

Fig. 2.6 Original and revisited one health concept. *Source* The original and the Revisited One-Health Concept, illustration by Maud Cassaignau and Markus Jung, 2024

ecological diversity and creating attachments to nature within unique spaces of more-than-human experience. The creation of these new types of urban ecosystems could support a shift of mindset away from human exceptionalism.

2.4.3 Designing and Realising Site-Specific Urban Animal Habitats

The previous concepts led to the question, how we may bring the concept of uncurated ecosystems to life in urban environments, as seen in Melbourne Cool Lines? Melbourne Cool Lines explored ways to create an urban strategy, identifying spaces for retrofitting and developing programs and general approaches. The goal of Wilding the Campus[4] was to develop a design strategy for untamed urban ecosystems in an urban context to address climate change. The term wild referred to the unrestrained nature of ecosystems rather than implying a flawed return to an unknowable original state. The approach was based on creating site-specific responses that suited the existing conditions. Questions remained about how to go about designing spontaneously developing urban landscapes with native species at a precinct scale with further details. What would be the steps to initiate the development of urban spaces where humans, animals, plants, and infrastructure coexist? The RMIT campus area next to Melbourne Cool Lines served as a real-life testing ground for the exploration of Wild Streets[5]prototypes, after developing a precinct strategy.

In Melbourne Cool Lines, found conditions had already shaped programs and the mix of urban form, landscape, and infrastructure through water sensitive and climate responsive design. In Wilding the Campus, site conditions also played a crucial role in shaping the specific design approaches towards an ecological city. Each urban environment had unique qualities that suited specific habitats and biomes. Certain spaces were both exposed and hot, while others were narrow and shaded or split by urban highways. The unique ecosystem designs were shaped by these morphologies. They resulted in the formation of ecosystems centred around specific endemic species and their unique habitats, taking into account factors such as shelter, food, predatory behavior, and evolutionary development. Examples demonstrate how these considerations led to very distinctive outcomes.

Carving Passages (Fig. 2.7) aimed to increase urban permeability for bats, birds and walkability for humans. By perforating and redesigning built volumes and façades, it would create vertical habitats for birds and bats. Refuge would be provided

[4] Design research led by Maud Cassaignau at RMIT alongside a studio with Daniel Spiteri, Jingtao Huang, Yang Xiao, Craig Danvers, Yilin Chen, Xiaomin Chen, Fangyu Chen, Ruisi Fu, Zixi Pan, Wanyi Fan, Yijia Yan.

[5] *Wild Streets* exhibition and prototypes were a collaboration between RMIT Landscape architecture (Maud Cassaignau, Heike Rahmann, Brent Greene) and Salad Dressing Landscape Architects Singapore (Yu Han Goh) supported by RMIT Place Lab (Kiri Delly, Brock Hogan, Chereese Sonkilla, Frances Gordon, Paloma Bugedo and Luke Gebert). The activity has been supported through RMIT PlaceLab funding, as part of the initiative's 'Cardigan Commons' Research Project.

by vegetated nooks, elevated crannies, and new passageways. Fungi, insects, and nocturnal creatures would find habitat in these obscured spaces. The implementation of these strategies would involve the planting of suitable trees to support the environment, food, and pollination needs of flight animals. Shaded passages and tree canopies would further increase heat sheltered spaces.

Green Bridges

The proposal for Green Bridges (Fig. 2.8) aimed to increase the habitats of urban possums, whose current territories are restricted by urban infrastructures. An urban arterial would have its tree canopies extended with a vegetated bridge. This would join habitats on both sides and provide a convenient way for pedestrians and cyclists to cross. Separate strips for pedestrians and non-humans would ensure engagement without interference. The new lush gateway would serve as a prominent symbol for the campus redesign, while providing safe shaded passage for pedestrians and cyclist.

Fig. 2.7 Carving Passages proposed a network of habitats for birds and bats. *Source* Green Passages, illustration by Yang Xiao and Jingtao Huang, permission obtained, 2023

Fig. 2.8 Green Bridges proposal extending possum habitats and pedestrian connection. *Source* Green Bridges, illustration by Craigh Danvers, permission obtained, 2023

Water Landscape

The Water Landscape (Fig. 2.9) suggested capturing rainwater from nearby building roofs and hard surfaces to create a wetlands area that would attract aquatic wildlife, insects, birds, and mammals. By leveraging the natural topography of the site, a uniquely 'wild' water-sensitive approach was implemented to retain rainwater through modulated ponds. Surrounding vegetation and water evaporation would cool the campus areas. Adjacent to the new Metro Station, a captivating and distinctive campus entrance would provide breathtaking views and study options on spacious shaded boardwalks.

2.5 Recommendations

2.5.1 Learning from Nature: Connection and Resilience

As part of the Wilding the Campus investigative process, we conducted in-depth site studies. A diverse range of site conditions and morphologies were exposed. Their capacity to support various ecosystems enabled us to develop a diverse geography of habitats. Each would offer humans unique experiences and create overlapping territories for different animal species. These varied retrofitting strategies resulted from opportunistically adjusting adaptation to specific conditions, while targeting different ecosystems, cool climate and biodiversity. The diversity of proposals reflects a new perspective on water and climate sensitive strategies. These often disregard local specificity, whereas our propositions acknowledged the importance of location-specific factors like urban morphology, micro-climate and ground conditions.

A site map collated (Fig. 2.10) these different ecosystems designed in response to found sites. It keyed those responses with the help of diagrammatic icons and two-word titles for ease of reading. The suggested ecosystem range presents opportunities for manifold explorations and wildlife encounters with different species and changing seasonal atmospheres. This providesa fertile ground for learning; witnessing how ecosystems develop and adapt to urban and climatic conditions highlights nature's dynamism. Such experiences promote climate and eco-conscious attitudes that embrace the dynamism of climate and the resilience of natural systems. Moving away from denial, rigid thinking, and paradigms of human control can thus encourage acceptance of our co-dependence with environmental and biotic systems, as illustrated in the Extended One-Health concept in the section' More than Human Model and Experiences'. It will help us embrace natural variability and change and find ways how we can harness knowledge from the plant and animal worlds to better adapt our cities to challenges of the present and future and gain true resilience in this process.

Fig. 2.9 The Water Landscape redesigns the site's topography to host aquatic species

2.5.2 Learning from and with Community: Immersion in Nature

During the process of Wilding the Campus, we thoroughly engaged with the community to test how to best communicate strategies around urban revegetation that involve the animal world. Wild Nights and Wilding Workshops were organized by Place Lab

Fig. 2.10 Map of Wilding the Campus including site locations for prototypical testing of Wild Streets

RMIT[6] to gauge appetite in the local community for campus greening to address climate and biodiversity challenges. People mapped their desired activities from reading under tree canopies, meeting outdoors, but also observing wildlife and listening to birdsong. Interestingly, they connected naturally to the idea of bringing back 'wild' ecosystems and animals to cities. Our framing and storytelling around iconic native animals supported their imagination and it turned out they were not concerned about the idea of critters or unruly vegetation.

This supported our assumption that framing ecosystems around animal habitats can form an enticing space for human connection. It helps people visualise the multiple functions vegetation can play in addressing climate and resilience and supports overcoming 'plant blindness' [28, 3–51] Further, it enticed community members to forward different ways of using urban park areas in consultation with the university [42], ways less framed around activity and more around observation, listening, being immersed, and connecting to place and other beings. This resonates with ideas of mindfulness, resilience building, forest bathing [36] and ecotherapy: calm approaches that promise improved health and wellbeing through immersion in the moment and in nature. In our hectic world, such occupations can complement more activity-based outdoor approaches to health that are often forwarded in outdoor design.

At the end of Wilding the Campus, RMIT Place Lab co-organised a public exhibition and talks, where we presented our works and ideas, but also launched four prototypes (Fig. 2.11) These tested how to initiate and plant four native ecosystems framed around a local mammal, bird, reptile and insect. Four habitats were incubated in mobile prototype boxes. Shrubland was destined for the marsupial bandicoot, a stepped micro-forest for the rainbow lorikeet, dry grassland with gabion wall for the skink, and an inclined flower meadow for the golden ray butterfly and its caterpillars. Here again, the local and professional community was invited to the process of planting, a further opportunity to discuss and engage with nature-based approaches. The event proved to be an activity for all. The four incubator boxes thus allowed us to test in reality, how a framing around wildlife enables people to better connect with urban greening, climate and resilience strategies, even when the animals are not yet present and just imagined through storytelling.

2.5.3 Benefits of Real-Life Concept Testing

During the following month, over the Australian summer, the prototypical boxes developed with minimal interaction apart from the occasional watering that the jardiniere planting demanded. It became interesting to observe that the plants thrived

[6] PlaceLabRMIT is an urban initiative from RMIT University to connect communities and shape places on RMIT's campuses. It operates as a spatial, digital, and social network. It incorporates a 'living lab' delivering through a focus on Places People Projects Platforms and Partnerships, working with local government, industry, and community, alongside academics and students [42].

Fig. 2.11 Wild Streets community planting event in front of exhibition launch

despite the less-than-ideal planting season that had resulted from the funding cycles, branching out beyond the boxes in fuzzy tangled vegetation. Birds and fluttering insects were attracted; we have yet to witness lizards or mammals. The growth of plants after a few months has become such that the native plant ecosystems need more space and relocation. We found a more permanent home, next to a Melbournian school, where they will potentially attract more extensive wildlife for the pleasure of the surrounding pupils and communities. Thus, we will be able to demonstrate how to close the circle between reinvigorated urban environments, ecosystems, and human health and education. Meanwhile, we were also able to observe how locals and students started to dwell much longer around our prototypes, seeing them as an agreeable backdrop for socialising, meeting, waiting, or other needs. This confirms for us the beneficial effect of such environments on the human psyche (Fig. 2.12).

2.6 Conclusion: Cities as Intertwined Ecosystems that Need Less Controlling, Holistic Approaches

The studies have provided a chance to view cities as ecosystems consisting of interconnected plants, animals, humans, and urban components. The Extended One Health model and more than human perspectives propose to review the human-centered approach in Water Sensitive and Climate Design, while recognising their immense role in highlighting the importance of natural systems in cities to address climate and biodiversity challenges with the goal of building resilience. These design practices have commodified ecosystems and waterways as infrastructures for solving manmade

Fig. 2.12 Wildly developing plants of the 4 prototypes attracting street dwellers

problems [56, 12–16). They have curated and controlled ecosystems. Through calculations, they have measured their effectiveness and transformed them into non-human capital and labour ([31], 364–365).

We suggest relinquishing this commodification and control while extending their unquestioned contributions to improving climate and biodiversity. What strategies can we implement to initiate and allow ecosystems to thrive and adapt in unpredictable urban and climatic conditions? How can we ensure autonomy for such open-ended processes? We need to be more accepting and acknowledge that our cities are connected to the ecosystems around and within, including the plant and animal life we impact. Eventually, their poor health affects us, polluting our urban systems and impacting our own well-being, as evident from pandemics, wildfires, droughts, and floods. By exploring ways to connect these diverse realms within an Extended One-Health Concept in symbiotic ways, we can cultivate flourishing life in plants, animals, and humans near and within cities. This might enable us to bridge the gap between the human and natural worlds and manmade systems and ecosystems to find more resilient ways to address climate and biodiversity crises. It might help us develop better modes of cohabitation [25, 237–260]. It might empower us to envision more inclusive models of planning, designing, and envisioning other species' value [1, 163–169] while refraining from instrumentalising them ([29, 476). Further, if we accept ecosystems have inherent intelligence to adapt and evolve, we may reframe our roles as initiators and enablers, designing frameworks and conditions for other beings to flourish amongst our cities. And we may redirect our controlling energies towards expanding areas for urban ecosystems, so that they form continuous networks and emerald oases[7] throughout our cities.

[7] In reference to Frederick Olmstead's "Emerald Necklace" in Boston, Emerald Necklace [17].

References

1. Barron E, Hess J (2020) Non-human labour: the work of Earth others. In: Gibson-Graham JK, Dombroski K (eds) The handbook of diverse economies. Edward Elgar Publishing, Cheltenham, pp 163–169
2. Cassaignau M, Jung M, Xue M (2018) Building Mixity! 37.83 S/144.993E. Melbourne: Monash University Publishing. Castree, N (2003) Environmental issues: relation alontologies and hybrid politics. Progress in Human Geography 27(2).https://doi.org/10.1191/0309132503ph422pr
3. Cassaignau M, Jung M, Morgan P (2016) The Kunshan Projects, Water responsive urbanism as driver for academic design innovation, Project to Practice: Innovating Architecture. Sydney: UTS, 56–64
4. City of Yarra (2019) Amendment C191-Swan Street, https://www.yarracity.vic.gov.au
5. CSRIO (2006) Water sensitive urban design, storm water management systems. https://www.publish.csiro.au/ebook/chapter/SA0601047. Accessed 3 April 2024
6. CABE (2024) Urban Green Nation: building the evidence base. https://www.designcouncil.org.uk/fileadmin/uploads/dc/Documents/urban-green-nation-summary1_0.pdf. Accessed 03 April 2024
7. CRC for water sensitive cities. Impacts of water sensitive urban design solutions on human thermal comfort green cities and microclimate. https://watersensitivecities.org.au/wp-content/uploads/2016/07/TMR_B3-1_WSUD_thermal_comfort_no2.pdf. Accessed 03 April 2024
8. City of Melbourne (2024) Elizabeth Street flooding is a serious issue. https://urbanwater.melbourne.vic.gov.au/industry/our-strategies/elizabeth-street-catchment-iwcm-plan/. Accessed 03 April 2024
9. City of Melbourne (2021) Good design guide for flood affected areas.https://participate.melbourne.vic.gov.au/amendment-c384/good-design-guide-flood-affected-areas (2021). Accessed 23 April 2024
10. City of Melbourne (2024) Cool Routes, accessed June 25, 24 https://www.melbourne.vic.gov.au/parking-and-transport/streets-and-pedestrians/Pages/cool-routes.aspx
11. City of Melbourne (2024) Cooling the City accessed June 25, 2024, https://www.melbourne.vic.gov.au/about-melbourne/sustainability/Pages/cooling-our-city.aspx#:~:text=Our%20goal%20is%20to%20plant,in%20a%20process%20called%20transpiration.
12. Creed B (2017) Stray, human-animal ethics in the Anthropocene. University of Sydney, Power Polemics Publications, Sydney
13. Chen Y, Jones CA, Dunse NA, Li E, Liu Y (2023) Housing prices and the characteristics of nearby greenspaces. Land 12(1):496.https://doi.org/10.3390/land12020496
14. Cocca C (2020) Creating the Emerald Necklace. Constructing nature: the Emerald Necklace. https://storymaps.arcgis.com/stories/7942e99f4f344ca083422b302e03f3ce
15. Deilami K, Kamruzzaman MD, Liu Y (2018) Urban heat island effect: a systematic review of spatio-temporal factors, data, methods, and mitigation measures. Int J Appl Earth Obs Geoinf 67:30–42. https://doi.org/10.1016/j.jag.2017.12.009
16. Department of Environment South Australia (2021) The benefits of native gardens. https://www.environment.sa.gov.au/goodliving/posts/2019/08/benefits-of-native-gardens. Accessed 26 April 2024
17. Emerald Necklace Conservancy (2024) The Emerald Necklace. https://www.emeraldnecklace.org/park-overview/emerald-necklace-map/. Accessed 29 April 2024
18. Environmental Protection Authority Victoria (2021) Current long term water qualities for the Yarra River and the Bay. https://www.epa.vic.gov.au/-/media/epa/files/publications/2036---long-term-recreational-water-quality---port-phillip-bay-beaches-and-yarra-river.pdf?la=en&hash=8FA9894DAC6B41C54CA824A9EC9F0655. Accessed 24 April 2024
19. Frank T (2019) After a $14-Billion Upgrade, New Orleans' Levees Are Sinking, Scientific American, New York, accessed June 25, 24 After a $14-Billion Upgrade, New Orleans' Levees Are Sinking | Scientific American
20. Fraser, M (2013), Introduction. Design research in architecture. Franham, Ashgate, 1

21. Frayling, C (1993) Research in art and design. Royal college of art research papers, vol 1 no.1 London, 1–5
22. Gruen L (2015) Entangled empathy: an alternate ethic for our relationships with animals. Lantern Books, New York. https://doi.org/10.1017/CBO9781139506755.023
23. Haraway D (2003) The companion species manifesto. University of Chicago Press, Chicago
24. Haraway D (2008) When species meet. University of Minnesota Press, Minneapolis
25. Hucy W, Mathieu N, Mazellier T et al (2005) L'habitabilité des milieux urbains. Un objet au croisement des disciplines (The live ability of urban environnements, a subject at the intersection of disciplines). In: Mathieu N, Guermond Y (eds) La Ville durable, du politique au scientifique (The sustainable city, from the politcial to the scientific). Cemagref/Cirad/Ifremer/Inra, Paris, pp 237–260
26. Jamei E and Tapper N (2019) Chapter 19—WSUD and urban heat island effect mitigation. In: Sharma AK, Gardner T, Begbie D (eds) Approaches to water sensitive urban design, Woodhead Publishing, https://doi.org/10.1016/B978-0-12-812843-5.00019-8, pp 381–407
27. Johar I (2003) Wild hope lectureRMIT. https://www.youtube.com/watch?v=zQJjfCSPvJI. Accessed 24 April 2024
28. Jose S, Wu CH, Kamoun S (2019) Overcoming plant blindness in science, education, and society. Plantspeopleplanet. https://doi.org/10.1002/ppp3.51
29. Kagan S (2011) Art and sustainability: connecting patterns for a culture of complexity, Verlag, Bielefeld
30. Latour B (2014) Agency at the time of the anthropocene. New Lit Hist 45(1):1–18
31. Legrand M (2015) La mise enordreecologique des parcs urbains: savoirs, pratiques et paysages: exemple d'un grand parc francilien (The ecological ordering of urban parks: knowledge,practices,landscapes:thecaseofalargeparkin Parisregion–France).Dissertation, MuseumNationald'HistoireNaturelle, Paris
32. Mau B (2011) You can DO better! Architect. https://www.architectmagazine.com/design/you-can-do-better_o. Accessed 25 April 2024
33. Mahon MB, Sack A, Aleuy AO et.al (2024) A meta-analysis on global change drivers and the risk of infectious disease. Nature.https://doi.org/10.1038/s41586-024-07380-6
34. Nesshöver C, Assmuth T, Irvine KN, et al. (2017) The science, policy and practice of nature-based solutions: an interdisciplinary perspective. Science of the Total Environment 579(1):1215–1227
35. Palazzo E, Rani WN, Mardiah WM (2017) Regenerating urban areas through climate sensitive urban design. Adv Sci Lett 23(7), 6394–6398. American Scientific Publishers. https://doi.org/10.1166/asl.2017.9277
36. Parks Victoria (2024) Forest bathing. https://www.parks.vic.gov.au/healthy-parks-healthy-people/nature-benefits/mental-health-and-wellbeing/forest-bathing. Accessed 29 April 2024
37. Puig de la Bellacasa M (2017) Matters of care: speculative ethics in more than human worlds. University of Minnesota Press, Minnesota
38. Rifkin J (2009) The empathic civilization: the race to global consciousness in a world in crisis. Penguin Press, Tarcher
39. Simard S (2016) How trees talk to each other. Ted Talk.https://www.ted.com/talks/suzanne_simard_how_trees_talk_to_each_other?language=. Accessed 26 April 2024
40. Simard S, Perry DA, Jones MD et al (1997) Net transfer of carbon between tree species with shared ectomycorrhizal fungi. Nature 388:579–582
41. Simard S, Martin K, Vyse A et al (2013) Meta-networks of fungi, fauna and flora as agents of complex adaptive systems. In: Puettmann K, Messier C, Coates KD (eds) Managing world forests as complex adaptive systems: building resilience to the challenge of global change. Routledge, New York, pp 133–164
42. Sonkkila C, Bugedo P, Gebert L (2023) Cardigan Commons wild streets zine. PlaceLab RMIT. https://placelab.rmit.edu.au/app/uploads/2023/12/CC-ZINE.pdf. Accessed 09 April 2024
43. Sonkkila C, Bugedo P, Gebert L (2023) Cardigan Commons research project summary", PlaceLab RMIT. https://placelab.rmit.edu.au/app/uploads/2024/02/Cardigan-Commons-Summary-Report-%E2%80%93-Digital.pdf. Accessed 09 April 2024

44. Steele W, Wiesel I, Maller M et al (2019) More than human cities: where the wild things are. Geo forum Special Issue106:411–415.https://doi.org/10.1016/j.geoforum.2019.04.007
45. Stormwater Solutions (2002) Managing storm flows in Boston and Brookline's Historic
46. Emerald Necklace, accessed May 25, 2024. https://www.stormwater.com/bmps/article/13001792/managing-storm-flows-in-boston-and-brooklines-historic-emerald-necklace
47. Van Heerden IL (2007) The failure of the New Orleans levee system following hurricane Katrina and the pathway forward. Public Administr Rev 67:24–35. https://doi.org/10.1111/j.1540-6210.2007.00810.x
48. Victorian Planning Authority (2019) Cremorne Place Implementation Plan. Accessed June 25, 24. https://vpa.vicgov.au/project/cremorne
49. Whatmore S (2006) Materialist returns: practicing cultural geography in and for a more-than-human world. Cultural Geographies 13(4):600–609
50. Wilson O, Hughes O (2011) Urban green space policy and discourse in England under New Labour from 1997 to 2010. Plan Pract Res 26(2):207–228
51. Willingham R (2022) As Melbourne clean-up begins, concerns raised over whether Flemington Racecourse wall exacerbated flooding, ABC, accessed June 25, 2024 As Melbourne clean-up begins, concerns raised over whether Flemington Racecourse wall exacerbated flooding—ABC News
52. WHO (2024) One-health. https://www.who.int/news-room/questions-and-answers/item/one-health. Accessed 25 April 2024
53. Wohlleben P (2015) The hidden life of trees, what they feel, how they communicate: discoveries from a secret world. Harper Collins, New York
54. Wolch J (1998) Zoöpolis. In: Wolch J, Emel J (eds) Animal geographies: place, politics and identity in the nature-culture borderlands. Verso, London, pp 275–301
55. Wolfe C (2010).What is post humanism? University of Minnesota Press, Minneapolis
56. Wong T, Eadie ML (2000) Water sensitive urban design: a paradigm shift in urban design. Proceedings of the 10th World Water Congress, pp 12–16
57. Wright L (2021) Architecture sharing space. Antennae: J Nat Vis Culture 56:46–54, https://www.academia.edu/68448550/Antennae_Issue_56_Species_and_Spaces
58. Yadav N, Rajendra K, Awasthi A, Singh C, Bhushan B (2023) Systematic exploration of heat wave impact on mortality and urban heat island: a review from 2000 to 2022, Urban Climate, vol 51.https://doi.org/10.1016/j.uclim.2023.101622

Chapter 3
Building Resilient Urban Futures: Adapting Cities to Climate Change Challenges

Anshul Jaiswal, R. Sagar, Aakansha Pandey, Deepankar Yadav, Mohd. Sirazuddin Ansari, and Richa Rawat

Abstract As urbanization continues to accelerate worldwide, cities face increasing challenges posed by climate change. Urban areas, as hubs of economic, social, and environmental activity, are particularly vulnerable to the impacts of climate change, including extreme weather events, rising temperatures, and sea-level rise. In response, the concept of urban resilience has emerged as a critical framework for understanding and addressing these challenges. This abstract explores the nexus of urban resilience and climate change, focusing on the strategies, challenges, and opportunities for building resilience in cities. It examines the various dimensions of urban resilience, including infrastructure, governance, social dynamics, and economic systems, and highlights the importance of integrating climate adaptation and mitigation efforts into urban planning and development processes. Key strategies for enhancing urban resilience include investments in green infrastructure, adoption of nature-based solutions, promotion of sustainable urban design and land-use planning, and strengthening of social cohesion and community engagement. However, numerous challenges hinder the implementation of these strategies, such as limited financial resources, institutional barriers, political constraints, and social inequalities. Nevertheless, amidst these challenges lie opportunities for innovation and transformation. Cities have the potential to leverage emerging technologies, harness data-driven approaches, and foster collaboration across sectors and stakeholders to enhance their resilience to climate change. Furthermore, by adopting a holistic and inclusive approach to urban resilience, cities can not only mitigate the impacts of climate change but also create more equitable, sustainable, and livable urban environments for all residents.

Keywords Climate change · Urban planning · Resilience · Nature-based solutions · Sustainable urban design · Adaptations and challenges

A. Jaiswal · R. Sagar (✉) · A. Pandey · D. Yadav · Mohd. S. Ansari · R. Rawat
Department of Botany, Banaras Hindu University, Varanasi 221005, India
e-mail: sagarbhu@gmail.com

S. Ghosh et al. (eds.), *Cities of Tomorrow: Urban Resilience and Climate Change Preparedness*, Urban Sustainability, https://doi.org/10.1007/978-981-97-9658-8_3

3.1 Introduction

Modern society is facing profound impacts from changing climatic conditions, including increasing global temperatures and more severe weather patterns [18]. These shifts are particularly taxing on urban areas, where dense populations and specific infrastructure make cities highly vulnerable to climate-related stresses. Changes in land and ocean temperatures have already inflicted significant damage on cities worldwide, with projections indicating more severe consequences ahead. Urban areas, due to their expanding populations, specific building materials, and land use patterns, face heightened risks from climate variability. Extreme weather events and gradual climate changes threaten urban livelihoods by damaging critical infrastructure and disrupting economic activities, potentially halting business operations for extended periods [28]. The economic toll of climate impacts is substantial, illustrated by events like the 1998 floods in Dhaka, Bangladesh, which cost firms millions and significantly impacted smaller sectors [4]. The urban poor are disproportionately affected by climate impacts, struggling to maintain livelihoods when disrupted by extreme weather events or slower-onset changes. These vulnerabilities vary by location and preparedness, with cities facing threats ranging from rising sea levels to infectious diseases and natural disasters. In response to these challenges, urban resilience has emerged as a critical concept. Resilient cities possess the ability to endure and recover swiftly from disruptions, maintaining essential functions during crises [6]. Strategies for building urban resilience involve learning from past experiences and integrating resilience into urban planning frameworks [87]. Urban resilience encompasses various approaches such as persistence, transformation, and adaptation, emphasizing cities' abilities to withstand and adapt to hazards from both human activities and natural disasters [21]. Discussions on urban resilience often adopt a systemic approach, highlighting the interconnected nature of urban ecosystems and the need for comprehensive risk management [10, 62].

As cities continue to grow and face increasing stresses from climate change and other risks, innovative urban planning strategies become imperative. Enhancing resilience through advancements in infrastructure, housing, transportation, and disaster management is crucial for cities to mitigate environmental challenges and ensure sustainable development. Failure to address these issues effectively could lead to severe economic and social repercussions for urban centers worldwide. Therefore, prioritizing resilience-building strategies is essential to safeguarding urban populations and infrastructure amidst a changing climate landscape.

3.2 Understanding Climate Change Challenges for Urban Areas

3.2.1 Climate Change Threats to Cities

Short-term fluctuations in temperature, precipitation, and wind within specific geographic areas are crucial in assessing urban vulnerability [14]. Extreme heat, a major environmental threat globally, is exacerbated by climate change, leading to more heat-related illnesses and deaths [43]. Temperature and precipitation variations also impact the spread of disease vectors like malaria-carrying mosquitoes, increasing infectious and diarrheal diseases. Heat reduces labor productivity, with the agriculture and construction industries being most affected. Construction is expected to lose about 19% of global working hours within the next decade due to heat stress, defined as temperatures above 35 °C, especially in humid areas. Extreme weather events, including severe temperatures, droughts, floods, wildfires, and storms, can severely disrupt urban economies and health. Climate change has increased the frequency and intensity of floods globally [9]. Urban floods, often from intense rainfall in poorly drained areas, are known as pluvial floods [76], while fluvial floods are caused by river overflows and are highly costly [27]. Compound flooding, resulting from fluvial floods and high coastal water levels, severely affects low-lying coastal cities [32]. By 2050, over 570 low-lying coastal communities are projected to experience sea level rises of at least 0.5 m, putting up to 800 million people at risk from storm surges and rising seas.

3.2.2 Impacts of Climate Change on Urban Infrastructure, Ecosystems, and Communities

Climate change significantly impacts city infrastructure, utilities, and overall quality of life, especially in low-lying coastal regions, which comprise just 2% of the Earth's surface but face immense challenges. These areas suffer from coastal erosion, saltwater intrusion, and construction damage, making some places uninhabitable and threatening buildings, pipelines, and infrastructure due to subsidence [64]. Rising temperatures, extreme weather, and sea level rise jeopardize roads, bridges, and buildings. Urban ecosystems are also at risk, with changes in temperature and precipitation disrupting local flora and fauna, leading to biodiversity loss and impaired ecosystem services like water purification and flood regulation. These shifts worsen urban heat islands, pollution, and reduce green spaces, adversely affecting residents' quality of life. Communities, especially in low-lying coastal areas, face degraded infrastructure and less resilient ecosystems, impacting well-being and livelihoods. Vulnerable populations, including low-income and marginalized groups, are disproportionately affected, facing greater health risks, displacement, and economic instability. This situation underscores the urgent need for comprehensive adaptation and mitigation strategies to build resilient cities [35]. Severe storms and heavy rainfall cause fatalities, injuries, and homelessness, with rising cases of PTSD and anxiety, as seen in recent floods in the Philippines, Pakistan, and Brazil. Climate change also exacerbates food and water shortages in cities like Delhi, Durban, and Mombasa, and increases hunger and water scarcity in Latin America, where income inequality is high [84]. Poor and disadvantaged populations are the most vulnerable and unprepared to deal with climate change effects, making it crucial to integrate climate change hazards into urban planning and infrastructure development to mitigate these impacts.

3.2.3 Vulnerable Urban Sectors

Cities are significantly impacted by climate change, with certain urban sectors being more vulnerable to various climate-related risks than others. For instance, while pluvial flooding might not directly affect public health, it can disrupt transportation systems and industrial activities. Heatwaves, one of the most significant health risks from climate change, also strain electricity and water supplies due to increased demand. Consequently, an assessment of potentially vulnerable urban sectors and activities should include industries, disaster risk management, public welfare, social harmony, urban design (including blue and green infrastructure), architecture, energy, transportation, biodiversity and conservation, higher education, tourism, insurance and the financial sector, and communication and information technology. Climate change exacerbates existing social and economic challenges in cities, such as hunger, poverty, resource demand, and economic instability. It deepens socioeconomic disparities, strains limited resources, prolongs poverty, complicates service delivery, and hinders the implementation of equitable and sustainable policies. Increasing urbanization increases the need for essential resources like food and drinking water, which could be reduced due to climate change. The combined social and economic impacts, including industrial damage, displacement, and loss of livelihoods, can progressively disrupt urban community functions [63].

3.3 Resilient Urban Planning

3.3.1 Definition and Its Significance

Resilience is crucial for urban sustainability, balancing human, economic, environmental, and social factors [70]. The United Nations describes resilience as the capacity for people, families, neighborhoods, urban centers, organizations, frameworks, and societies to thwart, endure, integrate, adjust to, tackle, and bounce back successfully from various threats. This capability is essential for preserving a functional state that does not hinder future objectives for lasting growth, harmony, safety, rights, and the well-being of everyone. Defined by [5] as a city's ability to withstand and recover from change, it includes adaptive abilities for maintaining efficiency post-disaster [20, 56]. Urban systems must quickly regain functionality after climate-related disruptions [46]. This resilience involves complex systems requiring strategic governance [53, 83] and is vital in managing socio-ecological systems sustainably amid climate change. Resilience is seen positively, contrasting with concepts like vulnerability [52]. It emphasizes systems' abilities for self-organization, regeneration, and development. According to the IPCC (2007), resilience is the ability of a social or environmental system to recover from disruptions while maintaining structure and function, organizing itself, and adjusting to stress and change. The IPCC's definition aligns with engineering resilience, focusing on a system's ability to return to equilibrium after disruption [38, 39]. Ecological resilience, however, emphasizes a system's capacity to learn, adapt, and organize itself while absorbing disturbances [2]. Shifting from innate robustness to planned resilience involves focusing on ecological resilience rather than engineering resilience [86]. [51] term this "bounce-forward," indicating transformative changes after catastrophes, assuming multiple system equilibria. Evolutionary resilience, emphasizing the system's changing nature over time regardless of external disturbances [26], focuses on dynamic processes within the system rather than a static state. This concept, termed "transform-forward," highlights transformation, regeneration, and development [30]. Systems with greater resilience are less vulnerable to shocks and stresses, as greater resilience decreases vulnerability [30]. Resilience and vulnerability originate from different perspectives: resilience from biophysical sciences, and vulnerability from structuralist social sciences and political environments [54]. Gallopin [31] argues that vulnerability is more inclusive than resilience, aligning with the idea of adaptive capacity.

3.3.2 Key Attributes Defining Resilient Cities

To achieve, retain, and strengthen resilience in urban systems, they should exhibit several key characteristics as identified by Sharifi and Yamagata [71]. Firstly, robustness and stability are essential, providing the ability to endure both short-term and long-term disturbances. Flexibility is also crucial, allowing urban systems to reorganize their structure and functions during such disturbances. Resourcefulness is another important trait, ensuring the availability of resources needed to enhance resilience. Coordination capacity plays a vital role in ensuring the optimal use of resources by residents, organizers, and decision-makers. Additionally, redundancy is necessary, with backup components available to replace any that fail. Diversity within the system is important, incorporating multiple components that can compensate for each other's limitations. Foresight capacity is required to prepare for unexpected disruptions, while independence ensures self-sufficiency to withstand adversity. Connectivity builds connections with larger-scale systems for rapid recovery, and collaboration emphasizes inclusive, grassroots urban management strategies. Agility is the ability to quickly recover functionality after disturbances, and adaptability involves incorporating "living with risk" into daily activities and planning. Self-organization through community-based activities focused on social networks is essential, as is creativity in finding novel approaches to tackle emerging challenges. Efficiency ensures that benefits are maximized within limited resources, and equity guarantees the fair distribution of benefits and impacts among societal groups. These characteristics involve trade-offs and synergies. For example, increasing redundancy might reduce efficiency, and balancing independence with connectivity can be challenging. Further research is needed to understand these trade-offs and synergies and clarify how these features support planning/preparation, absorption, recovery, and adaptation in resilient urban systems.

3.4 Strategies for Responding Cities to Climate Change Challenges

3.4.1 Green Infrastructure Development to Mitigate Urban Heat Island Effects and Manage Stormwater

Physical Infrastructure, Infrastructural Land Use, and Structural Design fall under physical resilience [58]. The National Infrastructure Plan addresses the economic, environmental, and social impacts of climate change on infrastructure, highlighting risks and financial consequences from recent floods and extreme weather events. Long-lasting urban infrastructure requires careful planning to avoid outdated systems. Redundancy allows for the replacement of damaged components, enhancing resilience. However, robust infrastructure can lead to complacency. Adaptability in urban environments and organizations boosts efficiency and diversity, crucial for absorbing shocks and promoting recovery. Regular monitoring alerts stakeholders to the need for redesign, renovation, and technological updates to improve infrastructure efficiency [71]. Communication and transportation networks are vital for resilience [56]. Investing in resilient infrastructure involves upgrading and redesigning critical systems, ensuring redundancy, and using natural solutions. Promoting green infrastructure, like gardens and urban forests, improves air quality, reduces heat, and lowers flood risk. Energy-efficient, sustainable practices should be incorporated into infrastructure planning, especially for transportation and buildings. Green roofs mitigate the heat island effect, enhance air quality, and often come with tax incentives. Vegetation, including trees and green roofs, provides shade, repels solar radiation, and releases moisture, reducing urban heat island effects.

3.4.2 *Enhancing Urban Mobility and Transportation Systems to Reduce Greenhouse Gas Emissions*

The transportation industry stands as a significant contributor to global greenhouse gas emissions, accounting for 57% of global oil demand and 28% of total energy consumption according to the UN. Between 2000 and 2019, CO2 emissions from transportation rose in most regions except Europe, where they declined by 2%. To combat climate change, integrating electric vehicles (EVs) is crucial. EVs produce zero tailpipe emissions, reducing greenhouse gases and improving air quality. Moreover, smart transportation leveraging ICTs has garnered substantial global investment. These technologies enhance road network efficiency, reduce emissions, and alleviate traffic congestion [50]. ICT solutions like computerized traffic lights and congestion pricing incentivize sustainable transport choices, influencing vehicle efficiency and driving behaviors [17, 25]. On-board vehicle systems such as pay-as-you-go driving and eco-driving measures further optimize energy use and reduce environmental impact [8, 36]. In conclusion, the transportation sector plays a pivotal role in addressing climate change. Through the adoption of EVs and smart transportation technologies, the industry can significantly promote energy conservation, minimize pollution, and contribute to a sustainable future.

3.4.3 *Promoting Sustainable Building Design and Energy Efficiency Measures*

Green or sustainable buildings are created with the least amount of negative environmental impact possible while utilising the most resources possible. These structures maintain environmental responsibility throughout their whole life cycle, from creation to operation to ultimate removal, and place a high priority on energy and water efficiency. They stand for the peaceful coexistence of nature and human endeavours. Energy efficiency is given top priority in sustainable buildings in order to reduce energy consumption and environmental effect. Building performance and occupant comfort can be improved by architects and designers through the use of various energy-efficient techniques and technology. The best possible temperature management and airflow are ensured by energy-efficient HVAC systems that include technologies like energy recovery ventilation and heat pumps. Energy-efficient fixtures, light sensors, and smart controllers maximise natural light while using less electricity in lighting design. To further increase energy efficiency, choose energy-efficient equipment and incorporate smart building management systems. Three essential policies are being promoted globally to ensure sustainable growth. Building Energy Performance Certification (BEPC) Programmes: intended to ensure the wider diffusion of novel innovations in combination with advanced energy technologies; Energy Efficiency Standard (EES): developed to boost energy efficiency; Feed-in-Tariff

(FiT): executed to motivate consumers to speed up the adoption of renewable energy [42].

3.4.4 Strengthening Social Resilience Through Community Engagement and Equitable Adaptation Strategies

The term "social resilience" describes the complex interactions between humans and their environment [30], impacting a society's ability to endure natural disasters like hurricanes and floods [59]. It encompasses the ecological, social, and psychological diversity within communities, allowing them to adapt and maintain essential functions under stress. Resource distribution plays a critical role in community vulnerability, uneven environmental damage can hinder effective hazard mitigation [33]. Disadvantaged communities face greater risks and post-disaster challenges. Calamities, defined by [45] as spatial and temporal events causing harm and disruption, can destabilize social structures, potentially leading to conflicts. Understanding social networks and organizations before, during, and after a crisis is vital. While all types of capital—human, physical, and social—are affected by disasters, social capital is the least impacted. Thus, rescue efforts should focus on enhancing social assets by involving residents as active participants [29]. Trust in official information and among citizens is crucial for rebuilding post-disaster [73]. Strong community relationships foster trust and cooperation before and after disasters, enhancing stability [16]. Social resilience is influenced by factors such as social networks, communication channels, learning ability, problem-solving skills, group activities, and flexibility [16, 56, 61]. Indicators of social resilience include collaborative culture, population distribution, intergenerational relationships, cultural diversity, communal unity, spontaneous organization, educational attainment, awareness levels, interpersonal engagement, poverty levels, income levels, community attachment, language proficiency, religious adherence, ethical behavior, healthcare system effectiveness, and healthcare availability [41]. Recognizing local cultures and traditions is crucial in enhancing adaptive resilience.

3.5 Tools and Technologies for Resilient Urban Development

3.5.1 Role of Data Analytics, Remote Sensing, and Geographic Information Systems (GIS) Inclimate Risk Assessment and Urban Planning

Experts can closely monitor climate change risks using Geographic Information Systems (GIS) tools and software. Detailed geospatial data and visualizations provide essential insights for institutions and government agencies preparing for future challenges. By combining GIS with meteorological studies, geospatial analysts can identify opportunities that benefit future generations. Additionally, technology for time-critical information significantly enhances disaster management by supporting vital decision-making during emergencies.

Traditional manual maps are often inefficient due to the labor, time, and effort required for updates and maintenance. In contrast, remote sensing is highly effective, enabling the creation, digitization, and storage of maps within a GIS environment. High-resolution satellite and aerial imagery can be utilized to detect wildfires, track floods, and assess disaster damage. Hazard mapping identifies risk and impact areas, while Digital Elevation Models (DEM) in geo-environmental research serve as foundational tools to create diverse maps. These layers can be incorporated into a GIS database for disaster planning, facilitating the identification of affected areas and the assessment of extensive damage to life and infrastructure. During extreme weather events, geospatial intelligence supports emergency responders and rescue workers. Risk maps from remote sensing data and satellite photography inform authorities of current conditions and highlight regions needing immediate attention. Response crews provide live reports and images, ensuring rapid and effective crisis management. After a large storm or fire, geographic information aids utility providers in restoring services and planning for crop regeneration or infrastructure reconstruction [13, 22].

3.5.2 Integration of Climate Modeling and Scenario Planning into Urban Development Processes

Utilizing predictive modeling for land changes to inform scenario-based planning has proven effective in enhancing preparedness for uncertainties associated with urban development, including urban sprawl and flood risk, compared to relying solely on comprehensive plans.

Scenario planning includes strategic planning that integrates technologies to navigate uncertain futures by developing alternative growth pathways. Scenarios have been extensively employed in urban settings for a variety of purposes, including land utilisation, transport, economic growth, ecological systems, resilience, and others [34]. [60] outlined a standard approach to scenario planning that involves identifying common predetermined elements and uncertainties, then assessing their implications and degrees of uncertainty [79]. They advocated enhancements to methodology in future scenario development, such as prioritising unpredictable or distinctive components. proposed that these scenarios might be enhanced by properly integrating them with computer tools within land-use simulations. Scenario planning has enabled the shaping of unpredictable futures into more predetermined situations through the use of Land Change Modeling (LCM). LCM, a valuable method for evaluating potential growth scenarios, has become increasingly prevalent in research on flood resilience and changing climates [12]. Potential land use forecasts are used by LCM, a planning support structure, to guide the planning and implementation procedures [11].

3.5.3 Use of Innovative Technologiestoenhance Urban Resilience

The twenty-first century faces two main challenges: meeting rising energy demand while minimizing environmental impact. Governments are developing policies to promote renewable energy, conservation measures, and technological innovations [48]. Sensor-connected vehicles, buildings, and infrastructure are crucial for collecting real-time data on vulnerabilities before disasters. Disaster volunteers can analyze scenarios and take necessary actions using sensors for earthquakes, floods, air quality, weather conditions, heating systems, and irradiation [7]. Wireless Sensor Web (WSW) technology aids in early detection and emergency preparedness [1]. Integrating wireless sensors with 3D visual technology creates near-real-time scenarios for improved decision-making [19]. Low-power, low-data-rate wireless sensor networks (WSNs) are vital for smart city management [72] and urban protection [44]. Renewable energy is transforming sustainable building design. Geothermal technologies, solar panels, and wind turbines provide reliable, eco-friendly power. These sources reduce fossil fuel reliance, enhance building self-sufficiency, and lower long-term energy costs. On-site generation lessens vulnerability to power outages, crucial in remote areas or during disasters. By integrating renewable energy, architects can design buildings that minimize fossil fuel use, reduce carbon emissions, and promote energy independence, combining creativity with environmental responsibility for a sustainable future.

3.6 Challenges and Barriers to Building Resilient Cities

The literature on urban resilience reveals a lack of quantitative assessment tools, with many existing frameworks remaining conceptual. Research often focuses on sustainability rather than resilience within urban systems [66, 67]. These are the following barriers against urban resilience [40, 77, 85].

3.6.1 Financial Constraints and Resource Limitations for Implementing Resilience Measures

Implementing resilience measures is challenging due to financial constraints and resource limitations across sectors like infrastructure, health systems, and community planning. High initial costs, such as flood defenses or retrofitting buildings, require significant investment that many communities, especially in developing countries, cannot afford [15]. Limited funding sources, strained government budgets, and difficulty securing external funding often leave long-term resilience measures underfunded. Municipal representatives in developing countries face additional challenges, including mismatched external financing cycles, difficulties in demonstrating and assessing adaptation, and meeting reporting requirements. Funding for urban adaptation should avoid a "one-size-fits-all" approach, recognizing the significant differences among developing country cities and their varied funding needs. Localities face varying levels and types of climate vulnerability, and existing funding guidelines may result in a "race to the bottom," where cities try to appear the most vulnerable to secure funding. Additionally, funders must consider each city's specific institutional and political contexts, as these factors influence how funds are used. Rigid funding stipulations can burden city governments, while more flexible funds allow accountable city governments to make effective decisions. Economic prioritization complicates matters, with immediate needs like healthcare and education taking precedence. Implementing resilience measures also requires technical expertise, which many regions lack, leading to delays and inefficiencies. Maintenance costs for inspections, repairs, and updates further strain local budgets. Social and political factors play a role, with political will often favoring short-term gains over long-term resilience investments [15].

3.6.2 Institutional and Governance Barriers to Coordinated Urban Adaptation Efforts

Institutional barriers exist at three stages, as identified by. These levels include "Individual," "Organizational," and "facilitating Environment." At the individual stage, barriers stem from the beliefs, attitudes, abilities, and personal knowledge of individuals inside an institution. Organizational-level obstacles pertain to the administrative and structural components of an organization. The "facilitating Environment" stage barriers involve how actors interact within the sociological, financial, and political context of climate mitigation. These barriers can impact a municipal capacity to develop and execute climate change mitigation plans. Challenges at the individual level include a lack of concern for climate adaptation and issues arising from leadership dynamics. Organizational-level challenges include inadequate scientific data and fragmented decision-making authority. Meanwhile, challenges at the enabling environment level may arise from inappropriate policies and regulations. Overall, overcoming these barriers requires effective coordination, communication, and adequate funding sources [57].

3.6.3 Socioeconomic Disparities and Inequities in Climate Change Vulnerability and Resilience

Socioeconomic disparities significantly influence how communities experience and respond to climate change impacts. Vulnerability to climate change is often heightened among low-income populations, marginalized groups, and developing regions due to limited resources, inadequate infrastructure, and restricted access to essential services. These groups often reside in areas more susceptible to climate hazards, such as floodplains, coastal zones, and urban slums, increasing their exposure to climate risks. Inequities in resilience are also evident, as wealthier communities can invest in more robust infrastructure, early warning systems, and adaptive technologies, while poorer communities may lack the financial capacity to do so. This gap exacerbates existing inequalities, as those with fewer resources face greater challenges in recovering from climate-related events. Additionally, social factors such as education, health, and political representation play crucial roles in determining a community's resilience, with disadvantaged groups often experiencing lower levels of these critical resources [3, 40].

3.7 Policy Recommendations and Future Directions

3.7.1 Public–Private-Civil Collaboration for Urban Resilience

Several cities have promoted community involvement in adaptation plans by engaging local organizations, residents, and leaders. Urban areas under the Asian Cities Climate Change Resilience Network (ACCCRN) involve the public and important stakeholders in vulnerability assessment and resilience strategy establishing through the implementation of the 'Shared Learning Dialogue' method. This approach ensures that plans go beyond mere documentation to actively test different resilience-building methods, aiming to understand what works, what is cost-effective, and what is equitable. Broad participation helps cities learn and adapt effectively. Additionally, it is essential to strengthen governance structures and processes to define roles, responsibilities, and coordination methods among various authorities and stakeholders. Forming alliances and partnerships with educational institutions, research centers, businesses, and community organizations enhances the use of knowledge, resources, and collaborative efforts. Securing a sufficient and diverse pool of funds, including seeking external financial sources and forming public–private partnerships, supports initiatives aimed at strengthening resilience [15].

3.7.2 Research Priorities and Emerging Trends

Research priorities and emerging trends in climate change adaptation for urban areas focus on enhancing resilience through cross-sectoral integration. Key strategies include addressing immediate policy relevance, identifying knowledge gaps, understanding adaptation limits, and mitigating maladaptation risks. Detailed risk assessments should consider natural disasters, climate change effects, and social vulnerabilities, integrating resilience into building codes, zoning laws, and land-use planning. Community engagement ensures diverse perspectives in decision-making, empowering local groups to create tailored resilience activities and addressing socioeconomic injustices. Economic resilience involves diversifying economies, promoting entrepreneurship, and investing in skills development to recover from economic shocks. Environmental sustainability focuses on renewable energy, energy efficiency, eco-friendly transportation, ecosystem protection, and improved water and resource management. Technology and innovation enhance monitoring, response times, and early warning systems, supporting resilient infrastructure and urban intelligence technology. These strategies enable cities to withstand shocks, adapt to climatic changes, and support equitable, sustainable development for their citizens.

3.8 Conclusion

Urban resilience, while beneficial, is challenging to define and implement. Research is expanding across various fields [81]. This chapter examines urban resilience definitions, traits, challenges, adaptations, and opportunities. Urban resilience refers to a city's ability to resist damage, minimize disruption, adapt to changes, and replace systems efficiently. Understanding this concept is key to building communities that withstand extreme events and disasters. Integrating diverse experiences is crucial for fostering resilient urban systems. Local management authorities need strategic leadership for effective implementation. The lack of assessment tools remains a significant challenge and opportunity for advancing research. New approaches to urban design are imperative in response to the increasing severity and frequency of climate-related events. It's essential to integrate information about rising risks into disaster management, community risk reduction, and other vital sectors. Designers can assist charitable organizations in adapting to climate change by implementing strategies such as conducting climate change risk assessments, designing inclusive infrastructure, and fostering partnerships between governments, businesses, and civil society. Advocacy efforts can also raise awareness about climate change's humanitarian impacts and encourage action among decision-makers and the public. Based on Coaffee's 2018 analysis, we identify four research targets to enhance urban resilience: First, document lines of responsibility for resilience implementation to clarify individual actor behavior and decision-making contexts, using extensive case studies to track resilience from concept to execution. Second, examine how resilience objectives are operationalized to ensure efforts improve adaptability and future outcomes rather than just maintaining the status quo. Third, identify conditions like "safe-to-fail" environments that encourage innovative solutions to acute shocks and chronic pressures, studying the institutionalization of various techniques to gain insights into alternative governance structures for resilience. Fourth, investigate the long-term viability of resilience as a policy objective through comparative research to show how practitioners balance immediate goals with promoting enduring resilience despite political and budgetary changes.

Acknowledgements AJ is thankful to the University Grant Commission New Delhi for providing Joint CSIR-UGC Fellowship. RS acknowledges Science and Engineering Research Board (SERB), New Delhi, is highly acknowledged for the financial support with file number: EEQ/2021/000356. We acknowledges IOE, Institute of Science, Banaras Hindu University for financial assistance.

References

1. Adeel A, Gogate M, Farooq S, Ieracitano C, Dashtipour K, Larijani H, Hussain A (2019) A survey on the role of wireless sensor networks and IoT in disaster management. Geological disaster monitoring based on sensor networks, 57–66
2. Adger WN, Brown K, Nelson DR, Berkes F, Eakin H, Folke C, ... Tompkins EL (2011) Resilience implications of policy responses to climate change. Wiley Interdisciplinary Rev Climate Change 2(5):757–766
3. Adger WN et al. (2006) Vulnerability. Global Environ Change. This article explores the concept of vulnerability in the context of climate change, focusing on how socioeconomic factors contribute to differential impacts and adaptive capacities 16(3):268–281
4. Alam M, Rabbani MG (2012) Vulnerabilities and responses to climate change for Dhaka. In: Adapting cities to climate change, pp 112–129. Routledge
5. Alberti M, Marzluff JM, Shulenberger E, Bradley G, Ryan C, Zumbrunnen C (2003) Integrating humans into ecology: opportunities and challenges for studying urban ecosystems. Bioscience 53(12):1169–1179
6. Al-Sayed MM, Alanizi A (2023) Resilient cities to mitigate effects of climate change. In: IOP conference series: earth and environmental science (Vol 1129, No 1, p 012029). IOP Publishing

7. Asimakopoulou E, Bessis N (2011) Buildings and crowds: forming smart cities for more effective disaster management. In: 2011 Fifth international conference on innovative mobile and internet services in ubiquitous computing, pp 229–234. IEEE
8. Axsen J, Plötz P, Wolinetz M (2020) Crafting strong, integrated policy mixes for deep CO_2 mitigation in road transport. Nat Clim Chang 10(9):809–818
9. Bai T, Mayer AL, Shuster WD, Tian G (2018) The hydrologic role of urban green space in mitigating flooding (Luohe, China). Sustainability 10(10):3584
10. Batty M (2008) The size, scale, and shape of cities. Science 319(5864):769–771
11. Berke P (2006) Urban land use planning
12. Brown D, Band LE, Green KO, Irwin EG, Jain A, Lambin EF, ... Verburg PH (2013) Advancing land change modeling: opportunities and research requirements. The National Research Council Press
13. Brumarová L, Kukuliač P, Brumar J (2021) Use of geographic information systems in crisis management. In: IOP conference series: earth and environmental science (Vol 900, No 1, p 012004). IOP Publishing
14. Cariolet JM, Vuillet M, Diab Y (2019) Mapping urban resilience to disasters–a review. Sustain Cities Soc 51:101746
15. Carmin J, Dodman D, Chu E (2013) Urban climate adaptation and leadership: from conceptual understanding to practical action
16. Chandra A, Acosta J, Howard S, Uscher-Pines L, Williams M, Yeung D, ... Meredith LS (2011) Building community resilience to disasters: a way forward to enhance national health security. Rand Health Quart 1(1)
17. Chan-Edmiston S, Fischer S, Sloan S, Wong M (2020) Intelligent transportation systems (its) joint program office: strategic plan 2020–2025 (No. FHWA-JPO-18–746). United States. Department of Transportation. Intelligent Transportation Systems Joint Program Office
18. Chapman S, Watson JE, Salazar A, Thatcher M, McAlpine CA (2017) The impact of urbanization and climate change on urban temperatures: a systematic review. Landscape Ecol 32:1921–1935
19. Cheikhrouhou O, Koubâa A, Zarrad A (2020) A cloud based disaster management system. J Sens Actuator Netw 9(1):6
20. Chelleri L (2012) From the «Resilient City» to urban resilience. a review essay on understanding and integrating the resilience perspective for urban systems. Documents d'anàlisi geogràfica, 58(2):287–306
21. Chelleri L, Waters JJ, Olazabal M, Minucci G (2015) Resilience trade-offs: addressing multiple scales and temporal aspects of urban resilience. Environ Urban 27(1):181–198
22. Chen Y (2022) Flood hazard zone mapping incorporating geographic information system (GIS) and multi-criteria analysis (MCA) techniques. J Hydrol 612:128268
23. Coaffee J, Therrien MC, Chelleri L, Henstra D, Aldrich DP, Mitchell CL, ... Participants (2018) Urban resilience implementation: a policy challenge and research agenda for the 21st century. J Contingencies Crisis Manag 26(3):403–410
24. Community and Regional Resilience Institute (2013) Building resilience in America's communities: Observations and implications of the CRS pilots
25. Creutzig F, Javaid A, Koch N, Knopf B, Mattioli G, Edenhofer O (2020) Adjust urban and rural road pricing for fair mobility. Nat Clim Chang 10(7):591–594
26. Davoudi S, Brooks E, Mehmood A (2013) Evolutionary resilience and strategies for climate adaptation. Plan Pract Res 28(3):307–322
27. Dottori F, Szewczyk W, Ciscar JC, Zhao F, Alfieri L, Hirabayashi Y, ... Feyen L (2018) Increased human and economic losses from river flooding with anthropogenic warming. Nature Climate Change 8(9):781–786
28. Douglas I, Alam K, Maghenda M, Mcdonnell Y, McLean L, Campbell J (2012) Unjust waters: climate change, flooding and the urban poor in Africa. In Adapting Cities to Climate Change, pp 201–223. Routledge
29. Dynes RR (2002) The importance of social capital in disaster response

30. Folke C (2006) Resilience: The emergence of a perspective for social–ecological systems analyses. Glob Environ Chang 16(3):253–267
31. Gallopín GC (2006) Linkages between vulnerability, resilience, and adaptive capacity. Glob Environ Chang 16(3):293–303
32. Ganguli P, Merz B (2019) Extreme coastal water levels exacerbate fluvial flood hazards in Northwestern Europe. Sci Rep 9(1):13165
33. Godschalk DR (2003) Urban hazard mitigation: creating resilient cities. Nat Hazard Rev 4(3):136–143
34. Goodspeed R (2017) An evaluation framework for the use of scenarios in urban planning (pp 7–8). Cambridge, MA, USA: Lincoln Institute of Land Policy
35. Gasper R, Blohm A, Ruth M (2011) Social and economic impacts of climate change on the urban environment. Curr Opin Environ Sustain 3(3):150–157
36. Haynes KE, Li M (2004) Analytical alternatives in intelligent transportation system (ITS) evaluation. Res Transp Econ 8:127–149
37. Hodgson GM (2006) What are institutions? J Econ Issues 40(1):1–25
38. Holling CS (1973) Resilience and stability of ecological systems. Annu Rev Ecol Syst 4(1):1–23
39. Holling CS (1996) Engineering resilience versus ecological resilience. Eng Ecol Constr 31(1996):32
40. Intergovernmental Panel on Climate Change (IPCC) (2014) Climate change 2014: impacts, adaptation, and vulnerability. This report provides an extensive review of the impacts of climate change and the challenges of adapting infrastructure and systems to be more resilient, including financial and resource constraints. Available at: IPCC Report
41. Irani M, Rahnamayiezekavat P (2021) An overview of urban resilience: dimensions, components, and approaches. Acta Scientiarum Polonorum Administratio Locorum 20(4):305–322
42. Jefferson M (2000) Energy policies for sustainable development. United Kingdom
43. Kalkstein L, Koppe C, Orlandini S, Sheridan S, Smoyer-Tomic K (2009) Health impacts of heat: present realities and potential impacts of a climate change. Distributional effects of climate change: social and economic implications (M. Ruth and M. Ibarraran, eds), Cheltenham, Edger Elgar Publishers
44. Khalifeh A, Darabkh KA, Khasawneh AM, Alqaisieh I, Salameh M, AlAbdala A, ... Rajendiran K (2021) Wireless sensor networks for smart cities: Network design, implementation and performance evaluation. Electronics 10(2):218
45. Kirschenbaum H (2004) Carl Rogers's life and work: an assessment on the 100th anniversary of his birth. J Couns Dev 82(1):116–124
46. Leichenko R (2011) Climate change and urban resilience. Curr Opin Environ Sustain 3(3):164–168
47. Levin, Simon. "ecological resilience". Encyclopedia Britannica, 28 Jan. 2024, https://www.britannica.com/science/ecological-resilience. Accessed 7 May 2024.
48. Lu Y, Khan ZA, Alvarez-Alvarado MS, Zhang Y, Huang Z, Imran M (2020) A critical review of sustainable energy policies for the promotion of renewable energy sources. Sustainability 12(12):5078
49. Lynch K (1981) Good city form The MIT Press. Massachusetts, Cambridge
50. Mäenpää H, Lobov A, Lastra JLM (2017) Travel mode estimation for multi-modal journey planner. Transp Res Part C: Emerg Technol 82:273–289
51. Manyena B, O'Brien G, O'Keefe P, Rose J (2011) Disaster resilience: a bounce back or bounce forward ability? Local Environ Int J Justice Sustain 16(5):417–424
52. McEvoy D, Fünfgeld H, Bosomworth K (2013) Resilience and climate change adaptation: the importance of framing. Plan Pract Res 28(3):280–293
53. Meerow S, Newell JP (2015) Resilience and complexity: A bibliometric review and prospects for industrial ecology. J Ind Ecol 19(2):236–251
54. Miller F, Osbahr H, Boyd E, Thomalla F, Bharwani S, Ziervogel G, ... Nelson D (2010) Resilience and vulnerability: complementary or conflicting concepts?. Ecol Soc 15(3)
55. Moudon AV (1986) Built for change: neighborhood architecture in San Francisco

56. Norris FH, Stevens SP, Pfefferbaum B, Wyche KF, Pfefferbaum RL (2008) Community resilience as a metaphor, theory, set of capacities, and strategy for disaster readiness. Am J Community Psychol 41:127–150
57. Oberlack C (2017) Diagnosing institutional barriers and opportunities for adaptation to climate change. Mitig Adapt Strat Glob Change 22:805–838
58. Ostadtaghizadeh A, Ardalan A, Paton D, Jabbari H, Khankeh HR (2015) Community disaster resilience: a systematic review on assessment models and tools. PLoS currents, 7
59. Paton D, Johnston D (2017) Disaster resilience: an integrated approach. Charles C Thomas Publisher
60. Postma TJ, Liebl F (2005) How to improve scenario analysis as a strategic management tool? Technol Forecast Soc Chang 72(2):161–173
61. Pfefferbaum B, Van Horn RL, Pfefferbaum RL (2017) A conceptual framework to enhance community resilience using social capital. Clin Soc Work J 45(2):102–110
62. Rogatka K, Starczewski T, Kowalski M (2021) Urban resilience in spatial planning of polish cities-True or false? Transformational perspective. Land Use Policy 101:105172
63. Ruth M (2010) Economic and social benefits of climate information: assessing the cost of inaction. Procedia Environ Sci 1:387–394
64. Ruth M, Gasper R (2008) Water in the urban environment: Meeting the challenges of a changing climate. In: OECD international conference: competitive cities in climate change, Milan, Italy
65. Ruth M, Ibarrarán ME (2009) Introduction: distributional effects of climate change–Social and economic implications. Edward Elgar Publishing, In Distributional Impacts of Climate Change and Disasters
66. Sharifi A, Murayama A (2013) A critical review of seven selected neighborhood sustainability assessment tools. Environ Impact Assess Rev 38:73–87
67. Sharifi A, Murayama A (2014) Neighborhood sustainability assessment in action: cross-evaluation of three assessment systems and their cases from the US, the UK, and Japan. Build Environ 72:243–258
68. Sharifi A, Yamagata Y (2014) Resilient urban planning: Major principles and criteria. Energy procedia 61:1491–1495
69. Sharifi A, Yamagata Y (2014) Major principles and criteria for development of an urban resilience assessment index. In: 2014 international conference and utility exhibition on green energy for sustainable development (ICUE), pp 1–5. IEEE
70. Sharifi A, Yamagata Y (2015) A conceptual framework for assessment of urban energy resilience. Energy Procedia 75:2904–2909
71. Sharifi A, Yamagata Y (2016) Principles and criteria for assessing urban energy resilience: a literature review. Renew Sustain Energy Rev 60:1654–1677
72. Sharma H, Haque A, Blaabjerg F (2021) Machine learning in wireless sensor networks for smart cities: a survey. Electronics 10(9):1012
73. Shimada G (2015) The role of social capital after disasters: An empirical study of Japan based on Time-Series-Cross-Section (TSCS) data from 1981 to 2012. Int J Disaster Risk Reduction 14:388–394
74. Stott P (2016) How climate change affects extreme weather events. Science 352(6293):1517–1518
75. Suárez M, Gómez-Baggethun E, Benayas J, Tilbury D (2016) Towards an urban resilience index: a case study in 50 Spanish cities. Sustainability 8(8):774
76. Torgersen G, Navrud S (2018) Singing in the rain: valuing the economic benefits of avoiding insecurity from urban flooding. J Flood Risk Manag 11(4):e12338
77. United Nations Office for Disaster Risk Reduction (UNDRR) (2015) Global assessment report on disaster risk reduction. This assessment covers the financial and economic constraints in disaster risk reduction and the importance of investing in resilience. Available at: UNDRR Report
78. United Nations Office for Disaster Risk Reduction (UNDRR) (2019) Global assessment report on disaster risk reduction 2019. This report emphasizes the importance of policy frameworks and regulatory measures in building urban resilience. Available at: UNDRR Report

79. Van der Heijden K, Bradfield R, Burt G, Cairns G, Wright G (2009) The sixth sense: accelerating organizational learning with scenarios. Wiley
80. Walker B, Holling CS, Carpenter SR, Kinzig A (2004) Resilience, adaptability and transformability in social–ecological systems. Ecol Soc 9(2)
81. Wang L, Xue X (2018) Exploring the evolution trends of urban resilience research. In: International conference on construction and real estate management 2018, pp 18–27. Reston, VA: American Society of Civil Engineers
82. Wang L, Xue X, Yang RJ, Luo X, Zhao H (2019) Built environment and management: exploring grand challenges and management issues in built environment. Frontiers Eng Manag 6(3):313–326
83. Wang L, Xue X, Zhou X, Wang Z, Liu R (2021) Analyzing the topology characteristic and effectiveness of the China city network. Environ Planning B: Urban Analyt City Sci 48(9):2554–2573
84. Winchester L, Szalachman R (2009) The urban poor's vulnerability to the impacts of climate change in Latin America and the Caribbean: a policy agenda. In: Fifth urban research symposium.
85. World Bank (2019) Lifelines: the resilient infrastructure opportunity. This report discusses the economic and financial aspects of investing in resilient infrastructure and highlights the challenges of funding such projects. Available at: World Bank Report
86. Wu J, Wu T (2012) Ecological resilience as a foundation for urban design and sustainability. Resilience in ecology and urban design: Linking theory and practice for sustainable cities. Springer, Netherlands, Dordrecht, pp 211–229
87. Xue X, Wang L, Yang RJ (2018) Exploring the science of resilience: critical review and bibliometric analysis. Nat Hazards 90:477–510

Part II
Integrating Resilience into Urban Planning

Chapter 4
Associating the Determinants of Urban Transformation in Mumbai Suburban District to the Status of Resilience to Climate Change

Sujayita Bhattacharjee

Abstract This chapter explores the linkages between the determinants of urban transformation and the status of resilience to climate change in the Mumbai Suburban District, which comprises a major portion of the metropolis of Mumbai. It aims to understand whether these determinants are influenced by the imperative to foster resilience to climate change and to what degree the transformation itself demonstrates climate change resilience. Here the locality of Mulund, located in the Mumbai Suburban district is chosen as the areal unit of investigation. Mulund also comprises the T-ward of the Municipal Corporation of Greater Mumbai (MCGM). The role of major determinants of Mulund's transformation, namely—the selling of the defunct industrial lands; government policies; public–private partnerships; gentrification and housing growth; and residents, retail, and businesses are considered. The study is based on both primary and secondary data. The methodology of the study involves the use of mixed methods involving both qualitative and quantitative analysis. The findings reveal the entwined nature of the role of the identified determinants and the limited consideration of climate change resilience in the transformation process, leading to the limited existence of climate change resilience. Based on the existing nature of the transformation and the role of the determinants contributing to it a set of recommendations are forwarded to develop climate change resilience in the study area. It is hoped that these recommendations will be implemented through necessary program interventions to upgrade the climate change resilience status of the area.

Keywords Climate change · Mumbai · Resilience · Transformation · Urban

Present Address:
S. Bhattacharjee (✉)
Assistant Professor, Department of Geography, D.G. Ruparel College of Arts, Science and Commerce, Mumbai, India
e-mail: sujayita100@gmail.com

S. Ghosh et al. (eds.), *Cities of Tomorrow: Urban Resilience and Climate Change Preparedness*, Urban Sustainability, https://doi.org/10.1007/978-981-97-9658-8_4

4.1 Introduction

The transformation of urban areas results from intricate interactions among numerous processes, influenced by various factors and stakeholders driving their occurrence and dynamics. It encompasses numerous decisions that integrate temporal and spatial actions across urban and architectural scales [4]. Concurrently, the looming threat of climate change necessitates urban areas to adapt and build resilience against its impacts [33, 37]. Here, it is essential to recognize that urban transformation can occur through both deliberate, planned initiatives and unplanned, haphazard occurrences [27, 43]. Moreover, the extent to which even planned initiatives consider the uncertainties of climate change is a question to ponder. Nonetheless, economic, demographic, and socio-cultural forces constantly interact with the urban spaces and create a context of change [21]. Consequently, urban areas evolve in response to a complex interplay of these multifaceted influences [32], shaping their physical, social, and economic landscapes. Therefore, understanding urban transformation requires a nuanced comprehension of the diverse processes and actors at play, including planned interventions, spontaneous developments, and the overarching influence of climate change.

The recent narratives surrounding urban transformation have been propelled by the acknowledgment of the imperative and potential for substantial change toward creating sustainable and resilient cities [18]. Besides, the existing research on the transformation of the urban spaces of contemporary times also suggests the functioning of urban processes such as gentrification [3, 39], suburbanization [23], redevelopment [13], and urban renewal [22, 24], without delving into the aspect of climate resilience. Nonetheless, what becomes evident here is the occurrence of urban transformation involving multiple stakeholders.

The current activities of urban development navigate multifaceted urban environments where globalization, technological innovation, urbanization, and various urban transitions intersect, involving a diverse array of governmental and non-governmental organizations, while local politicians and businesses actively seek development opportunities, prime locations, and suitable financing mechanisms to optimize land use [13]. This nexus is widely rampant across India where 'the growing urban population is further contributing to changing the urban dynamics' [1]. In India, urban areas are being created and transformed at a breakneck pace. Metropolitan cities in India are evolving into significant economic centers, marked by notable shifts in land use and a decline in environmental resources [7]. Many formerly suburban areas also have changed to become an essential component of the urban areas exhibiting the characteristics of the urban core. This is starkly clear in the instance of Mumbai, a major city whose suburban expansion has long been a defining attribute.

Suburban Mumbai or the Mumbai Suburban District no longer exhibits peri-urban characteristics. The polycentric growth of economic hubs across the Municipal Corporation of Greater Mumbai has diminished the difference between the Mumbai City District and the Mumbai Suburban District [12]. These economic hubs have developed into focal points for real estate growth, corporate sector businesses, and other privatized spaces. However, it is noteworthy that all these economic hubs did not emerge all at once. In fact, it was due to the saturation of the core of Mumbai, that the growth of suburban areas received a boost [9].

To explore the linkages between the determinants of urban transformation and the status of resilience to climate change in the suburban arena of Mumbai, the locality of Mulund has been selected as the study area. The urban transformation of Mulund and its growth as an economic hub has been a comparatively recent occurrence, making it conducive to exploring the determinants of urban transformation in the area and further searching for associations with the status of resilience to climate change.

4.1.1 *Objectives of the Study*

The main objectives of the study are as follows:

(a) To understand whether the determinants of urban transformation in Mulund are influenced by the imperative to foster resilience to climate change; and
(b) To examine how far the undergoing urban transformation itself demonstrates climate change resilience.

4.1.2 *Study Area*

The study area—Mulund comprises the T-ward of the Municipal Corporation of Greater Mumbai (MCGM). It is situated at the northeastern end of MCGM (Fig. 4.1). It is a part of the Mumbai Suburban District.[1] On the one hand, Mulund is a suburb of Mumbai, while on the other it is an integral part of Mumbai. Suburbs are essentially a continuum of the city [47]. Mulund too is a continuum of Mumbai, that has grown into a full-fledged urban entity following the expansion of the limits of the city [9]. Today it is a completely urban area.

4.1.3 *Research Methods*

The study utilizes a combination of primary and secondary data to provide a comprehensive understanding of the subject matter. Primary data is gathered firsthand from field observations and interactions, questionnaire surveys, interviews, and focus group discussions. On the other hand, secondary data is sourced from a detailed review of existing literature, incorporating findings from previous studies, reports,

[1] Both Mumbai City District and Mumbai Suburban District comprise the limits of the Municipal Corporation of Greater Mumbai (MCGM).

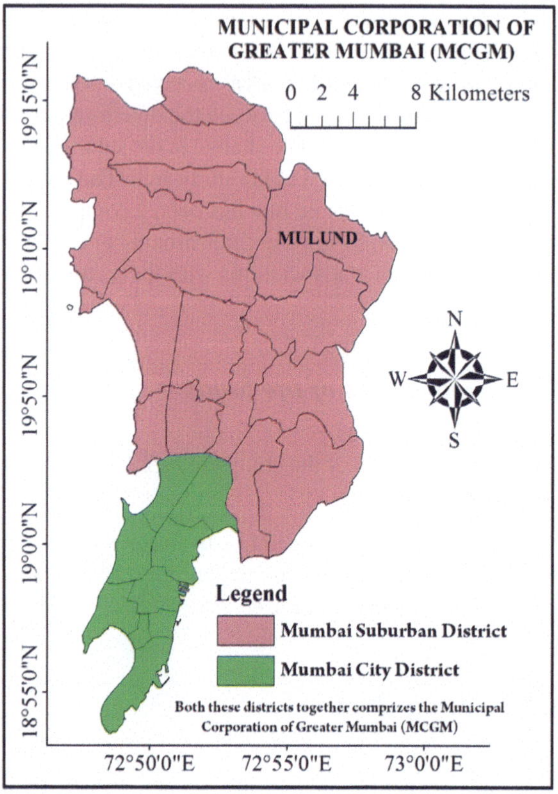

Fig. 4.1 The location of the study area (Mulund) in MCGM. *Source* Prepared by the author

and publications related to the topic. To achieve a comprehensive and detailed analysis, the research methodology employs a mixed-methods approach. This strategy blends qualitative methods, which delve into exploratory insights and provide contextual understanding, with quantitative methods that highlight statistical interpretation. In this study, the qualitative dimension takes precedence, while quantitative analysis serves to complement and support the qualitative findings. By merging these qualitative and quantitative approaches, the study seeks to present a well-rounded and thorough examination of the topic, effectively capturing its intricate and multifaceted aspects.

4.2 Determinants of Urban Transformation and Climate Change Resilience Consideration

Urban transformation can occur through planned regeneration efforts or spontaneously due to various processes and activities. Regardless of the approach, the role of determinants driving urban transformation cannot be ignored. The major determinants of urban transformation in Mulund have been identified as:

- Selling of the Defunct Industrial Lands
- Government Policies
- Public–Private Partnerships
- Gentrification and Housing Growth
- Residents, Retail and Business

The role of the determinants and the state of consideration of climate change resilience in the transformation process are examined below.

4.2.1 Selling of the Defunct Industrial Lands

Landowners are said to be the foremost entities initiating change, serving as primary decision-makers in determining the feasibility and extent of development, including the possibility of its occurrence [21]. In the case of Mulund, a pivotal action towards bringing about its transformation was taken by factory owners opting to sell their land to developers, marking the initial phase of transformation. According to the long-term residents of Mulund, the process of selling the industrial lands to builders/developers began with the initiation of Mumbai's deindustrialization phase. The decline of the textile mills located at the city core of Mumbai led to their eventual closure [20]. This was followed by the large-scale relocation of various industries beyond the limits of MCGM [6]. The industries of Mulund had the same fate, the area being an integral part of the MCGM. Faced with the shutdown of factories and industrial units in Mulund, most landowners opted to sell their properties to real estate builders/developers. Here it is evident that the motive behind selling the defunct industrial areas was purely driven by economic reasons without any consideration of the climatic aspect.

4.2.2 Government Policies

In recent times, government policies have become a major determining factor of urban transformation. Several policies such as those focused on the decongestion of Mumbai, liberalization of India, and imposition of regulations on the Mulund dumping ground, along with the integration of urban renewal schemes such as 'Jawaharlal Nehru National Urban Renewal Mission' (JNNURM), and 'Atal Mission for Rejuvenation and Urban Transformation' (AMRUT) in the area have significantly contributed to the transformation of Mulund [9].

At the same time around the decline of the cotton textile mills of Mumbai, the government laid the foundations for decongesting the city. In the 1980s, the government's policy towards decongesting the areal limits of MCGM led to the moving out of the industries from the metropolis of Mumbai [31]. With the implementation of this policy, the industries of Mulund, along with the rest of Mumbai underwent closure. Furthermore, the liberalization of the Indian economy in the early 1990s also contributed to the transformation of Mulund. Liberalization led to the pouring in of foreign direct investment (FDI) to several sectors of Mumbai [40], generating newer forms of employment in ICT, finance, and various other services, etc. This led to the congestion of the core and the escalation of prices too fast, directing growth opportunities toward the suburbs. With this, Mulund witnessed the growth of commercial real estate and office spaces to accommodate the service sector. Over the years, Mulund has also witnessed the coming up of many parks and gardens.

Mulund dumping ground, which is more than half a century old, has become a bone of contention for the residents of Mulund in recent times. Over decades of continuous garbage dumping, it transformed into a towering heap of waste, accumulating approximately 70 lakh metric tonnes of garbage, and reaching a height of roughly 36 m [34]. The dumping ground had way exceeded its capacity, becoming a more and more serious environmental and health hazard. Following years of contention and uncertainty regarding the status of the Mulund Dumping Ground, the MCGM decisively announced its closure on October 1st, 2018.

Mulund has seen significant transportation improvements, mostly due to the Mumbai Development Plan of 1991, 'aimed at easing congestion in the city centre and fostering transit-oriented development' [8]. These enhancements have positively impacted Mumbai's suburbs, including Mulund. Additionally, initiatives like the JNNURM have introduced Volvo Air-Conditioned Bus Services operated by entities like Brihanmumbai Electricity Supply and Transport (BEST), Navi Mumbai Municipal Transport (NMMT), and Thane Municipal Transport (TMT), improving Mulund's connectivity [5, 36, 38]. Furthermore, in alignment with AMRUT's objective of universal metering, MCGM implemented the 2019 metering policy, resulting in the installation of 12,635 water meters in Mulund [35].

Among these policies, some are loosely tied to the aspect of climate change. The decongestion of Mumbai was a mechanism to cope with the growing population. Here, it is noteworthy that population growth has connotations for the environment. Hence, the decongestion was in some way or another related to the aspects of the environment. Growing population densities and rise in impervious surfaces create the urban heat island effect [17], which compounds the effect of global warming. Reducing heat islands is, therefore, said to aid in slowing down climate change [46]. In addition to this, the imposition of restrictions on Mulund Dumping Ground is associated with the rise of environmental issues arising from the exceeding of its capacity [10]. Dumpsites, in general, are responsible for the emission of greenhouse gases causing climate change. Therefore, closing the dumpsites is one of the primary steps towards climate change resilience. Furthermore, the coming up of parks and gardens has boosted the status of Mulund on the front of climatic resilience.

4.2.3 Public–Private Partnerships (PPP)

Various PPP projects have been implemented in Mulund, particularly in housing infrastructure development. The suburb mainly comprises apartment living, with numerous multi-story housing projects underway. Based on the data collected on the nature of housing in Mulund, it is known that the ongoing projects are primarily private, while redevelopment projects involve both private and public–private partnership (PPP) arrangements, with the Slum Rehabilitation Authority (SRA) and Maharashtra Housing and Area Development Authority (MHADA) collaborating with builders. Additionally, the Central Railway has teamed up with Magicdil Health For All Private Ltd. to establish Emergency Medical Rooms (EMR) at various railway stations, including Mulund, to offer 24 × 7 medical services to commuters [15]. The PPP model has also come to be utilized for waste management in Mulund. The MCGM has engaged in PPP arrangements for the processing of waste at the Mulund dumping ground. The dumping ground is now being reclaimed through a five-year contract with Prakash Constrowell Ltd. for bio-mining 7 million tons of solid waste [30, 44].

Among the various PPP projects, the one concerning the processing of waste at the Mulund Dumping Ground is directly linked to environmental concerns. "Bio-mining, according to the guidelines of the Central Pollution Control Board (CPCB), is the scientific process of excavation, treatment, segregation and gainful utilization of aged municipal solid waste lying in dumpsites typically referred to as legacy waste" [41]. Although the project has been delayed due to various hurdles, the bio-mining of waste is expected to prove environmentally and climate-friendly, contributing to climate change resilience.

4.2.4 Gentrification and Housing Growth

The gentrification of Mulund began with the redevelopment of the old industrial landscape to high-rises, malls, and multiplexes, by the builders/developers. The process was mostly driven by economic motives. It was the occurrence of 'rent-gap' or "the disparity between the potential ground rent level and the actual ground rent capitalized under the present land use" [42] that benefited the builders and the once working-class neighborhood became gentrified. Although there are still certain remnants of the old industrial era, the process of gentrification has provided the area with a modern character. However, the gentrification process in Mulund caused a partial displacement of the factory workers and their families leading some of them

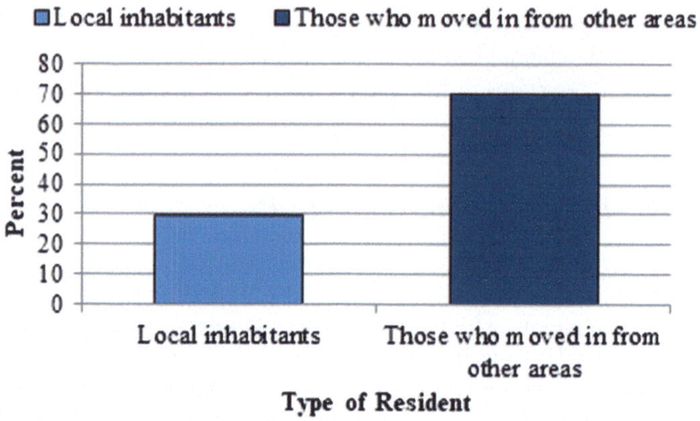

Fig. 4.2 Types of Residents in Mulund. *Source* Prepared by the author based on the data from the questionnaire survey

to leave Mulund, while many others eventually settled down in the slums of the area [11]. Thus, gentrification on the one hand led to the growth of upscale housing, and on the other hand, it also contributed to the settlement of the working class in the slums. Thus, it can be seen that the process of gentrification and housing growth was an economic occurrence unrelated to environmental considerations.

4.2.5 Residents, Retail, and Business

The urban transformation of Mulund has also been contributed by the people who have come to reside in the area. The sample data shows that 70.2% of Mulund's residents are migrants, while only 29.8% are locals (Fig. 4.2).

Among the migrants, 42.5% moved in within the last 30 years, and 57.5% before that (Fig. 4.3). After India's partition, many refugees settled here, with around 35,000 of their descendants now living in Mulund Colony [19]. Additionally, the area attracted migrants when its industries were operational in its western half. Mulund's evolving character has influenced migration patterns, attracting people from both Mumbai and beyond. Earlier (before 30 years), proximity to work in Mulund's industries drew many, who were housed in affordable *chawls* (Fig. 4.4). As industries declined, cheaper property became a draw, especially for those moving in 20–30 years ago. A growing family size drove relocation 10–20 years ago, reflecting a shift from joint to nuclear families. Recently, proximity to workplaces, including beyond Mulund, has become a key reason for moving, facilitated by improved transport links.

Over the years, there has been an increase in the number of slum residents too. Among the slum dwellers of Mulund, 69.0% are those who have moved to the area within the last 30 years (Fig. 4.5). The recent peopling of Mulund, on the one hand, is a product of the transformation of Mulund and on the other is itself a factor contributing

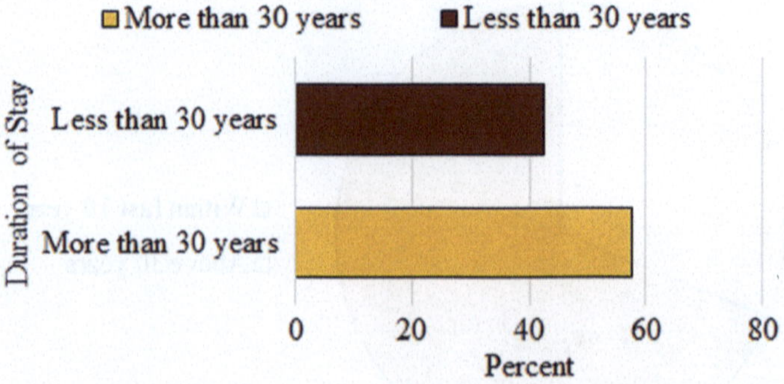

Fig. 4.3 Duration of Stay of the Residents in Mulund. *Source* Prepared by the author based on the data from the questionnaire survey

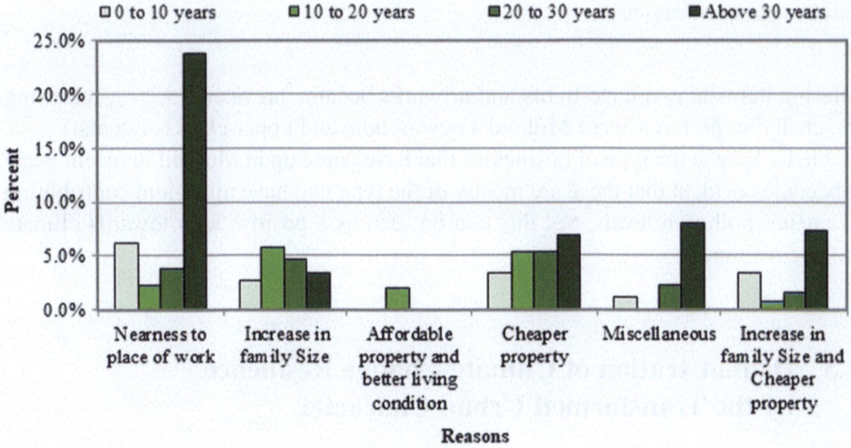

Fig. 4.4 Reasons for Migrating/Relocating to Mulund (Across Time Periods). *Source* Prepared by the author based on the data from the questionnaire survey

to the transformation of the area. However, it has not directly contributed to any form of climate change resilience in the area.

Along with the growth of residents, there has emerged immense retail and business sector activities that have further played a significant role in transforming the characteristics of Mulund. Over the years many retail and business activities have come up in Mulund, and most of it has been a recent addition (Fig. 4.6). Also, the variety of businesses has increased over the years. Businesses established before 30 years fall into three categories: real estate and travel agencies (2.6%), shops selling essentials like groceries and electronics (3.9%), and miscellaneous services like laundries and driving schools (3.9%). The latter two categories saw the highest growth (5.3% each) over the past 20–30 years. However, from 10–20 years ago, restaurants and cafes dominated new businesses (7.9%). In the last decade, decorative shops

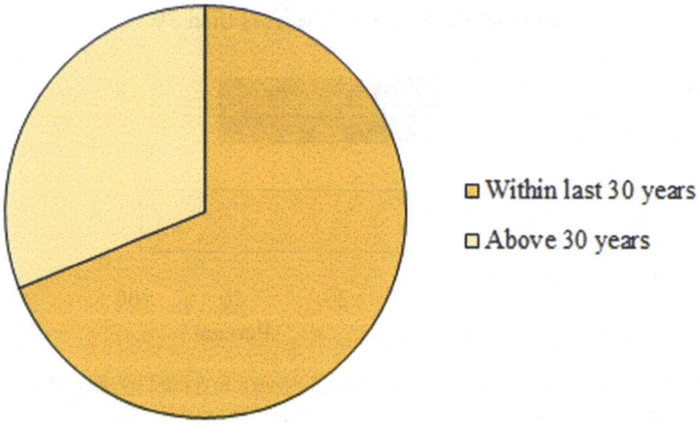

Fig. 4.5 Duration of stay of slum dwellers in Mulund. *Source* Prepared by the author based on the data from the questionnaire survey

offering items like vintage lights and artworks became prevalent (9.2%), reflecting the changing preferences of Mulund's new middle and upper-class residents.

On looking at the type of businesses that have come up in Mulund in recent years it becomes evident that these are mostly of the type that have minimum contribution to raising pollution levels. So, this can be seen as a positive step towards climate change resilience.

4.3 Demonstration of Climate Change Resilience by the Transformed Urban Character

From the preceding section, it is evident that not all the determinants of urban transformation in Mulund impact climate change resilience. Factors such as the selling of defunct industrial lands, gentrification, housing growth, and resident activities do not influence climate resilience. However, government policies, public–private partnerships (PPPs), and the retail and business sectors have had mixed impacts, with several of them contributing positively to climate resilience. Identifying and analyzing these determinants of urban transformation in Mulund posed several challenges such as the difficulties in comprehending the diverse stakeholder interests, navigating complex PPP structure, and exploring the dimension of public policy. Nonetheless, maneuvering through this complex dynamic enabled bringing together all the major determinants. As a result of these determinants, the urban transformation of Mulund has reached its present position.

Mulund today is more urbanized than it was ever before and in this growth trajectory, the level of urbanization at Mulund is expected to increase further. The selling of the defunct industrial lands; policies formulated by the government; PPP initiatives; the occurrence of gentrification and housing growth; and the rise of residents, retail,

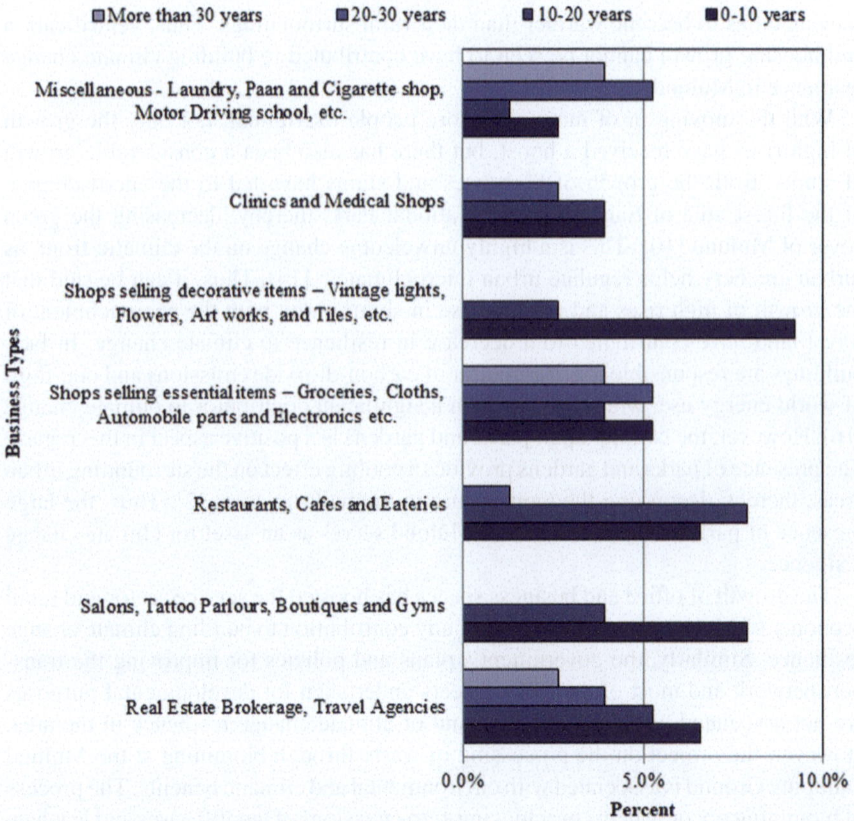

Fig. 4.6 Duration-wise Growth of Various Businesses in Mulund. *Source* Prepared by the author based on the data from the questionnaire survey

and business have transformed the character of Mulund drastically. However, it is important to understand how far this transformed urban character of Mulund exhibits resilience to climate change.

The selling of the defunct industrial lands to the builders/developers led to the construction of high-rises. Gentrification occurred and a new class of residents moved into the area, while, many of the displaced working-class families found refuge in the slums. As a result of gentrification, the area has seen the emergence of high-rise residential complexes equipped with various amenities. From a climate change perspective, these developments present challenges. High-rise buildings typically consume more energy for heating, cooling, and lighting, leading to higher greenhouse gas emissions [29]. The concentration of residents in a compact area may increase waste generation, potentially escalating landfill use and environmental pollution [25]. Gentrified and high-rise areas can intensify traffic congestion, contributing to elevated vehicle emissions. Additionally, the construction of high-rise buildings and the reduction of green spaces can exacerbate the urban heat island effect [26], causing cities to become warmer than their rural surroundings. Thus, gentrification and housing growth cannot be seen to have contributed to building climate change resilience in Mulund.

With the moving in of more and more people to Mulund, not only the growth of high-rises have received a boost, but there has also been a considerable growth of slums. Both the growth of high-rises and slums have led to the encroachment of the forest area of Sanjay Gandhi National Park, thereby, decreasing the green cover of Mulund [10]. This is a highly unwelcome change on the climatic front, as 'urban greenery helps regulate urban microclimates' [14]. Thus, it can be said that the growth of high rises and the increase in slums along with the encroachment of forest land have contributed to a decrease in resilience to climate change. In fact, buildings are responsible for one-quarter of carbon-di-oxide emissions and one-third of world energy use, which makes them a significant contributor to climate change [16]. However, the coming up of parks and gardens is a positive aspect in this regard. The presence of parks and gardens provides a cooling effect on the surrounding urban areas, thereby decreasing the temperature in the built-up areas [2]. Thus, the large presence of parks and gardens within Mulund serves as an asset for climate change resilience.

The growth of office and business spaces has boosted the service sector and retail economy of Mulund, without rendering any contribution to building climate change resilience. Similarly, the government's plans and policies for improving the transport network and most of the PPP projects undertaken for developmental purposes are not associated with building any kind of climate change resilience in the area. However, the project on the processing of waste through biomining at the Mulund Dumping Ground is associated with environmental and climatic benefits. The process of biomining not only helps in achieving zero emissions of landfill gases and leachate but also allows for the reuse of land with no maintenance required [28]. Thus, it can be expected that the project's completion will improve the climate change resilience status in Mulund.

Finally, the growth of residents, retail, and business has made Mulund populated, popular, and economically progressive. However, this has not led to any kind of contribution to climate change resilience. Rather, it can be argued that with the growth of residents and retail activities, the use of air conditioners, refrigerators, and similar other luxuries contributing to greenhouse gas emissions has also increased, thereby causing damage to the environment and climate. As hydrofluorocarbon refrigerants are powerful greenhouse gasses and require a lot of energy, a large portion of the cooling equipment currently in use contributes twice to climate change [45].

4.4 Policy Recommendations

Given the current dynamics of the transformation underway and the influential role of its contributing factors, a comprehensive set of recommendations is put forth to bolster climate change resilience within the study area. These recommendations encompass the following strategies:

(i) Urban planning strategies must be focused on minimizing the environmental impacts of urbanization. Accordingly, emphasis must be given to developing low-carbon and climate-resilient infrastructure, which may include green buildings and energy-efficient constructions.
(ii) Affordable housing options that incorporate energy-efficient design principles and renewable energy sources must be prioritized to reduce the emission of greenhouse gases from residential buildings.
(iii) The existing parks and gardens of Mulund are crucial green spaces and efforts must be made to protect them while creating more new such green areas to counteract the urban heat island effect and enhance resilience towards climate change by regulating local microclimates.
(iv) Generating awareness among local communities through education, awareness campaigns, etc., can help in engaging them in climate change resilience-building efforts through participatory decision-making.
(v) Incentives can be provided for the use of energy-efficient technologies and practices in residential as well as commercial buildings. This would make building climate change resilience more appealing to the people.
(vi) Climate change resilience considerations must be incorporated into the design and construction of the infrastructural projects. Emphasis must be given to the construction of flood-resistant buildings, stormwater management systems, and green roofs, to withstand extreme weather events and minimize climate-related risks.
(vii) Sustainable business practices must be promoted among local businesses too. Practices such as incorporating energy efficiency measures, waste reduction strategies, and sustainable sourcing practices, can reduce their carbon footprint and contribute to climate resilience.

4.5 Future Research Directions

Research is a continuous process. There is no end to it. All research opens doors to many more. As a continuum to this study, further research can be carried out to explore how the engagement of residents, retailers, and businesses in climate resilience initiatives can be improved. Similarly, longitudinal studies can be conducted to track changes in climate resilience over time in response to various interventions and policies. This study opens avenues for interdisciplinary research. The complexity and multifaceted nature of urban systems necessitates the integration of various disciplines to comprehensively address the challenges and opportunities associated with sustainable urban development. The future research path can be therefore navigated through interdisciplinary and collaborative initiatives.

4.6 Conclusion

From the above discussion, it is clear that the identified determinants of urban transformation in Mulund are intricately linked to each other, however, these determinants are loosely related to building climate change resilience in the area. In other words, the process of urban transformation in Mulund is associated with scarce initiatives geared towards building resilience to climate change. However, the coming up of parks and gardens in Mulund along with the closure of the Mulund dumping ground and implementation of biomining for waste processing are commendable steps concerning building climate change resilience in the area. Further, it is hoped that applications of the recommendations made in this chapter will help improve the area's position as a climate change-resilient area. Hence, it is envisaged that these recommendations be put into practice through the required program interventions.

Acknowledgements This chapter draws upon data gathered during my doctoral research undertaken at the Department of Geography, University of Mumbai, under the guidance of Dr. Sanjukta Sattar, whose mentorship I deeply appreciate. I am also thankful to the participants who generously participated in the study and provided their invaluable insights.

References

1. Abhishek N, Jenamani M, Mahanty B (2017) Urban growth in Indian cities: are the driving forces really changing? Habitat Int 69:48–57. https://doi.org/10.1016/j.habitatint.2017.08.002
2. Algretawee H, Rayburg S, Neave M (2019) Estimating the effect of park proximity to the central of Melbourne city on Urban Heat Island (UHI) relative to Land Surface Temperature (LST). Ecol Eng 138:374–390. https://doi.org/10.1016/j.ecoleng.2019.07.034
3. Andersen B, Eline Ander H, Skrede J (2020) The directors of urban transformation: the case of Oslo. Local Econ 35(7):695–713. https://doi.org/10.1177/0269094220988714
4. Ataman C, Tuncer B (2022) Urban interventions and participation tools in urban design processes: a systematic review and thematic analysis (1995–2021). Sustain Cities Soc 76:103462. https://doi.org/10.1016/j.scs.2021.103462
5. Banerjee S (2014) NMMT will run 2 routes from city to airports. The Times of India
6. Bhagat RB, Jones GW (2013) Demographic dynamics of mega-urban regions: the case of Mumbai
7. Bharath HA, Chandan MC, Vinay S, Ramachandra TV (2018) Modelling urban dynamics in rapidly urbanising Indian cities. Egypt J Remote Sens Space Sci 21(3):201–210. https://doi.org/10.1016/j.ejrs.2017.08.002
8. Bhatia N (2014) Mumbai's Eastern suburbs: Infrastructure leads development. The Indian Express
9. Bhattacharjee S (2021) The remaking of urban space: a study of transformation of Mulund. The University of Mumbai, Mumbai

10. Bhattacharjee S (2022) Degradation of urban environment: when residents' expectations and policy implementations do not coincide. The holistic approach to environment 12(2):79–85. https://doi.org/10.33765/thate.12.2.3
11. Bhattacharjee S (2021) Comprehending the gentrification of a suburb: the case of Mulund. Mumbai. GeoJournal 86(1):133–143. https://doi.org/10.1007/s10708-019-10067-5
12. Bhattacharjee S, Sharma M (2023) Polycentric Urbanism and the Growth of New Economic Hubs in Mumbai, India. In: Chatterjee U, Bandyopadhyay N, Setiawati MD, Sarkar S (eds) Urban Commons, Future Smart Cities and Sustainability. Springer International Publishing, Cham, pp 169–186
13. Chen Y (2022) Financialising urban redevelopment: transforming Shanghai's waterfront. Land Use Policy 112:105126. https://doi.org/10.1016/j.landusepol.2020.105126
14. Elmqvist T, Setälä H, Handel S, van der Ploeg S, Aronson J, Blignaut J, Gómez-Baggethun E, Nowak D, Kronenberg J, de Groot R (2015) Benefits of restoring ecosystem services in urban areas. Curr Opin Environ Sustain 14:101–108. https://doi.org/10.1016/j.cosust.2015.05.001
15. Ganapatye M (2017) One rupee clinics now open at 14 Mumbai railway stations—India Today. India Today
16. González-Torres M, Pérez-Lombard L, Coronel JF, Maestre IR, Yan D (2022) A review on buildings energy information: trends, end-uses, fuels and drivers. Energy Rep 8:626–637. https://doi.org/10.1016/j.egyr.2021.11.280
17. Guo A, Yang J, Xiao X, Xia (Cecilia) J, Jin C, Li X (2020) Influences of urban spatial form on urban heat island effects at the community level in China. Sustain Cities Soc 53:101972. https://doi.org/10.1016/j.scs.2019.101972
18. Hölscher K, Frantzeskaki N (2021) Perspectives on urban transformation research: transformations in, of, and by cities. Urban Transform 3(1):2. https://doi.org/10.1186/s42854-021-00019-z
19. Home Times (2018) State government announces to change leasehold refugee land to freehold 35000 Refugees at Mulund Colony, REJOICE!. Home Times
20. Khedkar AR (2018) Closing down of textile mills: a break-through in the real estate or a curse to Mumbai. Int J Curr Res 10(08):72641–72648
21. Knox P, Pinch S (2010) Urban social geography, 6th edn. Pearson Education Limited, Essex
22. Koramaz EK, Koramaz TK, Özer Ö (2018) Urban transformation through property-led regeneration: a case of building renewals in Istanbul. A|Z ITU J Faculty Archit 15(2):183–196. https://doi.org/10.5505/itujfa.2018.95815
23. Koumparelou A, Feiler A, Kühn L, Hübscher M (2023) Suburbanization Within the City? exploring suburban lifestyles in the inner-city of Leipzig (germany). Erdkunde 77(1):71–89
24. Lai Y, Chen K, Zhang J, Liu F (2020) Transformation of Industrial Land in Urban Renewal in Shenzhen. China Land 9(10):371. https://doi.org/10.3390/land9100371
25. Leao S, Bishop I, Evans D (2001) Assessing the demand of solid waste disposal in urban region by urban dynamics modelling in a GIS environment. Resour Conserv Recycl 33(4):289–313. https://doi.org/10.1016/S0921-3449(01)00090-8
26. Lin P, Lau SSY, Qin H, Gou Z (2017) Effects of urban planning indicators on urban heat island: a case study of pocket parks in high-rise high-density environment. Landsc Urban Plan 168:48–60. https://doi.org/10.1016/j.landurbplan.2017.09.024
27. Meral GİC, Özsoy DA (2021) Intersection of planned and unplanned residential patterns: a case from Istanbul. UIA 2021 RIO: 27th World Congress of Architects, 1022–1028
28. Mohan S, Joseph CP (2020) Biomining: an innovative and practical solution for reclamation of open dumpsite. In: Kalamdhad AS (ed) Recent Developments in Waste Management. Springer, Singapore, pp 167–178
29. Mostafavi F, Tahsildoost M, Zomorodian Z (2021) Energy efficiency and carbon emission in high-rise buildings: a review (2005–2020). Build Environ 206:108329. https://doi.org/10.1016/j.buildenv.2021.108329
30. Mumbai Mirror (2018) Mulund dumping ground, 2nd-largest in city, shut. Mumbai Mirror
31. Nainan N (2008) Building boomers and fragmentation of space in Mumbai. Econ Pol Wkly 43(21):29–34

32. Nijkamp P, Kourtit K (2013) The "New Urban Europe": global challenges and local responses in the urban century. Eur Plan Stud 21(3):291–315. https://doi.org/10.1080/09654313.2012.716243
33. Pamukcu-Albers P, Ugolini F, La Rosa D, Grădinaru SR, Azevedo JC, Wu J (2021) Building green infrastructure to enhance urban resilience to climate change and pandemics. Landscape Ecol 36(3):665–673. https://doi.org/10.1007/s10980-021-01212-y
34. Parab B (2017) Mulund dump closure cost doubles. The Asian Age
35. Praja Foundation (2020) Report on status of Civic Issues in Mumbai. Praja Foundation
36. Rao S (2010) 180 new BEST buses by year-end | Mumbai news—Hindustan Times. Hindustan Times
37. Salimi M, Al-Ghamdi SG (2020) Climate change impacts on critical urban infrastructure and urban resiliency strategies for the Middle East. Sustain Cities Soc 54:101948. https://doi.org/10.1016/j.scs.2019.101948
38. Sen S (2015) New AC buses connect south Mumbai to Thane via freeway. The Times of India
39. Sequera J, Nofre J (2020) Touristification, transnational gentrification and urban change in Lisbon: the neighbourhood of Alfama. Urban Studies 57(15):3169–3189. https://doi.org/10.1177/0042098019883734
40. Shaw A, Satish MK (2007) Metropolitan restructuring in post-liberalized India: separating the global and the local. Cities 24(2):148–163. https://doi.org/10.1016/j.cities.2006.02.001
41. Singh R (2022) Reuse of reclaimed land after biomining of legacy waste: what needs to change. https://www.downtoearth.org.in/blog/waste/reuse-of-reclaimed-land-after-biomining-of-legacy-waste-what-needs-to-change-82208. Accessed 12 Apr 2024
42. Smith N (1996) The new urban frontier: gentrification and the Revanchist City, 1st edn. Routledge, London
43. Solomou S, Sengupta U, Cheung E, Oredein O (2022) A strategic Planning strategic planning problem: the relationship between urban transformation transformations outcomes and the temporal order of planned projects. In: Iossifova D, Gasparatos A, Zavos S, Gamal Y, Long Y (eds) Urban infrastructuring: reconfigurations, transformations and sustainability in the global south. Springer Nature, Singapore, pp 241–264
44. The Indian Express (2018) Mulund: World's largest bio-mining project set to start this year. The Indian Express
45. UNEP (2023) Air conditioners fuel the climate crisis. Can nature help? In: United Nations Environment Programme. http://www.unep.org/news-and-stories/story/air-conditioners-fuel-climate-crisis-can-nature-help. Accessed 27 Apr 2024
46. US EPA (2023) Climate change and heat Islands. https://www.epa.gov/heatislands/climate-change-and-heat-islands. Accessed 12 Apr 2024
47. Vaughan L (2015) Introduction: suburbs are as old as the City Itself. In: Vaughan L (ed) Suburban urbanities: suburbs and the life of high street, 1st edn. UCL Press, London

Chapter 5
Assessing the Role of Urban Form in Ecosystem Service Provisioning as a Part of Urban Resilience: A Case Study on Kolkata

Swarnava Dey, Jayita Guha Niyogi, and Debashish Das

Abstract The concept of local ecosystem services is deeply intertwined with the urban microclimatic variability. This idea will be better applicable in urban planning when it can be delineated based on the urban form and its interrelationships. Therefore, we coupled a multivariate approach with urban landscape metrices as a proxy of landform characteristics to overcome existing limitations in interpreting the potential of ecosystem service in Kolkata Municipal Corporation. Five urban landscape metrices defined by size, density, complexity, and porosity of built-up patches were selected to comprehend the diverse dimensions of urban built-form, while five ecosystem performance indicators representing carbon fixation, hydrology, landscape-diversity, climatic regulation, and air filtration were used to define the spatial variability in ecosystem service potential. Several stepwise multiple linear regression models were used to investigate the influence of urban landscape metrics on ecosystem services. Results of the study indicated that areas characterized by large built-up patches, low fragmentation, and limited open spaces were less efficient in generating ecosystem services. In contrast, areas with smaller, fragmented built-up patches and higher porosity exhibited greater potential for ecosystem services. Porosity in built-up areas had the most significant impact on ecosystem service potential. This study not only has quantified the spatial intensity of ecosystem service potential in respect to different built-form but also introduced a methodology that might act as a comprehensive aid to policymakers and urban planners for a better understanding of sustainable and resilient urban form to maximize ecosystem services at the local administrative level.

S. Dey (✉)
School of Environmental Studies, Jadavpur University, Kolkata 700032, India
e-mail: swarnava65@gmail.com

J. G. Niyogi · D. Das
Department of Architecture, Jadavpur University, Kolkata 700032, India
e-mail: jayitaguha.niyogi@jadavpuruniversity.in

D. Das
e-mail: ddasju.arch@jadavpuruniversity.in

S. Ghosh et al. (eds.), *Cities of Tomorrow: Urban Resilience and Climate Change Preparedness*, Urban Sustainability, https://doi.org/10.1007/978-981-97-9658-8_5

Keywords Ecosystem services · Ecosystem performance indicators · Urban form · Landscape metrices · Urban Resilience

5.1 Introduction

As a major catalyst of urban sustainability and human well-being ecosystem services (ESs) encompass the provision of life-supporting products and services by natural landscapes [37, 39]. ESs bridge the gap between human well-being and the health of natural systems, offering valuable perspective for sustainable urban planning [7]. As a result, cities are increasingly embracing ecosystem-based adaptation (EBA) as a crucial strategy to combat climate change, enhance urban resilience, and foster sustainable urban transitions [38]. Studies often incorporated regulatory services like temperature regulation [48], carbon fixation [60, 61], runoff mitigation [18, 53], air filtration [3], biodiversity conservation [40, 44] as essential components of urban resilience. While numerous researches have incorporated the idea of EBA in urban planning, effective implementation across varied biophysical conditions and development contexts remains a complex challenge that demands further scientific research and experiential learning [25]. The potential of world ecosystem is significantly influenced by various urban growth patterns, each impacting the spatial heterogeneity of ecosystem functions differently [1]. Numerous studies have explored the link between urbanization and ecosystems [20, 50], but the impact of diverse urban designs on the distribution and flow of ESs remains relatively unexplored. However, there has been a notable increase in research examining the spatial configuration of urban forms and their relationship with ecosystem services (ESs) using landscape metrices [23], though the emphasis has mostly revolved around a restricted array of services and landscape characteristics. Apart from that, our comprehension of how centralized versus dispersed urban configuration, uniquely affect environmental status remains limited, as does our clarity regarding the ecological trade-offs associated with varying housing and infrastructure densities [1]. Minimizing the spatial footprint of urban areas by promoting more compact city designs is one method to lessen the effects of increasing urbanization. Although the "compact city" paradigm, has sparked considerable discussions about the possible long-term social and ecological advantages, but little is known about how this trend affects the unique features of urban resilience. Apart from that, previous studies also tried to incorporate the themes of garden city and smart city as a surrogate to urban resilience, which indirectly acknowledged the local ESs [11, 56]. The conventional description of urban resilience often encircles around the sustainable assimilation of natural environment, urban built-form, and local communities [52]. A comprehensive evaluation of these interconnected elements is vital for a resilient and sustainable urban future. Acceptable plans might moderate the dire impacts of urbanization and enhance the urban resilience, but in a complex urban setting like Kolkata, the relationship between the ecosystem and the city is still unclear. The initial phase in incorporating sustainable urban plans is to evaluate the ecological performance based on different urban configuration.

This study aims to addresses the interconnection of ecosystem, urban built-form, and urban resilience. In this context we analysed the ecosystem service potential in relation to spatial form and characteristics of the urban-built landscape and tried to suggest practical measures to enhance urban resilience.

5.2 Study Area

The study area (Fig. 5.1) Kolkata Municipal Corporation (KMC), located on the left (eastern) bank of the Hooghly River (22°28′00″–22°37′30″ N to 88°17′30″–88°25′00″ E). According to the Köppen climatic classification system Kolkata has a tropical wet-dry (Aw) climate [21]. Kolkata experiences an annual mean temperature of 26.8 °C, with monthly mean temperatures ranging from 19 °C to 30 °C, and receives approximately 1600 mm of annual precipitation [13]. KMC comprises 147 wards, divided into 16 boroughs, which serve as the territorial and functional core of the Kolkata Metropolitan Area [45]. According to the Kolkata Municipal Corporation, the city has approximately 4.5 million people, a population density of 24,252 individuals/km^2, and over one million buildings, of which 79.23% are reinforced cement concrete constructions.

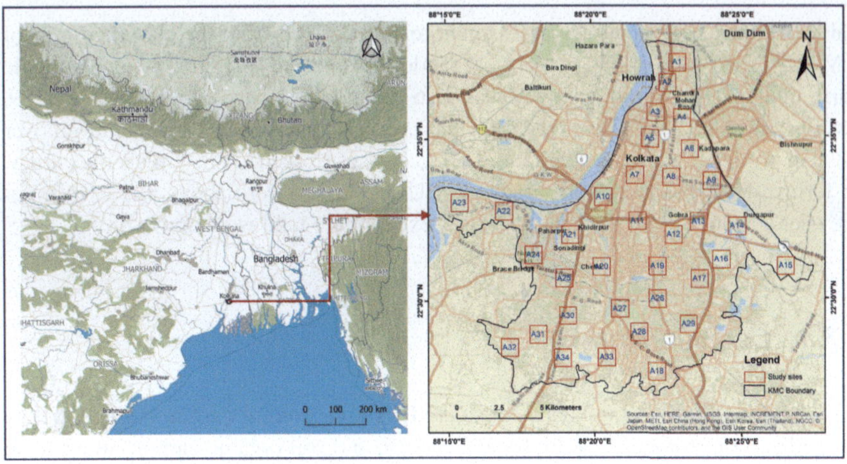

Fig. 5.1 Map showing study area (KMC) and sample sites

5.3 Materials and Methods

5.3.1 Data

Thirty-four sample sites (Fig. 5.1) ware selected based on built-up density, functional traits, and ratio of green spaces. Built-up density serves as a measure of intensity of urbanization, functional traits reflect socio-cultural aspects of an area [28], and the ratio of open space acts as a proxy for urban sustainability [35]. Different parts of the city were covered in the sampling process to maintain spatial variability. To create the land-use land-cover (LULC) map of KMC for 2022 (Fig. 5.2), two Sentinel 2A images (Date of Acquisition: 09/01/2022) were mosaicked together to delineate the desired study area, then supervised classification was performed to produce five LULC classes: built-up (represented settlements, roads, buildings, highways, foot-paths, construction sites, and metaled) surfaces; waterbody (represented inland water bodies, wetlands, canals, and ponds); bare/fallow (represented urban vacant land with unmetalled surface, fallow agricultural land); grass (represented land covered with grass, small bushes and shrubs, herbaceous plants); and tree cover (represented trees and large woody bushes and shrubs). Finally, the areas of every LULC class are calculated. All the processing of images is done by Arc GIS 10.3.1 software.

5.3.2 Accuracy Assessment

Table 5.1 shows the accuracy assessment results derived from the classified map of KMC, 2022. A total of 250 pixels were randomly selected and checked using google earth image. The overall accuracy was 91.67% and the Kappa coefficient was 0.89. Every Land Use Land Cover (LULC) class, except bare/fallow land, achieved over 88% in producer accuracy, while all LULC classes exceeded 88% in user accuracy. In terms of the producer's accuracy, all LULC classes were above 88%, while for the user's accuracy, all the LULC classes except bare land were above 88%.

5.3.3 Carbon Fixation

Four essential carbon pools are considered (Above ground carbon storage, below ground carbon storage, dead biological carbon, and soil organic carbon) to estimate the carbon storage by combining carbon densities assigned to every LULC class [59]. Carbon pool data for the vegetation classes was obtained from the India State of Forest Report, 2021. Field measurements offer precise biomass and carbon value estimation, but they are very labor-intensive and time-consuming [62]. So, carbon data for non-vegetation classes was acquired from the Intergovernmental Panel on

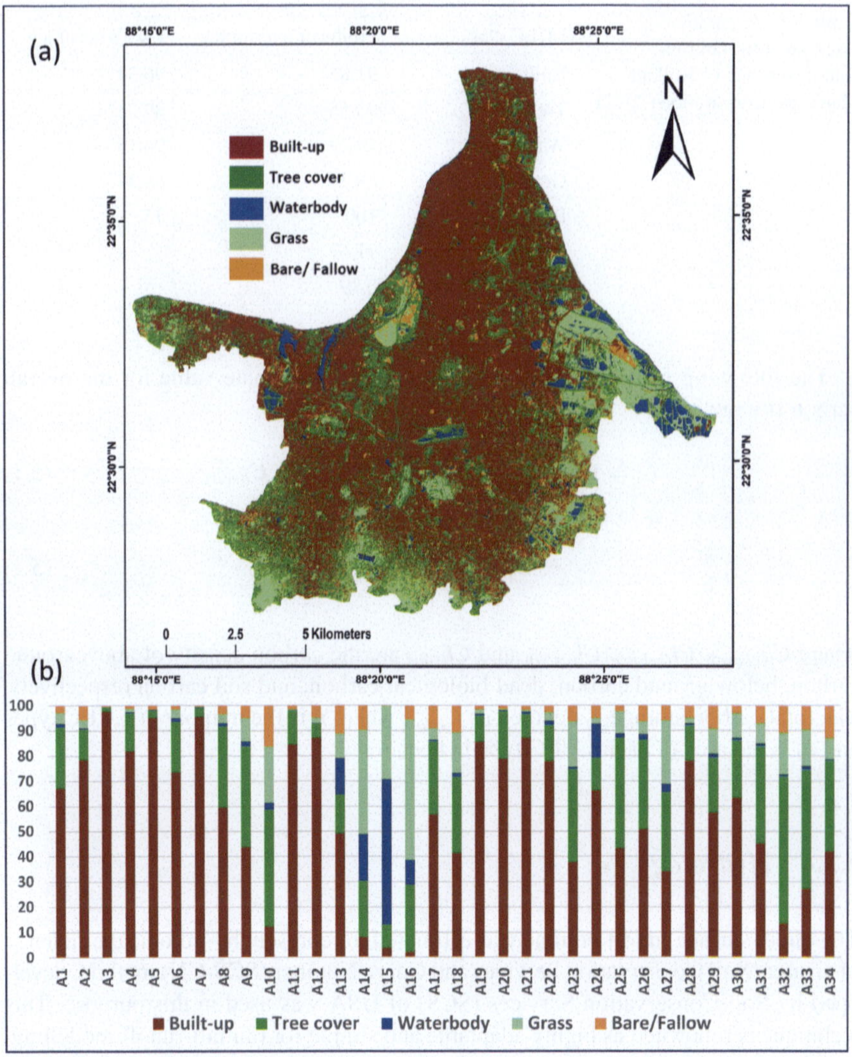

Fig. 5.2 **a** Land-use land-cover map of Kolkata Municipal Corporation for 2022, **b** Percentage of area covered by land-use land-cover class in each sample site

Climate Change's (IPCC) 2006 report, which is considered as highly reliable for carbon values globally [59].

Table 5.1 Accuracy assessment for land-use land-cover map of Kolkata Municipal Corporation (2022)

LULC class	Producer's accuracy	User's accuracy
Built-up	91.86	96.34
Tree cover	93.55	89.23
Water body	89.29	96.15
Grass	88.37	88.37
Bare/fallow land	100	75
Overall Accuracy	91.67	
Kappa Statistic	0.89	

The following formulae were used to get an approximate value for the overall carbon storage:

$$C_i = C_{i(above)} + C_{i(below)} + C_{i(dead)} + C_{i(soil)} \tag{5.1}$$

$$\mathrm{C_{total}} = \sum_{\mathrm{i}}^{\mathrm{n}} \mathrm{C_i} + \mathrm{A_i} \tag{5.2}$$

where $Ci_{(above)}$, $Ci_{(below)}$, $Ci_{(dead)}$, and $Ci_{(soil)}$ are the carbon density of aboveground carbon, below ground carbon, dead biological carbon, and soil carbon respectively. The total carbon storage is denoted as C_{total}, where n is the number of LULC types and A_i is the area of each LULC type in hectares.

5.3.4 Hydrology

The direct surface runoff volume was estimated to represent hydrological efficiency of an area. Soil Conservation Services and Curve Number (SCS-CN) method, developed by Soil Conservation Services (SCS) of USA was used in this purpose. This technique is renowned as highly adaptable and simple for rainfall-runoff modelling, widely adopted in numerous studies [6, 57, 65]. This method accounts for factors such as soil permeability, land-use land-cover, and antecedent moisture conditions to calculate thevolume of runoff [4]. The SCS-CN technique depicting the direct surface runoff can be expressed as.

$$Pe = \frac{(P - 0.2S)^2}{P + 0.8S} \tag{5.3}$$

P= rainfall and Pe= direct runoff, S= potential maximum retention

$$C_{total} = \sum_{i}^{n} C_i + A_i \tag{5.4}$$

Based on land-use land-cover, hydrological soil group (HSG), and antecedent moisture conditions (AMC), a table is used to generate the Curve Number (CN), a dimensionless number ranging from 0 to 100. AMC signifies the moisture level in the soil at the onset of the rainfall-runoff event under consideration. For modelling purposes, AMC II has been chosen as it indicates an average moisture condition. The hydrologic Soil Group (HSG) is typically categorized into four groups (A, B, C, and D), each representing different rates of soil infiltration. Group C has been selected for this study because it exhibits a moderately high runoff potential along with relatively low infiltration rates which resembles the characteristics of urban soil.

5.3.5 *Diversity*

Biodiversity is an important ecological performance indicator which ensures sustainability [55]. But biodiversity should encompass more than just species diversity, it should also consider various habitats or ecosystems [66]. Many studies incorporated landscape heterogeneity as a proxy to measure biodiversity [16, 36]. In this study we estimated landscape-diversity as a part of ecological performance, using Shannon-Weiner Index(D). This indicator was initially created by [26] as a measure that fuses the heterogeneity [67]. The following formula was used to derive the landscape diversity, where the index is applied to habitats rather than populations.

$$D = -\sum_{i=1}^{n}(p_i log_2 p_i) \tag{5.5}$$

where, $D =$ Diversity Index, $p_i =$ Proportion of Individual land use category. Urban land-use, vegetation, bare land, grass, and water were used as proxy habitats.

5.3.6 *Climate Regulation*

An effective measure of climate regulation may be depicted as the anomaly in air temperature distribution between the central metropolitan area and the outskirts or rural regions [64, 67]. Estimating air temperatures in a comprehensive and accurate manner is not straightforward, and establishing correlations between certain land use categories and air temperatures is difficult, requiring complex simulation and large amount of data [10]. This level of detail may not always be practical for daily planning. Hence, there is a need for proxy indicators that can effectively represent air temperature variations in urban areas. Urban heat studies often rely on satellite-derived Land Surface Temperature (LST) due to the limited availability of air temperature (Ta) observations [2, 12, 63]. This study also incorporated LST as a proxy to estimate the climatic regulation service.

5.3.7 Air Filtration

Urban vegetation often enhances air filtration by facilitating the deposition and retention of pollutants (Lindén et al. 2023). The rate at which air pollutants are removed from the atmosphere in the absence of precipitation is called the dry deposition rate, indicating the efficiency at which pollutants settle on surfaces under dry conditions [42]. Vegetation has much greater capacity for air pollution deposition than built-surfaces [31]. Therefore, this study calculated the air pollution removal (*Fi*) based on the dry deposition velocity (*Vdi*) and pollutant concentration (*Ci*). Pollutant concentration rate is calculated based on the compiled data from West Bengal pollution control board (wbpcb.gov.in) while due to absence of dry deposition related data we relied on an empirically established average dry deposition velocity rate [46].

$$F_i = Vd_i \times C_i \tag{5.6}$$

5.3.8 Defining the Urban Landscape Metrics

The term 'urban form' can be used in various ways. It can be quantified based on the physical structure of an urban region, such as based on LULC, or it might incorporate demographic or socioeconomic characteristics. In this study, six urban landscape metrics have been chosen to define urban form. All the metrics are quantitative measures of spatial heterogeneity employed based on physical characteristics of the urban region and determined on a class level approach. These metrics represent the size, density, complexity, and porosity of an urban area (Fig. 5.3) and calculated based on the urban land-use patches in ArcGIS 10.3.1 software. All the metrics of urban landscape are defined in Table 5.2.

Size: Mean patch size (MPS) indicates the average size of sealed urban patches. It indicates the average size of the patches comprising the landscape or class. A high urban MPS indicates a high scale of urbanized landscape, more consolidated and centralized city growth, whereas a lower MPS represents greater fragmentation [14].

Density: Mean patch density is calculated to estimate the density parameter of each selected site. It is calculated on the basis of per hundred sq. mt. area and indicated the number of urban patches. Patch Density can be considered as an indicator of urban fragmentation [51]. Patch density should be low in a region with a continuous stretch of urban land-use, but high in an area with small fragmented urban patches.

Complexity: The area weighted mean shape index (AWMSI), and the area weighted mean patch fractal dimension (AWMPFD) are used to define the complexity of an urban environment. These measurements mostly define the irregularity of an urban patch. The former denotes irregularity in the shape of an urban boundary; a greater value denotes an uneven shape, whereas a value of one denotes a perfectly square

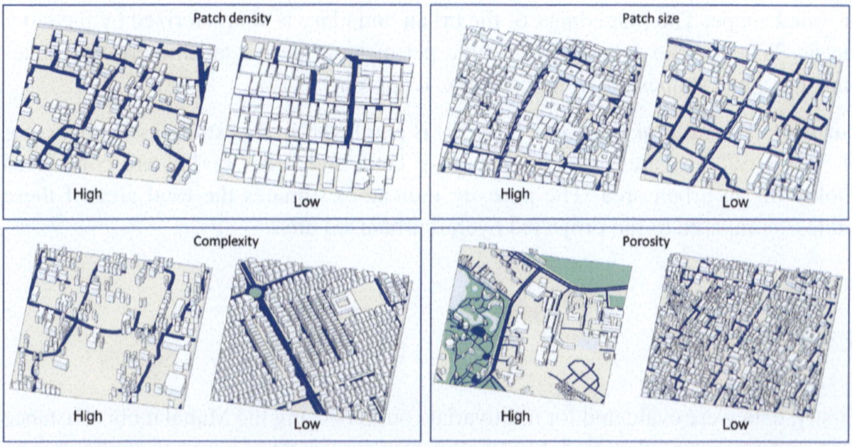

Fig. 5.3 Schematic map of urban landscape metrices

Table 5.2 Urban Landscape Metrics and their descriptions

Indicators	Abbreviation	Formula	Description	Source
Mean Patch Density	MPD	$\frac{n_i}{A*100}$	Where a_{ij} and p_{ij} are the area (sq. m) and perimeter of patch ij, n_i is the total number of patches and A is the landscape area (sq. m)	[41]
Mean Patch Size	MPS	$\frac{\sum_{j=1}^{n} a_{ij}}{n_i}\left(\frac{1}{10000}\right)$		
Area Weighted Mean Shape Index	AWMSI	$\sum_{j=1}^{n}\left[\left(\frac{p_{ij}}{2\sqrt{\pi a_{ij}}}\right)\left(\frac{a_{ij}}{\sum_{i=1}^{n} a_{ij}}\right)\right]$		
Area Weighted Mean Patch Fractal Dimension	AWMPFD	$\sum_{j=1}^{n}\left[\left(\frac{2 ln p_{ij}}{ln a_{ij}}\right)\left(\frac{a_{ij}}{\sum_{j=1}^{n} a_{ij}}\right)\right]$		
Ratio of Open Space	ROS	$\frac{s'}{s}(100)$	Where s' is the summarization area of all "holes" inside the extracted urban area, s is summarization area of all the patches	[29]

or round shape. The raggedness of the urban boundary is characterized by the latter metric. When urban forms have simple perimeters, the fractal dimension reaches one, and when shapes are more complex, it approaches two.

Porosity: The ratio of open space (ROS) is employed in this study as the indicator of "porosity." The areas of greenery, bare land, and water bodies are considered "holes" in an urban area. The porosity indicator estimates the total area of these "holes" compared to the projected overall urbanized area.

5.3.9 Analysis

Firstly, data were evaluated for multivariate outliers using the Mahalanobis Distance Test, and none were identified. Multicollinearity was checked between predictor variables, and no multicollinearity was found as the variance inflation factor (VIF) in all cases was <10 [24]. Pearson's product moment correlation was calculated between the urban landscape metrices (ULMs) and ecosystem performance indicators (EPIs) to determine the association between them. The stepwise multiple linear regression (SMLR) model was used to assess the relationship between dependent and all independent variables and their influence in regulation of ecosystem service potential. A regression normalized residual histogram was created to match the linear regression assumptions, which demonstrated that the data is normally distributed. Furthermore, the regression standardized residuals were normally distributed, with the actual and expected values falling along the line with no notable deviations. To test for homoscedasticity, scatter plots were created using the standardized Z scores of predictors and predictive variables, which estimated that there was no apparent relation between the residuals and the predicted value of the dependent variable, indicating that linearity between them. As a result, the assumptions of linearity and homoscedasticity in multivariate analysis were met. Finally, the Durbin Watson test determined that the observations were independent.

5.4 Result

5.4.1 Models of Ecological Performance Indicators

The comprehensive analysis (Table 5.3) delved into the intricate dynamics of urban landscape metrices (ULMs), revealing specific variables that exert a significant influence on ecological indicators. Among these variables, mean patch density (MPD) and the ratio of open space (ROS) emerged as pivotal predictors of carbon storage within urban environments. Notably, MPD exhibited a stronger predictive capability, emphasizing its crucial role in shaping urban carbon dynamics. Moreover, the analysis unveiled the multifaceted role of ROS in managing run-off dynamics, where higher ROS values were associated with reduced run-off rates. This finding underscores the invaluable function of open spaces in ameliorating run-off issues, a key consideration in sustainable urban water management strategies. Furthermore, UPD emerged as a fundamental determinant of pollutant removal rates, with higher MPD values correlating with enhanced pollutant removal capabilities. Interestingly, the fractal dimension of urban patches (AWMPFD) also contributed to pollutant removal, albeit exhibiting a negative relationship, highlighting the complex interplay between urban form characteristics and pollutant dynamics. In addition to these ecological implications, the study revealed significant impacts on landscape diversity, with both ROS and AWMPFD exerting a substantial influence on Shannon diversity in urban areas. Higher ROS values and more intricate urban patch structures were associated with greater diversity, emphasizing the need for urban planners and environmental managers to consider these factors in fostering urban ecosystems that support diverse flora and fauna. Furthermore, the study delved into the thermal dynamics of urban environments, showcasing the critical role of ROS and MPD in determining land surface temperature (LST). Higher ROS values were linked to cooler temperatures, reflecting the cooling effects of open spaces, while increased MPD was associated with higher LST due to the urban heat island phenomenon. These insights underscore the nuanced relationship between urban form variables and various ecological processes, highlighting the necessity of integrating these findings into urban planning and environmental management strategies for fostering sustainable and resilient urban ecosystems.

5.5 Discussion

The results showed that areas with dense urban growth tend to exhibit poor provision of ecosystem services (Fig. 5.4). High-density urban development denotes areas with a large continuous stretch of urban patches rather than multiple small patches (Fig. 5.3). In this study, more densely urbanized areas had less coverage by open spaces or holes (mostly blue and green spaces in this study). These areas displayed higher runoff coefficient and LST while stored less carbon, removed lesser pollutants, and had poorer landscape diversity. The majority of sample sites with high Mean Patch Density (MPD) exhibited lower levels of urbanization and greater vegetation cover. Along this line, the relation between MPD, carbon storage and pollutants removal are quite normal. The relationship between LST and the Ratio of Open Space (ROS) was robust, underscoring the crucial role of open spaces in moderating temperature dynamics within urban areas. Additionally, landscape diversity exhibited a significant association with ROS, indicative of the positive impact of diverse land uses on ecosystem resilience and habitat richness [33]. Furthermore, the study revealed a negative correlation between ROS and runoff, emphasizing the importance of non-sealed surfaces in mitigating stormwater runoff and maintaining hydrological balance in urban environments. Despite this general relationship, the ecological quality of urban areas with varying urban forms exhibited a high degree

Table 5.3 Stepwise regression models using urban landscape metrics to predict ecological performance indicators

Dependent variable	Model	Constant	R square	Adjusted R square	Unstandardized coefficients		Standardized Coefficients	Sig
					B	Std. Error	Beta	
Carbon storage	1	MPD	0.642	0.631	19.981	2.341	0.802	<0.001
					13.926	1.837		<0.001
	2	MPD ROS	0.739	0.722	14.675	2.566		<0.001
					8.725	2.214	0.502	<0.001
					0.235	0.069	0.431	0.002
	3	MPD ROS AWMSI	0.789	0.768	2.891	5.002		0.568
					8.141	2.035	0.469	<0.001
					0.301	0.068	0.554	<0.001
					1.782	0.668	0.246	0.012
Run-off	1	ROS	0.859	0.855	140.27	2.029		<0.001
					−0.552	0.039	−0.927	<0.001
Pollutant removal	1	MPD	0.635	0.623	11.791	2.289		<0.001
					13.394	1.796	0.797	<0.001
	2	MPD AWMPFD	0.722	0.704	−119.74	42.279		0.008
					9.542	2.016	0.568	<0.001
					91.497	29.376	0.374	0.004
Shannon diversity	1	ROS	0.635	0.623	0.452	0.062		<0.001
					0.011	0.001	0.848	<0.001
	2	ROS AWMPFD	0.722	0.704	−3.703	0.684		<0.001
					0.007	0.001	0.559	<0.001
					2.923	0.48	0.486	<0.001
LST	1	ROS	0.635	0.623	34.735	0.244		<0.001
					−0.038	0.005	−0.82	<0.001
	2	ROS MPD	0.722	0.704	34.738	0.219		<0.001
					−0.051	0.006	−1.081	<0.001
					0.565	0.189	0.377	0.005

of unpredictability. Notably, not all variables of urban form exerted the same level of impact on ecological indicators. Among these variables, Mean Patch Density (MPD), Ratio of Open Space (ROS), and Average Weighted Mean Patch Fractal Dimension (AWMPFD) emerged as significant influencers. Understanding the key drivers behind this variation is crucial for planning and designing cities that prioritize optimal ecological quality and abundant ecosystem services. Existing urban plans cannot be erased to build a new one; rather, they can be transformed in many ways.

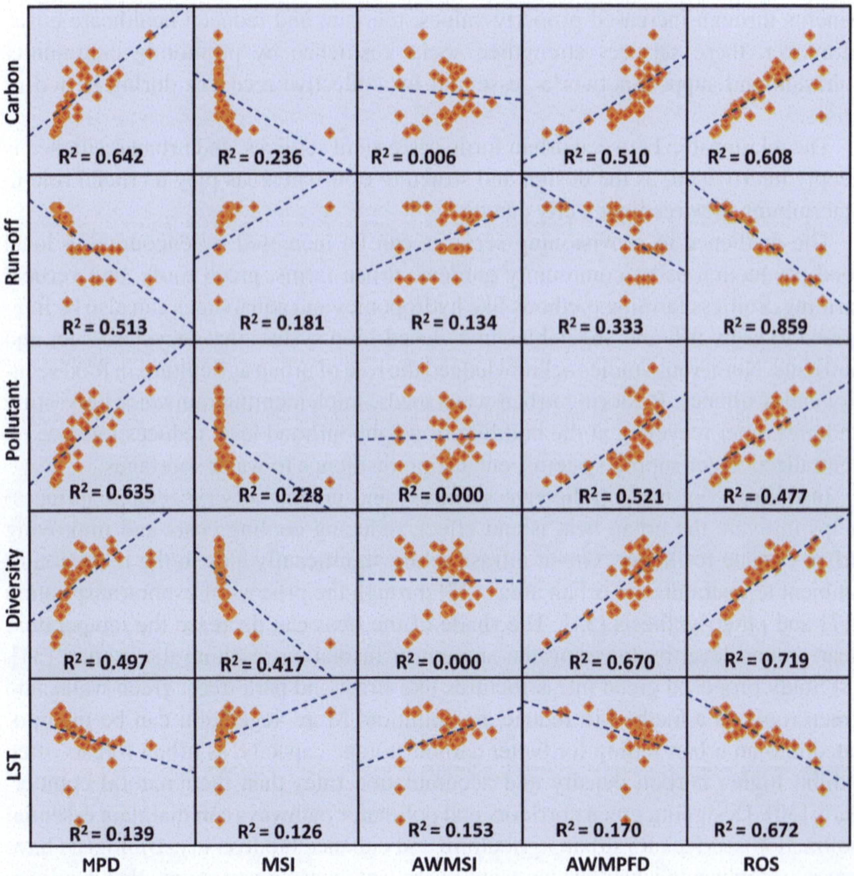

Fig. 5.4 Relationships between urban landscape metrics and ecological performance indicators across the 34 selected sites

Urban resilience refers to the capacity of urban systems to efficiently absorb, recover from, and adapt to a variety of disturbances and stresses, including environmental, economic, and social challenges. Numerous studies underscore the pivotal role of ecosystem services in boosting urban resilience [8, 9, 43]. Four major types of ecosystem services have their unique strengths in handling urban stresses and vulnerabilities [22]. Provisioning services, such as food production and fresh water supply, enhances food and water security, ensuring sustenance during crises. Regulating services including climate regulation and air filtration by green spaces and flood management by wetlands and green infrastructures help mitigate heatwaves and flooding. While supporting services like biodiversity maintenance ensure ecological stability and cultural services underpin recreational and aesthetic benefits, essential for mental health and social cohesion. Overall, urban ecosystem services provide natural buffers against disasters, aid in climate change adaptation, and offer economic benefits through increased property values, tourism, and reduced healthcare costs. Moreover, these services strengthen social resilience by promoting community cohesion and support networks, essential for collective recovery during period of crises.

The relationship between urban form, ecosystem services, and urban resilience is deeply intertwined, as the design and structure of urban areas play a crucial role in determining how resilient a city can be.

The resilience in provisioning services can be increased by encouraging local food production using community gardens, urban farms, green roofs, and vertical farming. Soilless farming methods like hydroponics and aquaponics can also be integrated to grow fish and vegetables in a closed-loop system that recycles water and nutrients. Numerous studies acknowledged the role of urban agriculture in food security and resilience. Restoring urban watersheds, implementing rainwater harvesting and greywater recycling at the building or neighbourhood level reduces reliance on centralized water supply systems, enhancing resilience to water shortages.

Installing cool roofs (reflective surfaces) and green roofs (vegetated surfaces) helps mitigate the urban heat island effect, reducing cooling costs and improving urban climate resilience. Green infrastructure significantly aids in the reduction of ambient temperatures in urban areas [27] through the process of evapotranspiration [17] and photosynthesis [54]. The shade of the tress can decrease the temperature near ground level by lowering the amount of insolation reaching the ground [34]. [5] Study proposed green infrastructures like street and park trees, green walls, and green roofs as a method to reduce air pollution. More vegetation can be incorporated into an urban setting for better carbon storage capacity, as urban forests often exhibit higher carbon density and accumulation rates than their natural counterparts [30]. Designing green corridors and pollinator pathways can maintain essential pollination services for urban agriculture and enhance biodiversity. Biologists have proposed enhancing habitat connectivity by safeguarding passages that facilitate species movement between existing green spaces to address issues related to habitat fragmentation [15, 32, 49]. Urban areas with permeable surfaces, retention basins, and natural drainage systems helps manage stormwater and reduces flood risks. In stormwater runoff management, trees in residential spaces can deliver several benefits [47]. A mature evergreen tree can intercept up to 4,000 gallons of water each year, while a deciduous tree is capable of intercepting between 500 and 700 gallons annually [58]. Interception of rainfall can reduce surface runoff rates, and rainwater moves through the tree canopy to trunks and the permeable areas, contributing to infiltration [19]. Application of permeable pavements, rain water harvesting, green roofs, rain gardens, bioswales, improving turf density and permeability, and application of organic mulches can decrease stormwater runoff and increase both surface water quality and infiltration [19].

Analyzing the relationship between urban form and ecosystem services is fraught with challenges due to the inherent complexity and multifaceted nature of both domains. Urban areas especially the complex urban fabrics of Kolkata exhibit a vast diversity in design, density, land use, and infrastructure, each uniquely influencing various ecosystem services. Relying on broad, generalized methods often oversimplify the analysis and become non conclusive. Therefore, longitudinal studies with context and scale specific application are necessary to capture temporal dynamics and complex feedback loops. Moreover, beyond ecological dynamics, the integration of ecosystem services in urban planning is heavily influenced by social perceptions, cultural norms, and traditions, which shape the supply, demand, and synergies of these services. On functional level fragmented jurisdictions, contrary policies and political red tapism hinder the implementation of integrated approaches to manage urban form and ecosystem services. Apart from individual service perspective urban resilience planning must address trade-offs between development and ecosystem services. High-density development can strain resources and reduce green spaces, while sprawling development can fragment habitats. Balancing these trade-offs involves smart growth strategies and compact development with integrated green infrastructure, while synergetic urban design can enhance resilience by simultaneously addressing multiple ESs. For example, urban parks not only provide recreational spaces but also improve air quality, support biodiversity, and reduce heat.

5.6 Conclusion

This study, across 34 sites in the KMC, showed that ecosystem services tend to decrease in urban forms where urban patches are large in size rather than number. A complex urban patch in terms of regularity in shape and size, with high raggedness and amount of open space, has better possibilities of creating ecosystem services than a compact urban patch with less open space. However, in sustainable urban planning compact, mixed-use developments with integrated green infrastructure can enhance urban resilience by reducing environmental impacts and improving resource efficiency. Conversely, poorly planned urban sprawl can exacerbate vulnerability to climate change and reduce the sustainability of ecosystem services. So, it can be estimated that urban form significantly shapes the provision and quality of ESs but it is sustainable planning which ensure the resilient flow of the ESs.

The 34 sample sites in this study cannot display a whole scenario of the larger KMC area, although the study sites represent variability in environmental conditions and urban forms that may be considered as a sample for the whole area. The simple framework of this study can enable us to compare the relative ecological performance of different urban settings effortlessly and also help in the assessment of future planning for the alteration of urban space based on ecological valuation and urban resilience. The ecological performance indicators allow us to decide the relative ecosystem execution of various urban structures with various environmental qualities and equally recommend that the indicators could be valuable in urban planning and future improvements. It additionally proposes an environmental approach as a driver for urban resilience, transcending a limited, superficial engineering-based mindset. This research could help nurture the urban ecosystem services in several ways: aiding in enhanced urban planning, fostering advancements in urban adaptability to prepare for uncertain climate impacts and other global environmental changes, operationalizing an integrated assessment of ecosystem services in sound urban decision-making, and strategizing in diverse scenarios.

5.7 Declaration of Competing Interest

The authors declare that they have no known competing financial interests or personal relationships that could have appeared to influence the work reported in this paper.

Author Contribution Swarnava Dey carried out the literature review, modelling, visualization, simulation, compilation, and interpretation of data, performed statistical analysis, drafted the manuscript and prepared the figures and maps. Jayita Guha Niyogi and Debashish Das conceptualized the study, participated in the modelling and statistical analysis, and helped to draft the manuscript. All the author reviewed and approved the final manuscript.

Funding Partial financial support was received from University Grants Commission (UGC), New Delhi, as Senior Research Fellowship.

Data Availability The authors confirm that the data supporting the findings of this study are available within the article and raw data are available from the corresponding author upon reasonable request.

References

1. Alberti M (2005) The effects of urban patterns on ecosystem function. Int Reg Sci Rev 28:168–192. https://doi.org/10.1177/0160017605275160
2. Amani-Beni M, Chen Y, Vasileva M, Zhang B (2022) Quantitative-spatial relationships between air and surface temperature, a proxy for microclimate studies in fine-scale intra-urban areas? Sustain Cities Soc 77:103584. https://doi.org/10.1016/j.scs.2021.103584
3. Anderson V, Zgela M, Gough WA (2023) Building urban resilience with nature-based solutions: a multi-scale case study of the atmospheric cleansing potential of green infrastructure in Southern Ontario. Canada. Sustainability 15:14146. https://doi.org/10.3390/su151914146
4. Bansode A, Patil KA (2014) Estimation of runoff by using SCS curve number method and arc GIS. Int J Sci Eng Res 5:1283–1287
5. Berardi U, GhaffarianHoseini A, GhaffarianHoseini A (2014) State-of-the-art analysis of the environmental benefits of green roofs. Appl Energy 115:411–428. https://doi.org/10.1016/j.apenergy.2013.10.047
6. Bhura CS, Singh NP, Mori PR, Prakashl MK (2015) Estimation of surface runoff for Ahmedabad urban area using SCS-CN method and GIS. Int J Sci Technol Eng 1:411–416
7. Biggs R, Schlüter M, Schoon ML (2015) Principles for building resilience: sustaining ecosystem services in social-ecological systems. University Press, Cambridge
8. Bush J, Doyon A (2019) Building urban resilience with nature-based solutions: how can urban planning contribute? Cities 95:102483. https://doi.org/10.1016/j.cities.2019.102483
9. Calderón-Contreras R, Quiroz-Rosas LE (2017) Analysing scale, quality and diversity of green infrastructure and the provision of Urban Ecosystem Services: a case from Mexico City. Ecosyst Serv 23:127–137. https://doi.org/10.1016/j.ecoser.2016.12.004
10. Chen YC, Chiu HW, Su YF, Wu YC, Cheng KS (2017) Does urbanization increase diurnal land surface temperature variation? Evidence and implications. Landsc Urban Plan 157:247–258. https://doi.org/10.1016/j.landurbplan.2016.06.014

11. Colding J (2011) The role of ecosystem services in contemporary urban planning. In: Niemelä J, Breuste JH, Elmqvist T, Guntenspergen G, James P, McIntyre NE (eds) Urban ecology: patterns, processes, and applications. Oxford University Press, pp 228–237. https://doi.org/10.1093/acprof:oso/9780199563562.003.0028
12. Cui YY, De, FoyB (2012) Seasonal variations of the urban heat island at the surface and the near-surface and reductions due to urban vegetation in Mexico City. J Appl Meteorol Climatol 51:855–868. https://doi.org/10.1175/JAMC-D-11-0104.1
13. Dasgupta S, Gosain AK, RaoS RS, Sarraf M (2013) A megacity in a changing climate: the case of Kolkata. Clim Change 116:747–766. https://doi.org/10.1007/s10584-012-0516-3
14. Das S, Angadi DP (2020) Assessment of urban sprawl using landscape metrics and Shannon's entropy model approach in town level of Barrackpore sub-divisional region, India. Model Earth Syst Environ 7:1071–1095. https://doi.org/10.1007/s40808-020-00990-9
15. Esbah H, Cook EA, EwanJ (2009) Effects of increasing urbanization on the ecological integrity of open space presDaserves. Environ Manag 43:846–862.https://doi.org/10.1007/s00267-009-9274-z
16. Ewers RM, DidhamRK WSD, Tylianakis JM (2005) Remotely sensed landscape heterogeneity as a rapid tool for assessing local biodiversity value in a highly modified New Zealand landscape. Biodivers Conserv 14:1469–1485. https://doi.org/10.1007/s10531-004-9786-z
17. Farrugia S, Hudson MD, McCulloch L (2013) An evaluation of flood control and urban cooling ecosystem services delivered by urban green infrastructure. Int J Biodiversity Sci Ecosyst Serv Manag 9:136–145. https://doi.org/10.1080/21513732.2013.782342
18. Freddy V, El Mehdi Saidi M, Douvinet J, Fehri N, Nasrallah W, MenadW, Mellas S (2016) Urbanization and land use as a driver of flood risk. In: The mediterranen region under climate change, IRD Éditions/Institut de Recherche Pour le Développement Marseille: Marseille, France, 2016, pp 563–575. ISBN: 978-2-7099-2219-7
19. Freeborn J (2011) Decreasing runoff and increasing stormwater infiltration. VTechWorks Home. https://vtechworks.lib.vt.edu/handle/10919/70585. Accessed 27 June 2023
20. García-NietoAP GIR, Baró F, Roche PK, Bondeau A, Cramer W (2018) Impacts of urbanization around Mediterranean cities: changes in ecosystem service supply. Ecol Ind 91:589–606. https://doi.org/10.1016/j.ecolind.2018.03.082
21. Ghosh D, Midya SK, Sarkar U, Mukherjee T (2015) Variability of surface ozone with cloud coverage over Kolkata, India. J Earth Syst Sci 124:303–319. https://doi.org/10.1007/s12040-015-0543-3
22. Gomez-Baggethun E, Gren Å, Barton DN, Langemeyer J, McPherson T, O'Farrell P, Andersson E, Hamsted Z et al. (2013) Urban ecosystem services. In: Elmqvist T, Fragkias TM, Goodness J, Güneralp B, Marcotullio PJ, McDonald RI, Parnell S, Schewenius M et al (eds) Urbanization, biodiversity and ecosystem services: challenges and opportunities: a global assessment. Dordrecht: Springer, pp 175–251
23. Grafius DR, Corstanje R, Harris JA (2018) Linking ecosystem services, urban form and green space configuration using multivariate landscape metric analysis. Landscape Ecol 33:557–573. https://doi.org/10.1007/s10980-018-0618-z
24. Hair J, Anderson R, Black B, Babin B (2016) Multivariate data analysis. Pearson Higher Ed
25. Hamel P, Tan L (2022) Blue–green infrastructure for flood and water quality management in Southeast Asia: evidence and knowledge gaps. Environ Manag 69:699–718. https://doi.org/10.1007/s00267-021-01467-w
26. Handley J (1988) Nature in the urban environment. City landscape: a contribution to the council of Europe's European campaign for urban renaissance, 47–59
27. Hardin PJ, Jensen RR (2007) The effect of urban leaf area on summertime urban surface kinetic temperatures: a terre haute case study. Urban Forestry Urban Greening 6:63–72. https://doi.org/10.1016/j.ufug.2007.01.005
28. Hu T, Yang J, Li X, Gong P (2016) Mapping urban land use by using landsat images and open social data. Remote Sensing 8:151. https://doi.org/10.3390/rs8020151
29. Huang J, Lu X, Sellers JM (2007) A global comparative analysis of urban form: applying spatial metrics and remote sensing. Landsc Urban Plan 82:184–197. https://doi.org/10.1016/j.landurbplan.2007.02.010

30. Hutyra LR, Yoonb, Alberti M (2011) Terrestrial carbon stocks across a gradient of urbanization: a study of the Seattle, WA region. Global Change Biology 17:783–797.https://doi.org/10.1111/j.1365-2486.2010.02238.x
31. Janhäll S (2015) Review on urban vegetation and particle air pollution–deposition and dispersion. Atmos Environ 105:130–137. https://doi.org/10.1016/j.atmosenv.2015.01.052
32. Jordán F, Báldi A, Orci K, Rácz I, Varga Z (2003) Characterizing the importance of habitat patches and corridors in maintaining the landscape connectivity of a Pholidoptera transsylvanica (Orthoptera) metapopulation. Landscape Ecol 18:83–92. https://doi.org/10.1023/a:1022958003528
33. Karuppannan S, Baharuddin ZM, Sivam A, Daniels CB (2013) Urban green space and urban biodiversity: Kuala Lumpur, Malaysia. J Sustain Dev 7(1). https://doi.org/10.5539/jsd.v7n1p1
34. Kawashima S (1990) Effect of vegetation on surface temperature in urban and suburban areas in winter. Energy Build 15:465–469. https://doi.org/10.1016/0378-7788(90)90022-b
35. Keeley M (2011) The Green Area Ratio: an urban site sustainability metric. J Environ Planning Manage 54:937–958. https://doi.org/10.1080/09640568.2010.547681
36. Li S, He Y, Xu H, Zhu C, Dong B, Lin Y, ... Wang K (2021) Impacts of urban expansion forms on ecosystem services in urban agglomerations: a case study of Shanghai-Hangzhou Bay urban agglomeration. Remote Sensing 13:1908.https://doi.org/10.3390/rs13101908
37. Liu Y, Lü Y, Fu B, Harris P, Wu L (2019) Quantifying the spatio-temporal drivers of planned vegetation restoration on ecosystem services at a regional scale. Sci Total Environ 650:1029–1040. https://doi.org/10.1016/j.scitotenv.2018.09.082
38. Ma Y, Jiang Y (2023) Ecosystem-based adaptation to address urbanization and climate change challenges: the case of China's sponge city initiative. Clim Policy 23(2):268–284. https://doi.org/10.1080/14693062.2022.2131503
39. Mashizi AK, Heshmati GA, Mahini ARS, Escobedo FJ (2019) Exploring management objectives and ecosystem service trade-offs in a semi-arid rangeland basin in southeast Iran. Ecol Indicators 98:794–803.https://doi.org/10.1016/j.ecolind.2018.11.065
40. McCloy MW, Andringa RK, Maness TJ, Smith JA, Grace JK (2024) Promoting urban ecological resilience through the lens of avian biodiversity. Front Ecol Evol 12:302002. https://doi.org/10.3389/fevo.2024.1302002
41. McGarigal K, Marks BJ (1995) Fragstats: Spatial pattern analysis program for quantifying landscape structure. https://doi.org/10.2737/pnw-gtr-351
42. McPherson EG (1998) Structure and sustainability of Sacramento's urban forest. Arboriculture & Urban Forestry 24:174–190. https://doi.org/10.48044/jauf.1998.023
43. McPhearson T, Andersson E, Elmqvist T, Frantzeskaki N (2015) Resilience of and through urban ecosystem services. Ecosyst Serv 12:152–156. https://doi.org/10.1016/j.ecoser.2014.07.012
44. Morelli F, ReifJ, Díaz M, Tryjanowski P, Ibáñez-Álamo JD, Suhonen J, ... Benedetti Y(2021) Top ten birds indicators of high environmental quality in European cities. Ecol Indicators 133:108397. https://doi.org/10.1016/j.ecolind.2021.108397
45. Mukherjee S, Bebermeier W, Schütt B (2018) An overview of the impacts of land use land cover changes (1980–2014) on urban water security of Kolkata. Land 7(3):91. https://doi.org/10.3390/land7030091
46. Nowak DJ, McHale PJ, Ibarra M, Crane D, Stevens JC, Luley CJ (1998) Modeling the effects of urban vegetation on air pollution. Air pollution modeling and its application XII399–407. https://doi.org/10.1007/978-1-4757-9128-0_41
47. Nowak DJ, Dwyer JF (2007) Understanding the benefits and costs of urban forest ecosystems. Urban and Community Forestry in the Northeast 25–46.https://doi.org/10.1007/978-1-4020-4289-8_2
48. Pallubinsky H, Kramer RP, van Marken Lichtenbelt WD (2023) Establishing resilience in times of climate change—a perspective on humans and buildings. Clim Change 176:135. https://doi.org/10.1007/s10584-023-03614-0
49. Parker K, Head L, Chisholm LA, Feneley N (2008) A conceptual model of ecological connectivity in the Shellharbour local government area, New South Wales, Australia. Landsc Urban Plan 86:47–59. https://doi.org/10.1016/j.landurbplan.2007.12.007

50. Peng J, Tian L, Liu Y, Zhao M, Wu J (2017) Ecosystem services response to urbanization in metropolitan areas: thresholds identification. Sci Total Environ 607:706–714. https://doi.org/10.1016/j.scitotenv.2017.06.218
51. Ramachandra TV, Aithal BH, Sowmyashree MV (2014) Urban structure in Kolkata: Metrics and modelling through geo-informatics. Appl Geomatics 6:229–244. https://doi.org/10.1007/s12518-014-0135-y
52. Ray B, ShawR (2018) Changing built form and implications on urban resilience: loss of climate responsive and socially interactive spaces. Procedia Eng 212:117–124.https://doi.org/10.1016/j.proeng.2018.01.016
53. Reinstaller S, Funke F, König AW, Pichler M, Kleidorfer M, Muschalla D (2024) Resilient urban flood management: a multi-objective assessment of mitigation strategies. Sustainability 16:4123. https://doi.org/10.3390/su16104123
54. Rosenzweig C, Solecki W, Parshall L, Gaffin S, Lynn B, Goldberg R, Hodges S (2006) New York City regional heat island initiative final report 06–06. Albany, NY: New York State Energy Research and Development Authority
55. Sachs J, Lange GM, Heal G, Small A (2005) Global initiative for environmental accounting: a proposal to build a comprehensive system of environmental and economic national accounts. Department of Economic and Social Affairs, Statistics Division, United Nations, New York
56. Salzman J, Arnold CA, Garcia R, Hirokawa KH, Jowers K, LeJava J, Pelosa M, Olander L (2014) The most important current research questions in urban ecosystem services. SSRN scholarly paper ID 2483455. Rochester, NY: Social Science Research Network. http://papers.ssrn.com/abstract=2483455
57. Satheeshkumar S, Venkateswaran S, Kannan R (2017) Rainfall–runoff estimation using SCS–CN and GIS approach in the Pappiredipatti watershed of the Vaniyar sub basin, South India. Modeling Earth Systems and Environment 3:1–8. https://doi.org/10.1007/s40808-017-0301-4
58. Seitz J, Escobedo F (2008) Urban forests in Florida: trees control stormwater runoff and improve water quality. EDIS 2008(5). https://doi.org/10.32473/edis-fr239-2008
59. Sharp R, Tallis HT, Ricketts T, Guerry AD, ... Douglass J (2018) InVEST 3.5.0.post284+ug.hf903e6636cca User's Guide. The Natural Capital Project, Stanford University, University of Minnesota, The Nature Conservancy, and World Wildlife Fund
60. Shi C, Guo N, Gao X, Wu F (2022) How carbon emission reduction is going to affect urban resilience. J Clean Prod 372:133737. https://doi.org/10.1016/j.jclepro.2022.133737
61. Shigeto S, Yamagata Y, Ii R, Hidaka M, Horio M (2012) An easily traceable scenario for 80% CO_2 emission reduction in Japan through the final consumption-based CO_2 emission approach: a case study of Kyoto-city. Appl Energy 90:201–205. https://doi.org/10.1016/j.apenergy.2011.03.049
62. Situmorang JP, Sugianto S (2016) Estimation of carbon stock stands using EVI and NDVI vegetation index in production forest of Lembah Seulawah sub-district, Aceh Indonesia. Aceh Int J Sci Technol 5:126–139
63. Soltani A, Sharifi E (2017) Daily variation of urban heat island effect and its correlations to urban greenery: a case study of Adelaide. Frontiers Archit Res 6:529–538. https://doi.org/10.1016/j.foar.2017.08.001
64. Tratalos J, Fuller RA, Warren PH, Davies RG, Gaston KJ (2007) Urban form, biodiversity potential and ecosystem services. Landscape and Urban Planning 83:308–317. https://doi.org/10.1016/j.landurbplan.2007.05.003
65. Vinithra R, Yeshodha L (2016) Rainfall–runoff modelling using SCS–CN method: a case study of Krishnagiri District, Tamilnadu. Int J Sci Res 5:2319–7064
66. Walz U, Syrbe RU (2013) Linking landscape structure and biodiversity. Ecol Ind 31:1–5. https://doi.org/10.1016/j.ecolind.2013.01.032
67. Whitford V (2001) "City form and natural process"—indicators for the ecological performance of urban areas and their application to Merseyside, UK. Landsc Urban Plan 57:91–103. https://doi.org/10.1016/s0169-2046(01)00192-x

Chapter 6
Modelling Urban Growth, Demand–Supply Dynamics, and Climate Resilience: A Study of a Class-I City in Eastern India

Pritiranjan Das, Somnath Rudra, Biswajit Maity, Suraj Kumar Mallick, Chumki Shikary, and Netrananda Sahu

Abstract The extensive transformation of natural landscapes into man-made landscapes places constraints on long-term urban sustainability. Therefore, it is important to analyse the issues of urban resilience relating to demand and supply for urban-related services in terms of human adaptation in a class-I town like Midnapore municipality in West Bengal's Paschim Medinipur district. To emphasize urban resilience, this study has been conducted to analyse the dynamics of the urban landscape from 1991 to 2021, predict the landscape for 2031 and 2041, and analyse the demand–supply chain of urban facilities to assess the nature of human adaptation in changing urban landscapes. The study used satellite images of 1991, 2001, 2011, and 2021, while information about the supply and demand chain, such as market, accessibility, medical and educational services, job facilities, drinking water services, and land availability was gathered from the field survey. This study revealed that urban expansion has encroached upon peripheral areas, replacing agricultural land and green spaces. The integrated demand and supply analysis indicated increasing demand in the periphery region while supply is concentrated only in the municipality's core. As urban land alteration will be a major issue for human adaptability soon, research into urban land dynamics and demand–supply analysis of urban amenities in the context

P. Das · S. Rudra (✉) · B. Maity · S. K. Mallick · C. Shikary
Department of Geography, Vidyasagar University, Midnapore 721102, West Bengal, India
e-mail: rudra347@rediffmail.com

P. Das · N. Sahu
Department of Geography, Delhi School of Economics, University of Delhi, New Delhi 110007, India

S. K. Mallick
Shaheed Bhagat Singh College (University of Delhi), Sheikh Sarai (Triveni), Phase-II, New Delhi 110017, India

S. Rudra
Department of Geography, School of Natural Resource Management, Central University of Jharkhand, Ranchi 835222, India

S. Ghosh et al. (eds.), *Cities of Tomorrow: Urban Resilience and Climate Change Preparedness*, Urban Sustainability, https://doi.org/10.1007/978-981-97-9658-8_6

of climate change and sustainable development will be extremely beneficial to local planners and administrators in managing the urban body for human living.

Keywords Urban expansion · Transition probability · CA-Markov chain · Demand–supply zones · Urban resilience

6.1 Introduction

Rapid urbanization and physical expansion towards the periphery have significantly altered the landscape of the urban core as well as rural surroundings through unplanned and unsystematic built-up area expansion particularly in developing countries like India [18, 19, 24]. Therefore, changes in the natural landscape, such as loss of farmland, water bodies, vegetation cover, and human adaptability over urban and rural surroundings, are becoming constraints for urban development in sustainable ways [6, 27]. Land utilization is transformed into many forms all over the world: from forest cover to agricultural land, from wasteland to developed areas etc. These changes can be evaluated by examining the urban land use pattern. Land use and land cover (LULC) maps have become increasingly important in providing critical information on landscape transformation, future development projections, and environmental impacts. As a result, analysing LULC has become vital for sustainable development and addressing climate change (Fikir et al. 2009). Consequently, LULC is undergoing rapid changes due to increasing urbanization [10], presenting both challenges and opportunities for urban populations [23].

Several urban studies have utilized remote sensing and geographic information systems (GIS) to assess LULC patterns and their implications [21, 28] (Lu et al. 2013). Furthermore, modelling is extremely beneficial for comprehending spatial growth and forecasting future urban development [24, 25]. Recently many studies have used a variety of models, including the Cellular Automata model, and the Markov Chain (CA-Markov) [20], regression models, like the logistic regression (LR) [15] and integrated models, like CLUE land alteration [26], (Guan et al. 2011). The Markov chain model predicts probable changes of land use in future, while cellular automata determine the specific locations of these changes (Jokar et al. 2013) [20]. The change simulation, on the other hand, is determined by an integrated CA–Markov model, which is best suited for this purpose.

Recent studies have found that the rapidly expanding built-up in urban fringe creates demand for food, water, and energy [8, 13]. Thus, the rapid expansion of the urban fringe raises concerns about urban resilience in terms of the supply and demand chain. Furthermore, newly established unplanned built-up areas are suffering from the imbalances in supply of services e.g., supply of water, road connectivity, water logging, health care facilities, education, street lighting, and so on. [13]. Moreover, the investigation of area-specific sustainable urban development about urban livelihood and resilience has become an essential component of urban landscape dynamics. Recent studies emphasized the role of horizontal expansion toward the outskirts in destroying croplands and natural habitats [1]. Although the effect of changing human demand and supply chain on human adaptability has been assessed for large cities like Siliguri [13], it was not explored earlier, particularly in medium-sized size city, as Midnapore, and West Bengal.

Thus, the main objectives of this study include assessing the changes in the urban landscape from 1991 to 2021, projecting future urban landscape dynamics using CA-Markov for 2031 and 2041, and analysing the demand–supply dynamics to evaluate urban resilience in both the core and peripheral areas of Midnapore City.

6.2 Study Area

Midnapore, founded in 1865, is one of West Bengal's oldest towns and the administrative centre of the Paschim Medinipur district. The Kangsabati river runs along the southern boundary of Midnapore. It falls within the Midnapore Kharagpur Development Authority (MKDA) Zone. The town experiences a hot tropical monsoon climate. Soils near Rangamati are typically lateritic, while those closer to the Kangsabati River consist mainly of alluvium, varying in composition from clay to sand. It has undergone significant urbanization, supported by robust infrastructure and enjoys excellent connectivity to neighbouring towns via well-developed highway and railway networks. Midnapore town consists of 25 wards, spreading over an area of 18.65 km^2. The population was 169,127 in 2011. In addition, this town's population is continuously growing. Because of this, the town is always growing in new directions with several informal settlements, and the built-up area is spreading beyond the municipality boundary (Fig. 6.1). The town has been chosen for this study due to its rapid urbanization over the past three decades.

6.3 Materials and Methods

6.3.1 Materials

Landsat imagery and ASTER DEM are utilized here to evaluate the spatiotemporal dynamics of the urban landscape. Consequently, cloud-free satellite images were acquired from the US Geological Survey (USGS) Earth Explorer platform (http://earthexplorer.usgs.gov/) for the pre-monsoon seasons of 1991, 2001, 2011, and 2021. During the field survey, ground control points (GCPs) and information related to human adaptation such as supply chain data based on market, accessibility, medical services, educational services, job facilities, land availability, and demand chain data based on drinking water, accessibility, market, educational services, and job facilities were collected. The integrated index score has been calculated to quantify the standard level performance of supply and demand of services. Moreover, the

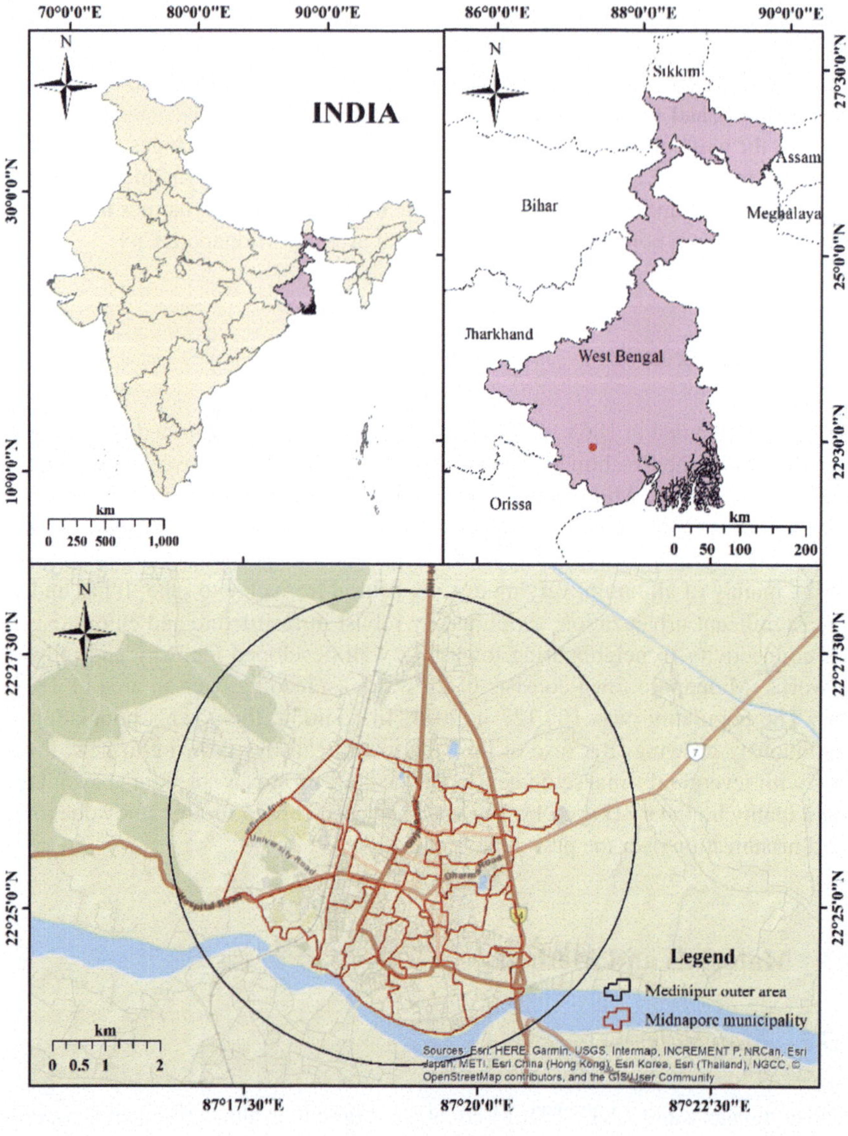

Fig. 6.1 Location map of Midnapore Municipality and outer boundary

information about the population is collected from the district census handbook (2011). The satellite data specifications are detailed in Table 6.1.

Table 6.1 Satellite data specification

Date of acquisition	Satellite data	Data type	Path/row	Spatial resolution	Reference system
23.10.1991	Landsat 5	TM_L1 TP	139–44/45	30 m × 30 m	UTM, WGS84, 45 N
26.10.2001	Landsat 5				
07.11.2011	Landsat 5				
23.11.2021	Landsat 8	OLI_TIRS_L1 TP			

6.3.2 Analysis of Land Use and Land Cover

The evaluation of LULC classification is crucial for studying urban landscapes and their planning. In this study, five predetermined land classes—vegetation cover, agricultural land, water bodies, built-up area, and open land—were identified using the supervised classification method within the GIS environment. The multi-temporal satellite images were rectified to generate standard false colour composites (FCCs) in ArcGIS 10.4. Subsequently, a 4.3 km buffer was delineated from the centre to indicate existing and potential urban expansion. Finally, LULC maps were prepared for the years 1991, 2001, 2011, and 2021 using supervised classification with the maximum likelihood method, and all maps were validated using the Kappa coefficient algorithm.

6.3.3 LULC Gain and Loss Analysis Using Land Change Modeler (LCM)

The LCM in IDRISI 18.3 Selva Edition is applied to depict the changing pattern of various land types. We are able to provide the date or gap between the current classified map and the previous classified map here [4]. This change analysis deals with the gains and losses of each land category and quantifies the net change analysis to indicate the contribution of numerous land classes to the net change of a specific land class. The area has been translated into square kilometres, and these values are presented on the diagram's horizontal scale by area in percentage.

6.3.4 Spatial Trend of Urban Expansion

The spatial pattern of urban growth has been used to ensure that all specified land types are converted to other feasible land categories [4]. In essence, the main objective is to depict how urban areas are changing and where urban expansion is detected for a specific year. After analysing the net changes in LULC using the IDRISI program, we generated spatial trend maps depicting urban expansion, along with transformation maps represented on a sixth-order polynomial scale. It is now useful for comprehending urban growth from the centre of the city towards rural surroundings of all expanding cities or urban bodies [5].

The spatial distribution of urban growth has been employed to ensure that all designated land types undergo conversion into other viable land categories [4]. Essentially, the primary objective is to illustrate how urban areas are evolving and where urban expansion is identified for a specific year. By tracking the net changes in LULC within the IDRISI program, we have efficiently generated spatial trend maps depicting urban expansion, alongside transformation maps on a sixth-order polynomial scale. This approach is valuable for understanding urban growth extending from the city centre towards the rural peripheries of all expanding cities or urban areas [5].

6.3.5 Transition Probability Matrix (TPM)

The TPM serves as a statistical tool, indicating the probability of future self-replacement for each land class and the anticipated alterations in the current land use scenario. Alternatively termed as a stochastic matrix, it delineates the transition within a Markov chain. Two types of stochastic matrices have been employed in our study: a right stochastic matrix, where each row of the real square matrix sums up to 1, and a left stochastic matrix, where each column totals 1. Equation 6.1 denotes the appropriate stochastic matrix utilized in this context (Thomas and Laurence 2006).

$$P = \begin{pmatrix} P_{1,1} & P_{2,1} & \cdots & P_{j,1} & \cdots & P_{s,1} \\ P_{1,2} & P_{2,2} & \ldots & P_{j,2} & \cdots & P_{s,2} \\ \vdots & \vdots & \ddots & \vdots & \ddots & \vdots \\ P_{1,i} & P_{2,i} & \ldots & P_{j,i} & \ldots & P_{s,i} \\ P_{1,s} & P_{2,s} & \ldots & P_{j,s} & \ldots & P_{s,s} \end{pmatrix} = \begin{pmatrix} P_{1,s} \\ P_{2,s} \\ \vdots \\ P_{i,s} \\ P_{s,s} \end{pmatrix}$$

$$\sum_{j=1}^{s} P_{j,i} = 1 \tag{6.1}$$

Here, Pr (j/i) = Pj,i represents the probability of transitioning from state j to state i within one time step, where the stochastic matrix P is constructed by assigning Pj,i as the element at the jth column and ith row. Consequently, the transition matrix is completed, ensuring that the sum of all states to the jth state equals one. This process yields a Markov chain transition matrix, illustrating the likelihood of transitioning from one land class to another across consecutive time intervals (Sarkar and Chouhan 2019).

6.3.6 Cellular Automata (CA) Model

We implemented a three-to-three transition rule, considering six neighbourhood cells, and utilized a 10-year interval based on the actual image of 2021. According to [9], the proximity of the functional cell of the filter impacts the cell values of the land classes depicted by the lattice state. This research classified the six functional cells in each community into five terrain classes: built-up areas, agricultural land, vegetation cover, open areas, and water bodies. Similarly, the future land use type of a cell is determined by one of its six surrounding cells, given the varying priorities of different land use categories. For instance, built-up unit does not transition into an open space, vegetated region, water body, or agricultural land (Thomas and Laurence 2006). Each land use type has its geographical allocations determined by significance (Guan et al. 2011). Consequently, cellular automata (CA) was consistently assigned using the spatial allocation method for the given time and remained unchanged during the simulation.

6.3.7 Combined C A-Markov Technique for Future Urban Growth

The CA-Markov technique has been applied to project the expansion of the city. These processes include three steps: (i) LULC map for the year 2021; (ii) Markov chain matrix from 1991 to 2001, 2001 to 2011, and 2011 to 2021; and (iii) simulated map of 2011, and 2021 based on 2001, 2011 actual image; and then integration of CA -Markov chain using the neighbourhood function to predict land use maps of 2031 and 2041. Land class gain and loss analysis is first performed on the data obtained from the creation of LULC maps through the use of geospatial tools. For every type of terrain, a transition probability matrix has been created based on the investigation. Then, the road map and DEM of the research region were utilized in the introduction of the CA-Markov technique. In order to maintain proper protocol, the projected map of 2031 was created by combining the road map, DEM map, and all five land classes from the LULC maps of 2011 and 2021, and the projected map of 2041 was created by combining the LULC maps of 2001 and 2021 with the DEM map.

6.3.8 Validation

Before investigating any errors in the geospatial-based model, it is imperative to first calibrate and validate the model. The Kappa Index is used as a key tool to provide spatial allocation information and validate the geographical model. This index is highly effective in validating both actual and expected images, as well as in the realm of simulation (Mondal et al. 2016). Furthermore, the integrated findings of CA-Markov are validated using area under curve (AUC) of receiver operating characteristic (ROC) curve, a well-established method for assessing the relationship between actual and simulated results. A value below 0.50 or equal to 0.50 indicates a random and insignificant distribution, while a value of 1 suggests a perfect match (Pontius et al. 2004).

6.3.9 Changing Supply and Demandzones with the Land Use Transformation

The adjustment of urban land use involves assigning appropriate weights based on spatial location. As demand or supply increases, corresponding weights also increase. Consequently, two separate maps—one combining supply maps and the other combining demand maps—are necessary to facilitate a qualitative spatial assessment based on performance. These maps illustrate the changes in the urban environment over time. Subsequently, the combined demand -supply zone maps are categorized into various groups based on index values (Table 6.2). Each zone or cluster signifies its performance regarding urban change. Consequently, a demand and supply map organized by zones or clusters is generated. This outlines the process for computing the integrated demand and supply zone.

$$ISZ_a = \left\{ \frac{1}{n} \sum_{0}^{s} ifd \leq d\_max \right\} \tag{6.2}$$

$$IDZ_a = pop * gcap \tag{6.3}$$

where ISZ_a indicates the integrated supply zone area; d denotes the Euclidean distance between two areas (a); d_max is the maximum radius to provide the supply services.

IDZ_a stands for the integrated demand zone area; *pop* indicates population of that area; *gcap* is the demanded area of required per person for services (km^2 /cap).

Table 6.2 Integrated demand and supply index

Integrated index score	Standard level of supply performance	Standard level of demand performance
0–1	Very Low	Very High
1–2	Low	High
2–3	Moderate	Moderate
3–4	High	Low
4–5	Very High	Very Low

To achieve this, the ArcGIS analysis tool Arc-hotspot (Getis-Ord Gi*) is utilized, which is a spatial statistical method. A hotspot is generally characterized by a high clustered value of the z-score, although it must be statistically significant [11]. The formula equivalent to the Getis-Ord Gi* (Gi*) local statistics is:

$$G_i^* = \frac{\sum_{j=1}^{n} w_{i,j}x_j - \overline{X}\sum_{j=1}^{n} w_{i,j}}{s\sqrt{\frac{n\sum_{j=1}^{n} w_{i,j}^2 - \left(\sum_{j=1}^{n} w_{i,j}\right)x^2}{n-1}}} \tag{6.4}$$

where, xj indicates the attribute value for the feature j; w_i, j indicates spatial weight between features i and j, n is equal to the total number of the features in s area.

$$\overline{X} = \frac{\sum_{j=1}^{n} x_j}{n} \tag{6.5}$$

$$S = \sqrt{\frac{\sum_{j=1}^{n} x^2{}_j}{n} - (\overline{X})^2} \tag{6.6}$$

In Getis-Ord Gi* statistics, the greater z-score is used to indicate statistically significant positive results. According to [13], it denotes hotspots or areas with high demand or supply. The cold spot or zone with lower demand or supply is shown by the z-bigger score's negative values.

6.4 Result

6.4.1 Changing Land Use with Urban Growth

Table 6.3 displays the results of the five classified land use classes. In 1991, farmland and vegetation cover were the dominant land classes. Subsequently, the built-up area increased significantly encroaching upon farmland and vegetative cover. Rapid development was particularly evident in the northern, southern, and southwestern parts of the study area. The northern portion experienced growth facilitated by improved connectivity via National Highway 60 and proximity to the Keranichoti region. The southern portion expanded rapidly due to lower land prices in the Palbari area. Similarly, the southwest portion witnessed rapid growth owing to its proximity to the state highway and railroad station (Fig. 6.2). The maps of 2011 and 2021 depict a substantial increase in water surface compared to those of 1991 and 2001, attributed to the damming of the Kangsabati River. However, over the last 20 years, open land has decreased, while vegetation cover and agricultural land have followed a similar pattern. To assess the accuracy of both classified maps, Kappa coefficient values are applied. The accuracy assessment results for the years 1991, 2001, 2011, and

2021 are 0.85, 0.90, 0.89, and 0.92, respectively, with corresponding Kappa values of 0.81, 0.88, 0.86, and 0.90. Consequently, the significantly higher accuracy level of the classified maps, exceeding 0.8, effectively validates the model [17].

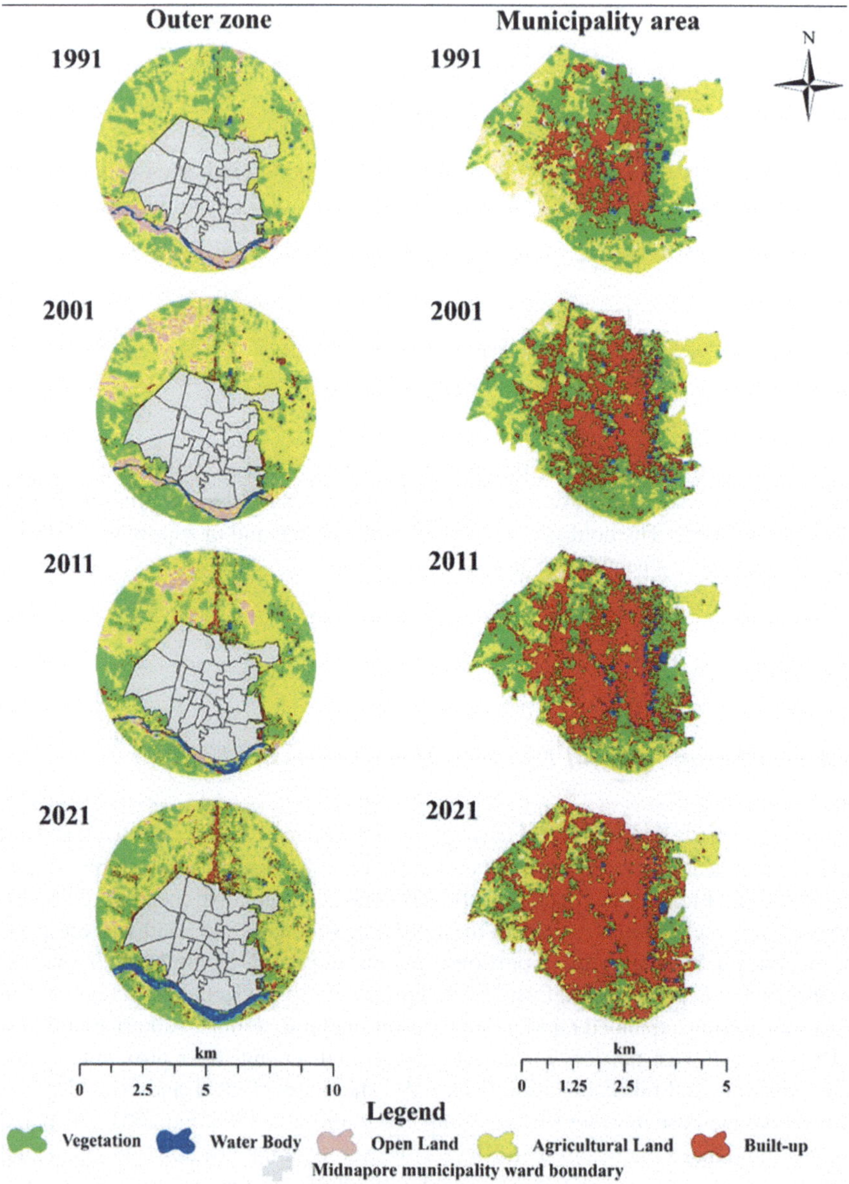

Fig. 6.2 Spatiotemporal changes of LULC from 1991 to 2021

Table 6.3 Distribution of area shared by LULC in 1991, 2001, 2011 and 2021

Area (km^2)				
LULC	1991	2001	2011	2021
Vegetation	15.78	17.33	18.44	17.35
Water body	1.21	1.16	1.65	2.49
Open land	5.47	5.08	2.49	2.48
Agriculture land	31.03	27.15	24.85	23.74
Built-up area	4.72	7.49	10.78	12.14

6.4.2 Land Transformation to Built-Up Land

Between 1991 and 2021, the urban built-up land area expanded from 4.72 to 12.14 km^2, nearly tripling in size. The built-up area transition map (Fig. 6.3) indicates that vegetation cover has a higher propensity to transition into built-up areas compared to other land types. The overall outcome of net land changes reveals that population growth has been the predominant factor exerting tremendous pressure on land from 1991 to 2021. The construction of dams on the Kangsabati River has led to an increase in water bodies and vegetation cover, while open land and agricultural land have experienced a decline in recent decades. However, built-up areas have emerged as the dominant land use since 1991.

6.4.3 LULC Loss and Gain and Net Change Analysis

Figures 6.4 and 6.5 illustrate the regional and temporal variations of LULC across different time intervals, highlighting land use gains and losses. The findings indicate a proportional decline in agricultural land over time, with the built-up area experiencing the dominant increase between 1991 and 2021. Notably, built-up areas have predominantly replaced open land between 2011 and 2021, while agricultural land has witnessed the most substantial decrease over the past 30 years, followed by vegetation cover. This trend is attributed to an influx of people after 2011, as reported by local surveys, consequently necessitating adjustments in other land classes.

6.4.4 Built-Up Area Expansion

Ca-Markov approach for future growth simulation

The transition matrices for the periods 1991–2001 (Table 6.4), 2001–2011 (Table 6.5), 2011–2021 (Table 6.6), and 2001–2021 (Table 6.7) were computed using the Markov chain. In these matrices, the likelihood of self-replacement is denoted

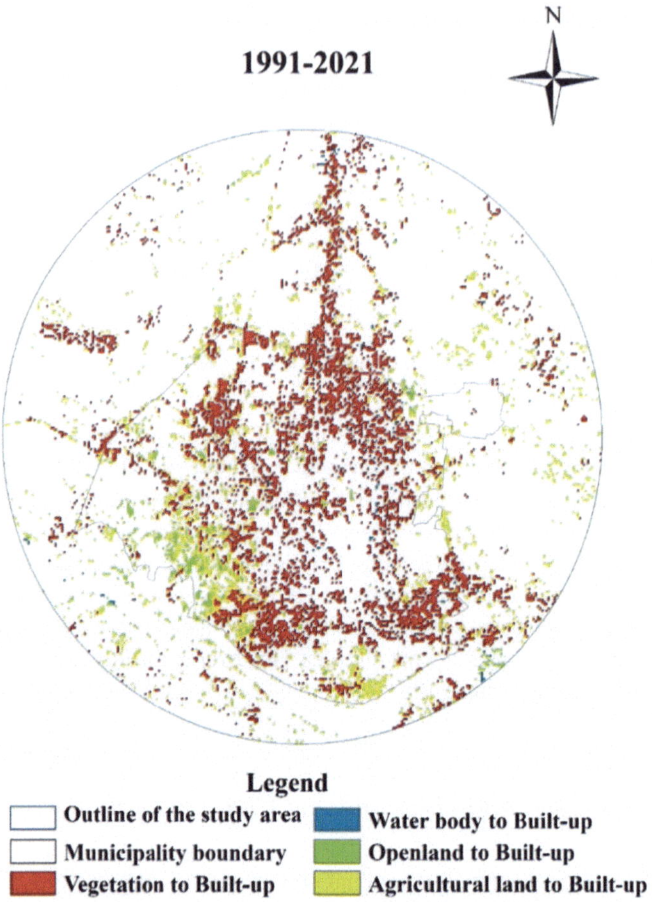

Fig. 6.3 Different land use transformation to built-up land between 1991 and 2021

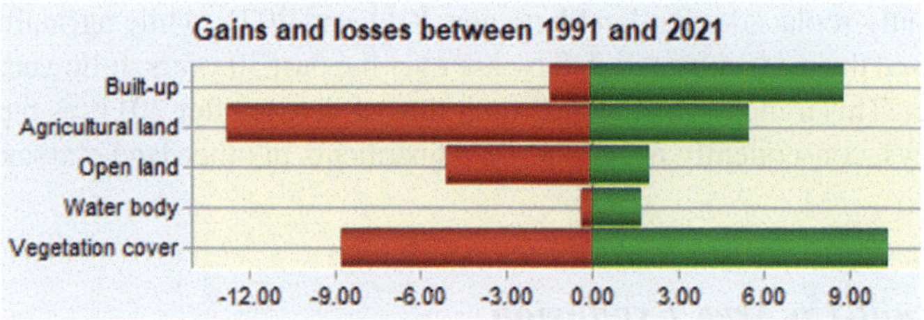

Fig. 6.4 Gain and loss of LULC between 1991 and 2021

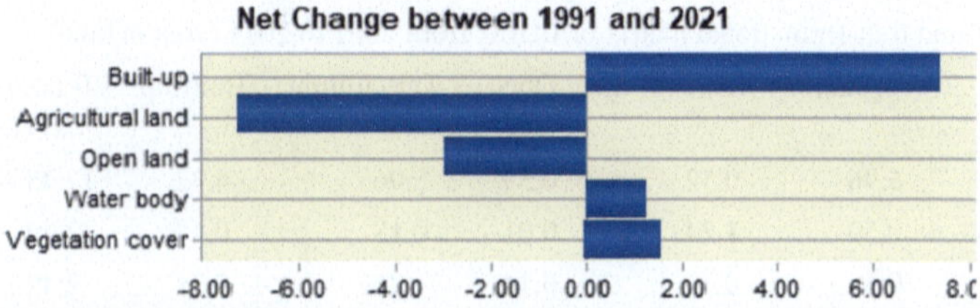

Fig. 6.5 Net changes of LULC between 1991 and 2021

by the diagonal values, while the off-diagonal values represent the probability of land transition (Keshtkar and Voigt, 2016). The findings show a notable expansion of the built-up area from 1991 to 2001, primarily at the expense of vegetation and agricultural land. Open land was transformed into built-up areas, agricultural fields, and to some degree, vegetative cover. The conversion of other land classes occurred in proportion to the remaining land classes. Furthermore, the findings from CA-Markov suggest that agricultural fields and vegetation areas have experienced rapid encroachment, with the built-up area emerging not only as a source of income but also as a significant employment hub.

Validation of the simulated model

The CA-Markov techniques have been employed to predict future urban expansion, facilitating urban planning and development. Consequently, simulated maps of urban development have been generated for 2031 and 2041 (Fig. 6.8). The results predict that by 2031 and 2041, the built-up area will expand to 15.33 and 16.52 km^2, respectively. These anticipated outcomes suggest that the peripheral regions will serve as the future urban development area. The town is expected to continue its expansion, though at the cost of reduced agricultural land and vegetation cover due to the increasing demand for land driven by population growth. For the validation of the predicted maps of 2011 and 2021, the Kappa Index Agreement was utilized for error

Table 6.4 Land transformational matrix of LULC from 1991 to 2001 (area in km^2)

LULC	Vegetation	Waterbody	Open land	Agricultural land	Built-up	Total (1991)	Loss
Vegetation	**10.34**	0.13	0.04	3.58	4.35	18.44	8.10
Water body	0.19	**1.19**	0.03	0.06	0.18	1.65	0.46
Open land	0.16	0.13	**0.76**	1.39	0.05	2.49	1.73
Agricultural land	6.19	0.14	0.81	**16.85**	0.86	24.85	8.00
Built-up	2.16	0.37	0.02	0.29	**7.94**	10.78	2.84
Total (2001)	19.04	1.97	1.65	22.17	13.38	**58.20**	
Gain	8.70	0.78	0.89	5.32	5.44		

Note Bold diagonal values denote the self-replacement probability

Table 6.5 Land transformational matrix of LULC from 2001 to 2011 (area in km^2)

LULC	Vegetation	Waterbody	Open land	Agricultural land	Built-up	Total (2011)	Loss
Vegetation	**5.96**	0.52	0.59	5.96	4.32	17.35	11.39
Water body	0.59	**1.24**	0.01	0.18	0.47	2.49	1.25
Open land	0.58	0.30	**0.14**	1.26	0.21	2.49	2.35
Agricultural land	6.37	0.92	1.36	**12.43**	2.66	23.74	11.31
Built-up	3.47	0.50	0.22	2.40	**5.55**	12.14	6.59
Total (2021)	16.97	3.48	2.31	22.24	13.20	**58.20**	
Gain	11.01	2.24	2.17	9.81	7.65		

Note Bold diagonal values denote the self-replacement probability

Table 6.6 Land transformational matrix of LULC from 2011 to 2021 (area in km^2)

LULC	Vegetation	Waterbody	Open land	Agricultural land	Built-up	Total (2011)	Loss
Vegetation	**8.32**	0.24	0.40	4.76	3.64	17.36	9.04
Waterbody	0.44	**1.75**	0.00	0.02	0.28	2.49	0.74
Open land	0.28	0.33	**0.25**	1.58	0.04	2.48	2.23
Agricultural land	5.20	0.44	1.65	**15.28**	1.17	23.74	8.46
Built-up	2.82	0.32	0.09	1.23	**7.68**	12.14	4.46
Total (2021)	17.06	3.08	2.39	22.87	12.82	**58.20**	
Gain	8.74	1.33	2.14	7.59	5.14		

Note Bold diagonal values denote the self-replacement probability

Table 6.7 Land transformational matrix of LULC from 2001 to 2021 (area in km^2)

LULC	Vegetation	Waterbody	Open land	Agricultural land	Built-up	Total (1991)	Loss
Vegetation	**7.88**	0.16	0.3	4	5.01	17.35	9.47
Waterbody	0.49	**1.72**	0	0.05	0.22	2.48	0.76
Open land	0.33	0.4	**0.24**	1.46	0.05	2.48	2.24
Agricultural land	6.1	0.44	1.44	**14.09**	1.66	23.73	9.64
Built-up	2.59	0.34	0.07	1	**8.14**	12.14	4
Total (2001)	17.39	3.06	2.05	20.6	15.08	**58.20**	
Gain	9.51	1.34	0.24	6.51	6.94		

Note Bold diagonal values denote the self-replacement probability

checking (Fig. 6.6). The 'K-no' values for the years 2011 and 2021 were found to be 0.8680 and 0.8596, respectively, indicating satisfactory error verification. Additionally, the 'K-location' values were determined to be 0.9128 and 0.8670, while the 'K-location Strata' values were 0.9128 and 0.8671, indicating accurate distribution of grid values. The 'K-standard' values for the years 2011 and 2021 were calculated as 0.8556 and 0.8481. Subsequently, the integration of CA-Markov was validated using the ROC curve for the generation of future urban development zones in 2031 and 2041. The ROC curve revealed AUC values of 83.4 and 84.1 for 2011 and 2021, respectively (Fig. 6.7), thus validating the future growth predictions.

The last 20 years' geographical patterns of LULC transformation are utilized to show how LULC has changed. The Kasai River is becoming contaminated due to the constant discharge of municipal and hospital waste into the waterway. If the authorities do not take immediate action, the water quality is expected to deteriorate by 2030. As a result, a perspective city planning can be promoted this historical city to a modern and appealing future city.

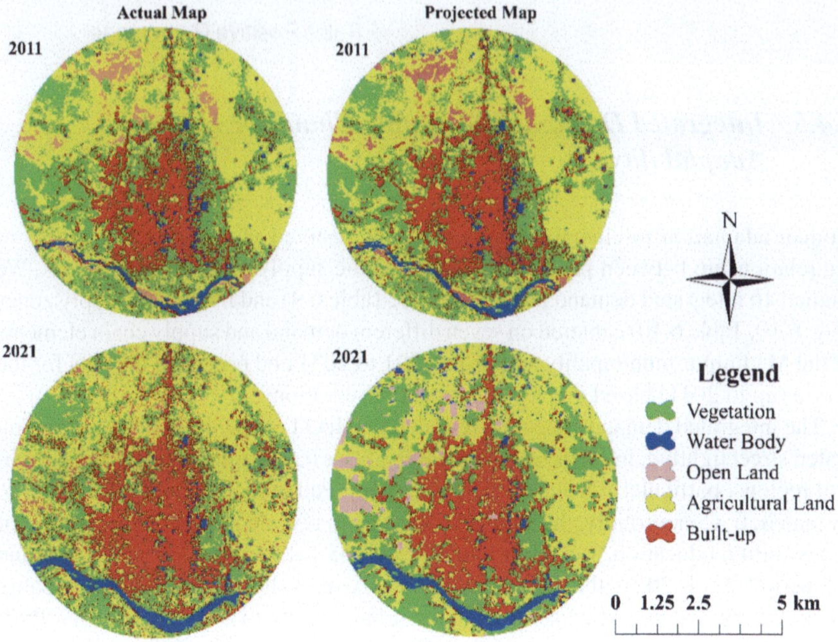

Fig. 6.6 Actual and projected LULC map of 2011 and 2021

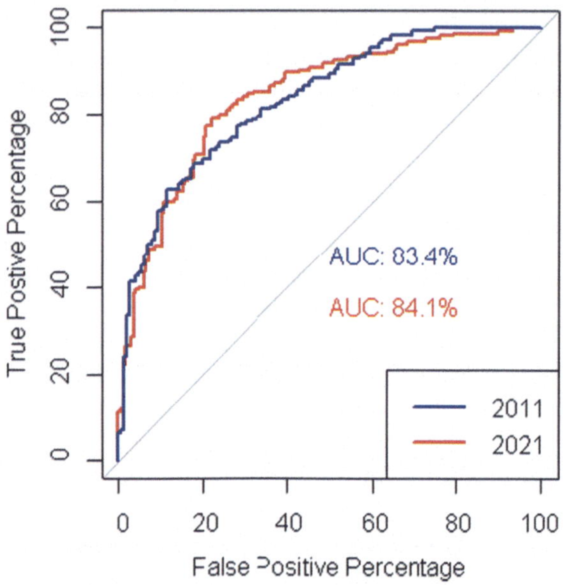

Fig. 6.7 ROC curve for validation of predicted map of 2011 and 2021

6.4.5 *Integrated Demand and Supply Chain for Human Adaptability*

Human adaptation to changing landscapes has resulted in an important aspect of the relationship between people's demand and the supply chain in urban areas. We studied 16 integrated demand zones (Fig. 6.9, Table 6.9) and integrated supply zones (Fig. 6.10, Table 6.10) cantered on seven different demand and supply chain elements of the Medinipur municipality area from 1991 to 2021 and predicted the data for the future (up to 2041) based on past and recent observations.

The integrated demand zone map revealed that local residents' needs for drinking water, street lighting, markets, and medical facilities have shifted towards the peripheral regions, particularly from Palbari, Najarganj, Patherghata, and Satyajitnagar to Keranichoti. Concurrently, the supply chain system revealed a changing situation of accessibility, education, job opportunities, and market facilities over time. During the period 2011–2021, the supply of services within the Medinipur municipality advanced rapidly, providing services to the urban society in a variety of ways. Prior to the recent decade, the entire municipality was concentrated within a small area, and the supply chain system was negligible. As a result, development activities can only address demand issues and provide various facilities to those wards that are still experiencing water problems, streetlight issues, accessibility issues, and market facilities. Increasing urbanization and land alteration will become critical issues for human adaptability and urban resilience in terms of living shortly. As a result, the predicted results for 2021–2031 and 2031–2041 can be useful for further research to elucidate the existing problems related to demand and supply chain within and

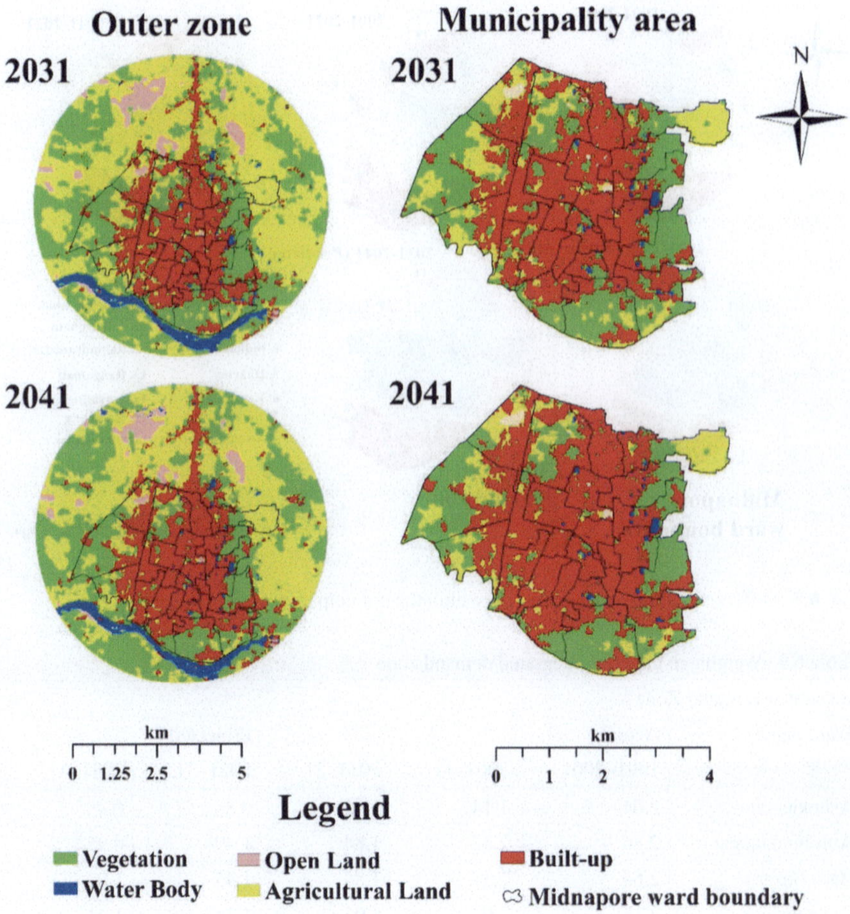

Fig. 6.8 Projected map (outer zone and municipality area) of built-up growth and different land uses of 2031 and 2041

adjacent to the Midnapore municipality and thus, make this urban body suitable for living by cooperating with current demands.

6.5 Discussion

This study integrates CA-Markov models to predict the expansion of the Midnapore municipality region. Despite the limitations associated with the CA-Markov approach, this research thoroughly examines the level of urbanization in the Midnapore municipality region, which has historically held a prominent position in the Paschim Medinipur District. The spatiotemporal scenario was projected for 2031 and

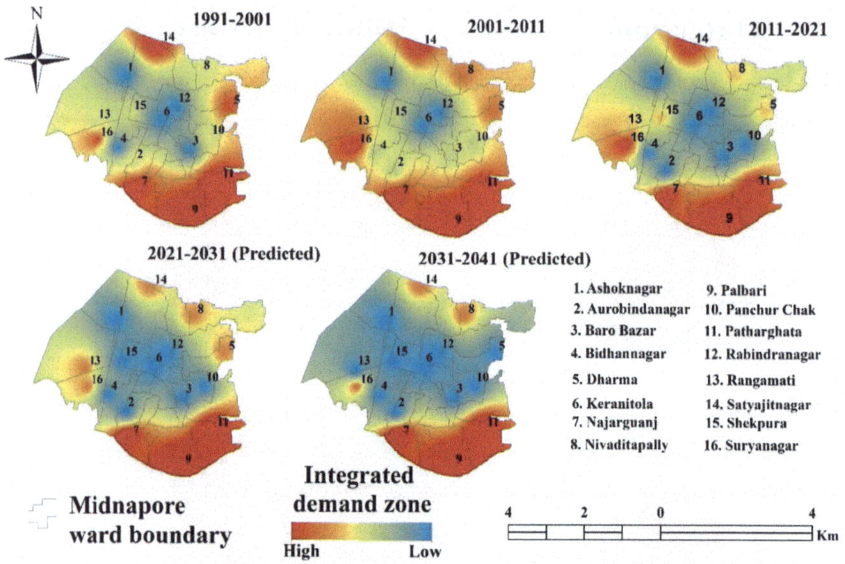

Fig. 6.9 Different demand zone scenarios within the municipality ward boundary

Table 6.9 Weightage index of integrated demand zone

Integrated Demand Zone					
Ward name	Actual			Projected	
	1991–2001	2001–11	2011–21	2021–31	2031–41
Ashoknagar	2.04	1.14	1.02	0.86	0.45
Aurobindanagar	2.86	2.42	1.84	1.49	0.95
Baro Bazar	2.64	2.31	1.86	1.45	1.02
Bidhannagar	2.15	2.03	1.95	1.46	1.21
Dharma	3.73	3.45	2.89	2.65	1.89
Keranitola	2.09	1.89	1.64	1.23	0.84
Najarguanj	4.81	4.46	3.76	3.24	3.13
Nivaditapally	3.18	3.02	2.75	2.42	2.14
Palbari	4.39	4.02	3.84	3.43	3.08
Panchur Chak	3.34	2.88	1.76	1.54	1.19
Patharghata	4.23	4.07	3.72	3.25	3.07
Rabindranagar	2.23	1.79	1.56	1.42	1.19
Rangamati	2.76	2.45	2.24	2.07	1.47
Satyajitnagar	4.89	4.23	3.78	2.73	2.28
Shekpura	3.27	2.87	2.45	1.28	0.89
Suryanagar	4.13	4.04	3.74	2.79	2.54

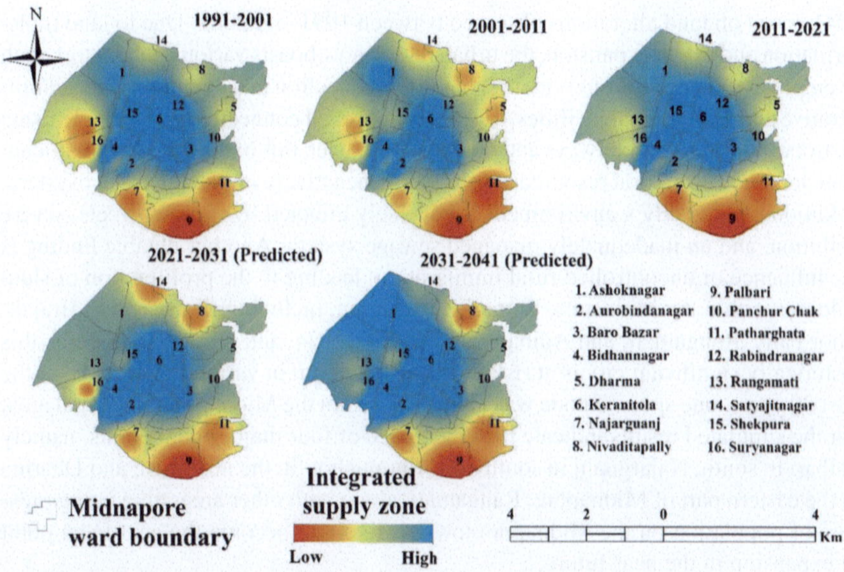

Fig. 6.10 Different supply zone scenarios within the municipality ward boundary

Table 6.10 Weightage index of integrated supply zone

Integrated supply zone					
Ward name	Actual			Projected	
	1991–2001	2001–11	2011–21	2021–31	2031–41
Ashoknagar	3.89	4.23	4.58	4.78	4.98
Aurobindanagar	3.84	4.24	4.64	4.78	4.89
Baro Bazar	4.25	4.56	4.68	4.87	4.96
Bidhannagar	4.06	4.29	4.67	4.83	4.97
Dharma	2.29	2.78	3.48	4.49	4.86
Keranitola	4.07	4.23	4.65	4.73	4.98
Najarguanj	1.64	2.75	2.98	3.64	3.98
Nivaditapally	2.35	2.87	3.48	3.79	3.98
Palbari	0	1.23	1.48	2.48	2.79
Panchur Chak	3.27	3.64	4.19	4.49	4.87
Patharghata	1.24	1.68	2.84	3.78	3.97
Rabindranagar	4.19	4.28	4.56	4.76	4.94
Rangamati	1.24	2.46	3.75	4.43	4.83
Satyajitnagar	2.78	2.89	3.48	3.79	4.48
Shekpura	3.29	3.49	4.24	4.85	4.97
Suryanagar	1.28	2.67	3.73	3.97	4.89

2041 based on land alterations observed between 1991 and 2021. Due to land transformation and urban expansion, the urban region now boasts various pull factors such as employment opportunities, market facilities, education and health service, administrative and recreational facilities, as well as improved connectivity with the Kolkata metropolitan area via roadways and railways. However, this has resulted in significant land demand and natural resource consumption, negatively impacting the ecosystem. Additionally, the city's environment is adversely affected by narrow streets, severe pollution, and an inadequately managed sewage system. Another notable finding is the influence of uncontrolled rural immigration, leading to the proliferation of slum colonies within the Midnapore municipality region, including areas such as Brajakishor Pally, Rangamati, and Adibasipara (Wards 17, 24, and 18). Consequently, this historically significant city is at risk of losing its charm in various aspects. Over the past 30 years, one specific node was identified within the Midnapore municipal area, but the simulated results indicate the emergence of four major nodal points, namely Palbari in south, Najarguanj in southwest, Keranichoti in the northeast, and Dharma in the eastern part of Midnapore. Rangamati, along with other areas alleviating pressure of population on the Midnapore town, is likely to become the next focal point of expansion in the near future.

The changing patterns of urban land use have profound impact on the urban eco-environment (Nelson et al. 2010). Over the past three decades, this study has identified a notable degree of urban sprawl, particularly in areas such as Ashoknagar, Suryanagar, Rangamati, Palbari, Rabindranagar, Patharghata, and Dharma on the outskirts of the Midnapore municipality. These regions have emerged as significant demand zones for urban services. Additionally, the spreading of built-up areas has followed a ribbon-shaped development along the Bhadutala-Panchkhuri roadside in the northern part of the Midnapore municipality, resembling growth patterns observed in the urban agglomeration of Kolkata [2]. Notably, the growth in this research area has been characterized by horizontal expansion instead of vertical development, a distinction from larger cities, driven by population growth [22] (Li and He 2008). As highlighted by previous researchers, the increase in built-up areas has often occurred at the cost of agricultural land, with vegetation cover being another vulnerable terrain encroached upon [3, 16]. Agricultural fields and vegetation areas, being crucial resources for urban dwellers, are particularly susceptible to urban expansion [14]. Moreover, factors such as a decent residential environment, access to drinking water facilities, and low land prices have fuelled the development of built-up areas over agricultural fields, with this trend expected to persist in the future. Consequently, the Midnapore municipality and its peripheral areas face immense pressure on urban facilities and services, posing challenges to urban resilience and adaptability.

6.6 Policy Recommendations for Climate Resilient Urban Planning

Resilience in urban planning often faces challenges in linking local adaptation efforts with environmental changes over time [11]. Climate resilience urban planning integrates sustainable practices and policies aimed at reducing greenhouse gas emissions, enhancing energy efficiency, and promoting the use of renewable resources. It needs to design green infrastructure, such as parks and green roofs, which help mitigate urban heat islands and manage stormwater. Additionally, it involves implementing zoning regulations of demand and supply in urban areas that discourage development in high-risk areas, thereby protecting vulnerable populations (Mallick et al. 2021). Thus, effective governance is crucial to manage urban development; without it, planning efforts will falter due to the public's lack of understanding and commitment to the concept of future urban resilience and adaptability.

Robust governance frameworks are essential for regulating urban growth and ensuring that policies promote climate-resilient infrastructure and sustainable development practices. Inclusive planning processes engaging local communities, businesses and other stakeholders can incorporate their adaptive capacities and needs into resilience strategies, fostering a sense of ownership and commitment. To manage the influx of rural urban migrants, it is necessary to develop policies that create affordable housing options and integrate slum areas into the formal urban fabric to prevent unregulated settlements. Implementing slum upgradation programs can improve living conditions and equip these areas to handle climate-related stresses. Economic diversification is very essential to reduce dependency on vulnerable sectors and enhance the overall economic resilience of urban areas.

Promoting sustainable practices such as the adoption of renewable energy sources, including solar, can reduce dependency on fossil fuels and enhance energy resilience. Developing and promoting public transit, cycling, and walking infrastructure can reduce urban carbon footprints and improve air quality. Investing in resilient infrastructure capable of withstanding climate impacts, including upgrading water supply systems, drainage and sewerage, and transportation networks, is crucial. Promoting energy-efficient building designs and materials ensures resilient to extreme weather events.

Utilizing various demand and supply indicators can guide urban planning in achieving Sustainable Development Goals (SDGs) such as inclusive and sustainable economic growth (SDG 8), access to clean and healthy environments (SDG 11.1), and sustainable consumption and production patterns (SDG 12.a). Implementing these policy recommendations for resilient urban planning will help Midnapore better prepare for and adapt to the impacts of climate change.

6.7 Conclusion

The urban expansion of the Midnapore municipality area is represented by a compact pattern of settlement, and it has various opportunities to expand in all directions. Furthermore, high traffic, human strain on land, pollution, sewage issues, and so on can have a negative impact on appropriate urban development. This area requires immediate urban development planning through the establishment of public–private partnerships, diversified economic activities, using renewable energy etc. to make the town more resilient. Similarly, increasing green space along roadsides could help combat pollution while also making the city more walkable. To fully develop the future urban growth zone and solve the urban problem, more diligent work is needed. The demand–supply gap for urban facilities and amenities is created by rural-to-urban migration, pollution, meteorological conditions, or vehicle numbers. If we had access to all this information, we would be better prepared to deal with the environmental consequences of future city growth. By focusing on integrating climate resilience into urban planning, enhancing infrastructure resilience, promoting sustainable practices, and fostering inclusive governance, the city can ensure a sustainable and resilient future for its inhabitants.

Acknowledgements The authors are thankful to the DST-FIST sponsored Department of Geography, Vidyasagar University, Midnapore for conducting this research. Authors are also thankful to the residents for providing valuable information during this research work.

Conflict of Interest Statement: On behalf of all authors, the corresponding author states that there is no conflict of interest.

References

1. Afriyie K, Abass K, Adjei P-W (2019). Urban sprawl and agricultural livelihood response in peri-urban Ghana. Int J Urban Sustain Dev.https://doi.org/10.1080/19463138.2019.1691560
2. Bhatta B (2009) Analysis of urban growth pattern using remote sensing and GIS: a case study of Kolkata. India. Int J Remote Sens 30(18):4733–4746
3. Chen R, Ye C, Cai Y, Xing X, Chen Q (2014) The impact of rural out-migration on land use transition in China: past, present and trend. Land Use Policy 40:101–110
4. Clarke KC, Gaydos LJ (1998) Loose-coupling a cellular automaton model and GIS: long-term urban growth prediction for San Francisco and Washington/Baltimore. Int J Geogr Inf Sci 12(7):699–714
5. Clarke KC (2018) Land use change modeling with SLEUTH: improving calibration with a genetic algorithm. In: Geomatic approaches for modeling land change scenarios, pp 139–161. Springer, Cham
6. Dubovyk O, Sliuzas R, Flacke J (2011) Spatio-temporal modelling of informal settlement development in Sancaktepe district, Istanbul, Turkey. ISPRS J Photogramm Remote Sens 66(2):235–246

7. Keshtkar H, Voigt W (2016) A spatiotemporal analysis of landscape change using an integrated Markov chain and cellular automata models. Modeling Earth Syst Environ 2(1):1–13
8. Ma Q, He C, Wu J (2016) Behind the rapid expansion of urban impervious surfaces in China: major influencing factors revealed by a hierarchical multiscale analysis. Land Use Policy 59:434–445
9. Maithani S (2010) Cellular automata-based model of urban spatial growth. J Indian Soc Remote Sensing 38(4):604–610
10. Maity B, Mallick SK, Rudra S (2020) Spatiotemporal dynamics of urban landscape in Asansol Municipal Corporation, West Bengal, India: a geospatial analysis. GeoJournal 87,https://doi.org/10.1007/s10708-020-10315-z
11. Mallick SK, Rudra S (2021) Land use changes and its impact on biophysical environment: study on a river bank. Egypt J Remote Sens Space Sci 24(3):1037–1049
12. Mallick SK, Das P, Maity B, Rudra S, Pramanik M, Pradhan B, Sahana M (2021) Understanding future urban growth, urban resilience and sustainable development of small cities using prediction-adaptation-resilience (PAR) approach. Sustain Cities Soc. https://doi.org/10.1016/j.scs.2021.103196
13. Mallick SK, Rudra S, Maity B (2023) Unplanned urban built-up growth creates problem in human adaptability: evidence from a growing up city in eastern Himalayan foothills. Appl Geogr.https://doi.org/10.1016/j.apgepg.2022.102842
14. Mohan M, Pathan SK, Narendrareddy K, Kandya A, Pandey S (2011) Dynamics of urbanization and its impact on land-use/land-cover: a case study of megacity Delhi. J Environ Prot 2(09):1274
15. Mozaffaree Pour N, Oja T (2021) Prediction power of logistic regression (LR) and multi-layer perceptron (MLP) models in exploring driving forces of urban expansion to be sustainable in Estonia. Sustainability 14(1):160. https://doi.org/10.3390/su14010160
16. Naikoo MW, Rihan M, Ishtiaque M (2020) Analyses of LULC (LULC) change and built-up expansion in the suburb of a metropolitan city: Spatio-temporal analysis of Delhi NCR using Landsat datasets. J Urban Manag 9(3):347–359
17. Pal S, Ziaul S (2017) Detection of land use and land cover change and land surface temperature in English Bazar urban centre. Egypt J Remote Sens Space Sci 20(1):125–145
18. Patra S, Sahoo S, Mishra P, Mahapatra SC (2018) Impacts of urbanization on land use/cover changes and its probable implications on local climate and groundwater level. J Urban Manag 7(2):70–84
19. Ramachandra TV, Aithal BH, Sowmyashree MV (2014) Urban structure in Kolkata: metrics and modelling through geo-informatics. Appl Geomatics 6(4):229–244
20. Rimal B, Zhang L, Keshtkar H, Haack BN, Rijal S, Zhang P (2018) Land use/land cover dynamics and modeling of urban land expansion by the integration of cellular automata and Markov chain. ISPRS Int J Geo Inf 7(4):154
21. Shikary C, Rudra S (2020) Measuring urban land use change and sprawl using geospatial techniques: a study on Purulia Municipality, West Bengal, India. J Indian Soc Remote Sens 48(11), https://doi.org/10.1007/s12524-020-01212-6
22. Shikary C, Rudra S (2022) Urban growth Prediction for sustainable urban management using Markov Chain Model: a study on Purulia Municipality, West Bengal, India. J Indian Soc Remote Sens. https://doi.org/10.1007/s12524-022-01596-7
23. Shikary C, Rudra S (2023) Identifying obstacles and opportunities of urbanization in the district primate town of Purulia, West Bengal. GeoJournal. https://doi.org/10.1007/s10708-023-10878-7
24. Siddiqui A, Siddiqui A, Maithani S, Jha AK, Kumar P, Srivastav SK (2018) Urban growth dynamics of an Indian metropolitan using CA Markov and logistic regression. Egypt J Remote Sens Space Sci 21(3):229–236
25. Thapa RB, Murayama Y (2010) Drivers of urban growth in the Kathmandu valley, Nepal: examining the efficacy of the analytic hierarchy process. Appl Geogr 30(1):70–83
26. Verburg PH, Veldkamp A (2004) Projecting land use transitions at forest fringes in the Philippines at two spatial scales. Landscape Ecol 19(1):77–98

27. Wu KY, Zhang H (2012) Land use dynamics, built-up land expansion patterns, and driving forces analysis of the fast-growing Hangzhou metropolitan area, eastern China (1978–2008). Appl Geogr 34:137–145
28. Zeshan MT, Mustafa MRU, Baig MF (2021) Monitoring land use changes and their future prospects using GIS and ANN-CA for Perak River Basin. Malaysia. Water 13(16):2286. https://doi.org/10.3390/w13162286

Part III
Green Infrastructure and Nature-Based Solutions

Chapter 7
Disaster Resilience in Fiji Through Urban Green Infrastructure: A Review

Mumtaz Alam, Kunal Avishek Gounder, and Radhisha Shivnal Nath

Abstract Fiji, a South Pacific country, is known for its frequent natural disasters, including cyclones and flooding. This chapter explores the role of Urban Green Infrastructure (UGI) in enhancing disaster resilience in urban areas of Fiji. UGI focuses on strategic planning and development of green spaces, parks, trees, and flora, providing environmental, social, and economic benefits. UGI is emphasized for flood mitigation, cyclone protection, biodiversity promotion, ecosystem services, urban heat island reduction, community space exploitation, economic benefits, public health enhancement, and climate adaptation. The chapter emphasizes the importance of integrated planning, community involvement, regulatory interventions, continuous monitoring, skill development, and resilient landscaping techniques. The chapter asserts that UGI embodies a comprehensive methodology that enhances disaster resilience and sustainability in urban regions. By implementing UGI principles, Fiji has enhanced its capacity to mitigate natural catastrophe impacts and ensures the welfare of its inhabitants in a more unpredictable environment.

Keywords Disaster resilience · Urban green infrastructure · SDG · Cyclone protection and climate adaptation

7.1 Introduction

The latest phantom haunting the globe is confronting modern societies: the shifting global climate. For almost a century, the concept of the "greenhouse effect" [26] has been firmly established in theory. Data gathered from a measurement station built atop Hawaii's Mauna Loa volcano in 1959 to track CO_2 and other pollutants in the natural environment indicate that there are elevated amounts of "greenhouse gases"

M. Alam (✉)
Department of Social Sciences, College of Humanities and Education, Fiji National University, Suva, Fiji
e-mail: mumtaz.alam@fnu.ac.fj

K. A. Gounder · R. S. Nath
College of Humanities and Education, Fiji National University, Suva, Fiji

S. Ghosh et al. (eds.), *Cities of Tomorrow: Urban Resilience and Climate Change Preparedness*, Urban Sustainability, https://doi.org/10.1007/978-981-97-9658-8_7

in the atmosphere. These measurements demonstrated that CO_2 was rising unabated because it was not being entirely utilized by the oceans. Scholze et al. [19] claim that temperature variations impact "photosynthesis, plant respiration", and the breaking down of biological material, all of which impact the global carbon flux from land to atmosphere. In order to "achieve stabilization of greenhouse gas concentrations in the atmosphere at a level that would prevent dangerous anthropogenic interference with the climate system," the United Nations Framework Convention on Climate Change was put into effect in 1994 [25]. Urry [26] states that this stage should "allow ecosystems to adapt naturally to climate change."

7.1.1 Climate Crisis in the Pacific

Pacific Island nations are extremely susceptible to the consequences of global warming. It does exist that empirical evidence is only one component of a robust risk-management technique; however, it ignores the "dynamic psychological, cultural, social, economic, and political systems that affect climate impacts" [8]. Despite this, policymakers continue to turn to science to help them decide how communities should respond to climate challenges. Since Pacific communities, infrastructures, and habitats in the Pacific area are already feeling the effects of climate change, it is imperative that research be incorporated and interpreted into society. In order to formulate a comprehensive policy strategy that includes ocean-focused and climate-responsive urban planning methods adapted to island systems, the UNFCCC and UNESCAP formed an innovative collaboration with Pacific Island developing states during the 2017 Conference of Parties [34]. The UNFCCC and UNESCAP pledged to advocate for enhancing resiliency, reducing greenhouse gas emissions, and safeguarding the wellness of the marine environment in coastal island cities. The Pacific Ocean's "Ocean Cities" [34] are areas where manmade and natural habitats converge, where human behaviour and urbanization significantly impact aquatic and terrestrial ecosystems. Ocean Cities are particularly vulnerable to the effects of climate change, the problems associated with urbanization, and other development pressures.

7.1.2 Urban Green Infrastructure in Fiji for Disaster Resilience

Fiji comprises approximately 332 tropical islands spread across more than 18,000 square kilometres in the South Pacific Ocean and is home to approximately 881,000 people. Of those, only one-third are habitable. As shown below in Fig. 7.1, Viti Levu and Vanua Levu, the two largest islands in Fiji, are home to 70% of the nation's population.

Fig. 7.1 Map of Fiji. *Source* Google Maps - https://satellites.pro/Google/Fiji_map

Due to the climate crisis, the low-lying island territories of the Republic of Fiji experience frequent disasters. Fiji has emerged as a global leader in emphasizing the difficulties faced by low-lying island countries due to the serious threats that the changing climate creates to individuals and businesses, including rising sea levels, severe weather, coastal flooding, and degradation of ecosystems [4, 20]. The Fijian people's health, society, and personality are integrally related to their oceanic and coastal surroundings. The fishery industry contributes approximately 1.8% of the GDP and 7% of total export earnings, whereas prior to the COVID-19 pandemic, tourism accounted for 38.9% of the GDP and 35.5% of employment [10, 15]. Both tourism and the fishing industry are important components of the Fijian economy. The ability of the aquatic and coastal areas to maintain livelihoods, safeguard coastlines, and advance Fiji's economy is in jeopardy due to the growing effects of climate change.

Climate change is affecting numerous industries in Fiji, including forestry, urban development, health, agriculture, water and fisheries, and tourism. Chaudhary [12] highlights the informal and informal responses to environmental issues, such as seawall construction and mangrove replanting. Despite increased global climate financing, the amount available for adaptation is insufficient to sustain threatened communities along the coastline [12, 20]. Over a billion people live in unofficial housing, exacerbated by global urbanization and scarcity of affordable accommodation [17]. Fiji faces challenges due to rural communities being disconnected from essential services and built near hazardous areas. Informal inhabitants often choose risky locations due to lack of legal residency and official tenure status. The growing urban population presents issues like poor service, squatter colonies, joblessness, impoverishment, and connectivity issues. Understanding these challenges requires understanding land rights in Fiji, which are divided into state, freehold, and i-Taukei land categories.

The GSUA, located in Suva, has seen a surge in sugarcane farming families, leading to the establishment of squatter colonies due to inadequate infrastructure, lack of affordable housing, and the influx of migrants seeking better lives. These settlements also face environmental and health issues due to inadequate water supplies, sewerage, and sanitation services.

Furthermore, plans and programs for minimizing disaster risk and addressing climate change begun to combine from 2007 [12]. The government of Fiji is dedicated to incorporating climate change and disaster risk mitigation into the national budgeting and development process, as seen by the shift from merely disaster management to including climate variation in the Strategic Development Plan 2007–2011 (SDP), aiming to embrace disaster risk reduction by encouraging to alert public prior to a natural disaster or flooding.

Chaudhary [22] states that some of the projects underway are:

1. Mangrove Ecosystem and Climate Adaptation to Livelihood (USAID)
2. Climate Change and Health Adaptation (GEF, UNDP, WHO, MOH)
3. Community Adaptation to Climate Change and Land Use Planning (AusAID, WWF)
4. Ridge to Reef (GEF, UNDP)
5. Integrated Flood Management—NadiBasin Pilot (World Bank)

Chaudhary [22] and [17] highlight the challenges in installing natural infrastructure in informal settlements to mitigate climate change effects. Space limitations, financial resources, and governance issues hinder the installation process. However, the Fijian government has made strides in climate change adaptation, but still needs to address vulnerabilities and broaden the scope of adaptation initiatives. New vulnerability assessments should consider ecological, physical, and socioeconomic vulnerabilities. Diversifying projects to include tourism, food production, medical buildings, and forestry is crucial for making industries environmentally friendly and ensuring the livelihoods of farmers and fishers.

7.2 Materials and Methods

This section provides a comprehensive explanation of the methodology employed to assess the capacity of Urban Green Infrastructure (UGI) in bolstering catastrophe resilience in urban regions of Fiji. The study employs secondary data sources, such as official papers, academic literature, and relevant case studies from comparable locations. This project seeks to provide a realistic framework for implementing UGI (Urban Green Infrastructure) in Fiji. The framework will be developed using a comprehensive approach that involves analysing secondary data and official reports. The goal is to enhance disaster resilience and promote sustainable urban development in Fiji.

7.3 Results and Discussion

7.3.1 Urban Green Infrastructure and Sustainability

Town and city green areas and infrastructure provide sustainable living, urban expansion, and ecological, economic, and social benefits [14]. Urban woods, wetlands, parks, and green roofs increase air purification, water quality and quantity, carbon storage, and ecosystem protection. UGI might be crucial to Fiji's urban growth. The urban heat island effect is threatened by increasing infrastructure development in metropolitan Fiji. APN Global Change Research says urban heat concerns boost air-conditioning and fan energy use. Residents also face health concerns [2]. Thus, UGI promotes social sustainability by creating recreational parks to boost mental health, relieve stress, and find serenity. Moreira et al. [16] found that dense greenery with diverse plant compositions improves urban inhabitants' mental health. Alam et al. [1] note that conventional urban expansion has experienced several obstacles. These include rising traffic, inadequate amenities, poor air quality, a lack of place, land use segregation, and other rural features. These concerns harm people. Figure 7.2 depicts Fiji's top 10 killers, largely non-communicable illnesses. Lifestyle, especially urban lifestyle, is the major cause. Lack of parks and sidewalks hinders physical activity. North Carolina State University research found that everyday activities can alter moderate-to-intense exercise. Walkable, green locations have high levels of moderate-to-intense exercises, researchers found. The study found that greener home, work, and other settings boost walking opportunities. The researchers suggest lawmakers, landscape architects, and urban planners to plant greenery in high-walkability zones to improve exercise [13].

From an economic perspective, UGI attracts investment and tourism and creates jobs for new sectors like landscaping. UGI provides affordable ways to manage stormwater, reduce energy consumption, and mitigate urban heat islands, saving cities and their residents money in the long run. Thus, every dollar invested in UGI is a dollar saved in future disasters. Every economy needs infrastructure, and Fiji's economic progress is no exception. The UGI option offers exceptional returns on investment in Fijian infrastructure. UGI contributes to cost reduction; a study shows a 67% reduction in the cost of development [30]. Devon County Council states that many visitors are drawn to the area for its distinctive natural environment. Thus, it's important to preserve and enhance these green gems. Enterprises looking for a new workplace are often drawn to the scenic backdrop of many residential and commercial spaces. Green infrastructure can increase worker productivity and is essential for drawing in new investment. In addition, green infrastructure can create

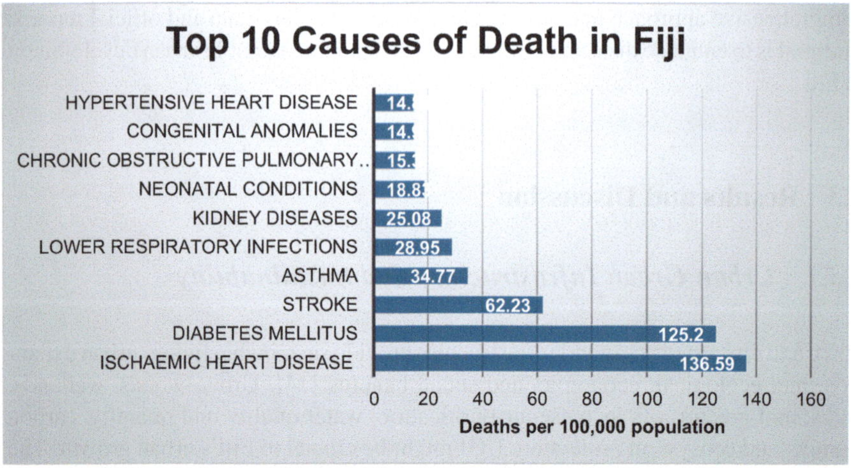

Fig. 7.2 Causes of Death in Fiji (*Source* World Health Organizations)

jobs by offering resources for products and crafts, renewable energy projects, and leisure activities.

7.3.2 Green Infrastructure for Climate Resiliency

Under the climate change framework, UGI is essential for mitigation and adaptation. By improving resistance to extreme weather events like heatwaves, droughts and floods, green spaces and natural places can help in climate change adaptation. Urban forests and green areas will absorb carbon dioxide and lower the energy needed to cool buildings, which helps mitigate the effects of climate change. Fiji's major urban centres like Nausori, Nadi, Ba and Navua are the latest victims of flash flooding in 2024. Flash flooding has become frequent in urban centres of Fiji, mainly due to poor drainage, urban planning, and high stormwater runoff.

UGI helps benefit from both physical and social perspectives. Many scholars have proven the benefits of UGI, pushing government agencies to adopt and implement it into their policies. But practically, it has not been implemented in developing countries, including Fiji. Figure 7.3 shows the summary of benefits that could be achieved through UGI.

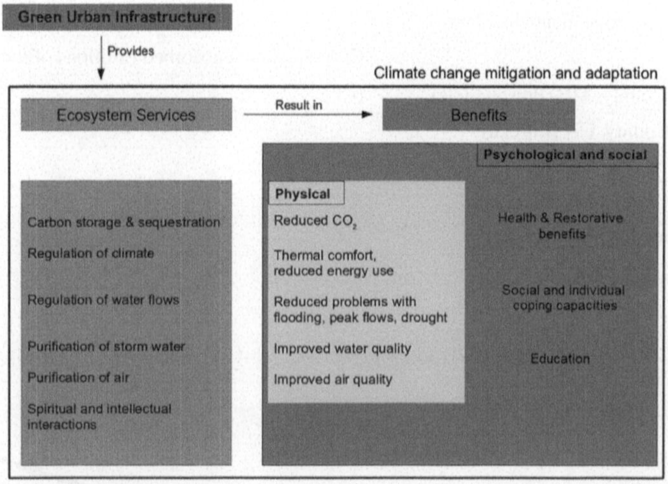

Fig. 7.3 Green Urban Infrastructure (*Source* [5])

7.3.3 *Physical Benefits*

7.3.3.1 Reduction of Carbon Dioxide

[29] discussed that vehicle exhaust and domestic heating are the major causes of urban heat issues with limited vegetation to balance the scale with carbon dioxide. UGI reduces carbon dioxide in urban centres by removing carbon dioxide through photosynthesis in daylight and releases through respiration at night [5]. Fares et al. [6] state that photosynthesis increases carbon dioxide sequestration in the urban forest during the growing season but not in the residential sites. This is mainly because of more carbon dioxide present in urban compared to residential areas. Many scholars have also proven this (Davies et al., Zhao et al., and Hutyra et al.), where carbon storage by vegetation is greater in urban areas. Therefore, more vegetation in urban Fiji could help achieve the same result. Also, due to the high population and less availability of vegetation, it increases the quantity of carbon dioxide released.

Research by U.S. Forest Services further adds that urban trees absorb approximately 150 million tons of carbon dioxide annually. Table 7.1shows the results from some of the 242 cities they studied.

7.3.3.2 Thermal Reduction and Reduced Energy Consumption

UGI development in Fiji's urban countries will reduce air and surface temperature by improving evapotranspiration and providing more shade. This will reduce the urban thermal island issue since the environment will become cooler, and less air-conditioning and fans will be used, thus reducing energy consumption.

Table 7.1 CO_2 equivalent absorbed

County	CO_2 equivalent absorbed (millions of tons) *
Aroostook County, ME (Presque Isle)	6.2
Humboldt County, CA (Eureka)	5.5
Lane County, OR (Eugene)	4.1
Penobscot County, ME (Bangor)	3.8
Jackson County, OR (Medford)	2.4

Source CentralClimate.org

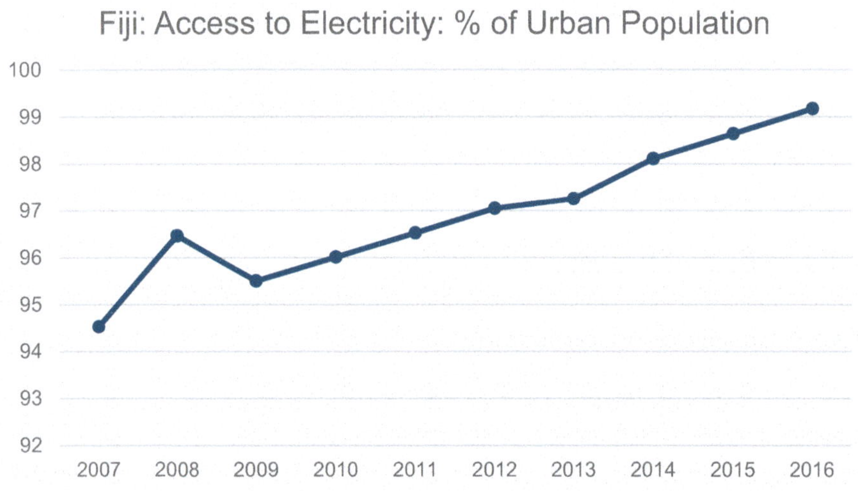

Fig. 7.4 Electricity in the urban population of Fiji (*Source* CEICDATA.COM; World Bank)

Currently, 40% of Fiji's energy is produced through non-renewable sources, which is contributing to climate change [11]. A study by [33] in Singapore has shown that the temperature in urban areas increases as they move away from green parks, and the energy consumption of buildings near the parks is 10% less. Also, green roofs can reduce direct evapotranspiration, hence reducing the heating of buildings.

Figure 7.4 shows the accessibility of electricity in the urban population of Fiji. The graph has been increasing steadily, showing increased demand for electricity.

7.3.3.3 Improved Air Quality

Plants have the capability to absorb and filter air pollutants, such as particulate matter, nitrogen oxides, Sulphur dioxide and carbon monoxide, through leaves and bark. Brantley et al. [3] found that there is a decrease in black carbon, signifying vehicle emissions, behind vegetative barriers. Cooler temperature reduces the formation of

Table 7.2 Air pollution

County	Air pollution reduced (millions of pounds) *
Aroostook County, ME (Presque Isle)	185
Lane County, OR (Eugene)	150
Humboldt County, CA (Eureka)	125
Penobscot County, ME (Bangor)	106
St. Louis County, MN (Duluth)	71

Source CentralClimate.org

secondary pollutants, ensuring air quality improves. Moreover, UGI includes more footpaths and cycling transportation. Residents can choose healthy activities like afternoon or morning walks and cycling. This will reduce the use of vehicles, hence reducing carbon emissions. Xing and Brimblcombe [31] further add that parks are the lungs of cities as they provide healthier air; nevertheless, attributing this to vegetation impedes the development of better park designs and adaptation of horticulture idea that reduces exposure to air pollutants. A study by the U.S. Forest Service in an estimated 242 cities has proven that urban trees reduce almost 4 billion pounds of air pollution. Table 7.2 shows some of the study results by U.S. Forest Services.

7.3.4 Green Infrastructure to Improve Disaster Resiliency

UGI offers significant advantages in terms of the environment, society, and economy in addition to reducing the disaster impacts such as treating stormwater at its source. UGI will improve Fiji's financial line and encourage urban livability through appropriate introduction, repair, and operation.

7.3.5 Flooding

Grey infrastructure, which is being historically used, is primarily intended to divert stormwater from the built environment. Stormwater runoff dumps hazardous contaminants, including bacteria, fertilizers, silt and heavy metals, into Fiji's lakes, streams and oceans. Grey infrastructure swiftly releases stormwater into our rivers and streams, increasing peak flows and the risk of flooding; this does not completely reduce the risk of flooding. A well-planned green infrastructure system can lessen the risk of flooding by decreasing and slowing the release of stormwater. Rainwater collection and infiltration are examples of green infrastructure that improves water supply efficiency. The constantly increasing need for water supplies and wastewater treatment can be greatly eased by using the water collected in rainwater harvesting systems for indoor and outdoor irrigation and other purposes.

Table 7.3 Stormwater runoff avoided (millions of gallons)

County	Stormwater runoff avoided (millions of gallons) *
Harris County, TX (Houston)	5.9
Allegheny County, PA (Pittsburgh)	4.6
Cook County, IL (Chicago)	4.6
Multnomah County, OR (Portland)	4.0
Jefferson County, AL (Birmingham)	3.3

Source CentralClimate.org

Restoring wetlands and planting trees is a practical substitute for constructing a new water treatment facility. Restoring floodplains would effectively avoid floods and cost less than constructing a new, taller dike. Furthermore, in addition to helping combat erosion and climate change, a restored forest, wetland, or floodplain will offer some rare and endangered animal and plant species with needed habitat. It is well-recognized that forests, wetlands, and floodplains act as peak flow buffers and purify water by removing pollutants. These services are important for urban areas to adjust to shifting weather patterns and dynamics of human needs [7]. Yeo [32] concluded that most of the urban development has occurred in Nausori and Nadi, which is prone to flooding.

Urban forest and their benefits in the improvement of stormwater runoff was studied by the U.S. Forest Service in the estimated 242 cities (counties) in the United States, and it was discovered that urban trees in these cities prevent roughly 110 billion gallons of stormwater annually, which is equivalent to 167,000 Olympic sized swimming pools. Table 7.3 shows some of the results discovered by the U.S. Forest Service.

7.3.6 Tropical Cyclone

Tropical cyclones are the most aggressive natural disaster experienced in Pacific Island Countries. With damage costing nearly billions to small island nations, the frequency and intensity are proven to increase due to climate change. Historically, tall trees have been used in the farms to protect the agricultural farms from strong winds, acting as the windbreaker. Similarly, in the era of urbanization where buildings in urban areas are becoming vulnerable to tropical cyclones, trees in urban can act as a wind breaker. Hence reducing the impacts of strong winds of tropical cyclones. A study by [27] supports the idea that urban tree cover can serve as a "debris filter", collecting debris velocity and material and mitigating damage if the trees are not a threat. The trees might become risky if they are not well maintained and trimmed on time. Old and bulky trees are very fragile causing more damage rather than protecting the urban environment. Therefore, UGI requires cross-disciplines, planners, and other professionals to undertake the best practices to ensure sustainable approaches to reduce the impacts of tropical cyclones. Thompson et al. [23] pointed out that trees that are 30 m high will generate debris due to a decrease in the strength of the trees. Turton [24] discovered that a smaller scale terrain, layout and structure influence the cyclone's impacts at local scales among forestry systems. This demonstrates the complexity of the interactions between tree behaviour and cyclones. This is made more difficult in urban settings by altered natural landscapes, a wide range of species selection, active tree management, and the built environment, all of which patterns are strongly impacted by socio-cultural, economic and driving forces.

7.3.7 Fiji's Vision for Urban Green Infrastructure

Like many other Pacific Island Countries, Fiji is the champion of fighting the climate change crisis. In 2015, Fiji was part of producing the Suva Declaration on Climate Change to increase the support for adaptation actions in the Pacific region. In 2016, Fiji became the first country to ratify the Paris Agreement to limit the global average temperature to below 1.5 degrees Celsius. Fiji was also the president of the 23rd session of the Conference of Parties (COP 23) to the United Nations Framework Convention on Climate Change. In 2021, Fiji government passed its first Climate Change Act to ensure that developments prioritize climate change and Fiji's National Climate Change Policy (NCCP) was also introduced.

Fiji has pledged to promote and achieve climate-resilient green growth domestically and internationally by implementing creative policies and initiatives. The five-year partnership between Fiji and the Global Green Growth Institute (GGGI) to meet these aims is outlined in the Country Planning Framework (CPF). The year 2014 saw Fiji join GGGI. In 2015, the Fijian government and GGGI established a Host Country Agreement. To integrate the needs for climate change adaptation across various domains, the Fiji 2019–2023 CPF places a strong emphasis on growing the capacity for producing renewable energy, decreasing the reliance on fossil fuels in the transportation sector, and creating inclusive green communities and cities [9].

The National Development Plan (NDP) has identified urban sustainability as a priority area, with the objective of creating dynamic, environmentally sustainable metropolitan centers that are resilient to the effects of climate change.

The government of Fiji wants to create town and city centers that support green growth initiatives, such as green energy, efficient and environmentally friendly transportation systems, the construction of green buildings, and technological advancement [9]. The country is rapidly becoming more urbanized. NDP has proposed several policies and initiatives aimed at bolstering urban planning. These include enhancing Fiji's ability to conduct climate risk assessments and creating climate-resilient infrastructure.

NDP seeks to improve the environment and mobilize investment to establish sustainable urban centres in Fiji. The development of green cities and towns, as well as employment generation, will depend heavily on the productivity and climate resilience of the private sector. To support these initiatives, GGGI helps improve sub-national planning, evaluations and analyses of options for green cities and towns, gives local governments technical support and capacity building, and examines ways to raise green finance in the tourism, manufacturing, and commercial sectors. The NDP also emphasizes cooperative ventures between the public and private sectors and the development of waste, water, and sanitation initiatives. Along with developing and supporting the financing of green city initiatives, GGGI evaluates investment potential in land and oceans capes as well as in cities, towns, and islands. Through acknowledging Fiji's towns and cities as both landscapes and oceans capes, GGGI seeks to promote sustainable urban expansion and the creation of green jobs.

Fiji's challenge to achieving UGI is the financial constraints and limited international investments. Fiji has been seeking expressions of interest from many stakeholders in green investment, but the aging infrastructure is a concerning issue arising at the moment. Urbanization pressure to increasing population in the urban area has been a major factor for the unsustainable urban developments with limited planning.

7.3.8 Urban Green Infrastructure and SDG

By developing UGI, Fiji will be able to achieve some of the Sustainable Development Goals. UGI enables the creation of more green spaces such as parks, gardens and natural elements that increase the natural environment rather than surrounded by concrete buildings. A green environment will create a better environment for residents to take advantage of, such as exercise, walking, strolling and many more, motivating people to choose healthy habits. Green vegetation will also increase the supply of fresh oxygen, improving air quality. Incorporating UGI will help Fiji to achieve SDG 11: Sustainable Cities and Communities. Moreover, UGI will increase the vegetation, preventingexcessive evapotranspiration, maintaining the soil's moisture and preventing excessive drought in the long period without rain. UGI also improves stormwater runoff, preventing frequent flooding. In addition, an increase in vegetation will make the environment much cooler due to an increase in the supply of wind and oxygen, preventing urban heat issues. It will also absorb excessive carbon emissions released from vehicles and industries, escaping to the atmosphere and damaging the ozone layer. All this contributes to SDG 13, taking "Climate Action". From a human perspective, UGI improves human health and well-being by providing provisions to exercise in parks, supply of fresh oxygen, and improve air quality, contributing to SDG 3: Good Health and Well-being. UGI can play a pivotal role in achieving SDG6: Clean Water and Sanitation. Creating wetlands and vegetation will contribute to water treatment and improve the water quality in urban areas. Incorporating these elements and measures to reduce the stormwater runoff in the urban design will ensure that cities have access to quality clean water and enough quantity to increase the urban population.

Fiji has the potential of adopting UGI due to the availability of land and a developing economy, hence allowing Fiji to achieve SDG simultaneously. Integrating nature into urban development will increase the beauty of urban Fiji, attracting more tourism and economic activity. However, this requires the collaboration of the government and many stakeholders to ensure that UGI development's outcome is sustainable and efficient.

7.3.9 *Effectiveness of UGI in Enhancing Urban Resilience to Climate Change*

UGI has proven its effectiveness in enhancing urban resilience to many cities across the world. UGI strategies have enabled cities to attract investments and tourism activity. In Singapore, "Garden City" approach as adopted in the late 1960's, giving the opportunities to become one of the most resilient green cities today [21]. Programs such as "Community in Bloom", has increased from 150 gardens in 2005 to 1700 gardens in 2021, enabling the well-being of the residents [21]. Figure 7.5 shows the study in Singapore about mental resilience of people who engages themselves in gardening. Gardening in urban areas has positive effect as people tend to be happier hen their plant grows [22].

The green urban strategy also assists in combating urban heat island issues. In Portland, USA, they reduce stormwater runoff, protect water quality, and enhance the general conditions of watersheds by using green infrastructure such as trees, green roofs, and green roadways. 51% of the impermeable surfaces under city control are presently managed by the integrated green stormwater infrastructure (Stormwater Strategies, n.d). Including UGI has helped to reduce the cost of Tabor to the River Program in Portland from $144 million to $81 million with multiplying benefits to the community. In China, the "Sponge City" initiative was launched in 2015 has increased greenery by 27% and enabled the city of Wuhan to retain 70% of rainwater [36]. These examples from major cities around the world shows the effectiveness of UGI in enhancing urban resilience to climate change.

7.3.10 *Challenges and Limitations in Implementing UGI*

Implementing UGI could be a challenge for some cities especially if they are implementing for the first time. One of the biggest obstacles to the implementation of UGI is the lack of political effort in providing adequate policies and incentives that would encourage and increase the effort of practice of UGI actions [35]. Policies are key players in promoting the application of UGI and incentives such as tax-free solar panels can boost people purchasing it. Secondly, the accurate and practical deployment of UGI components of buildings is impeded by the lack of building codes and technical support in developing countries. This makes hesitations with contractors and builders in planning and installing strategies like green roof systems. A lack of unified guidelines and standards for planning and implementing green roof systems

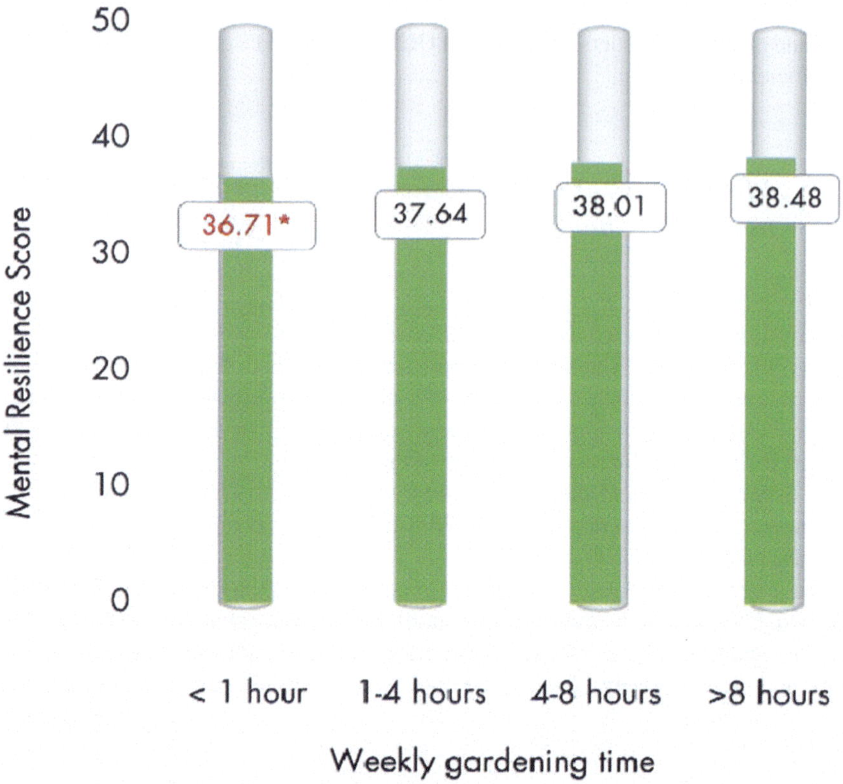

Fig. 7.5 Mental Resilience Score by Weekly Gardening Time (*Source* [21])

is linked to the limitation of research and analysis in this field [35].Weak collaboration between important agents and knowledge systems, as well as procedural hurdles resulting from low sectoral and disciplinary integrity, point to deficiencies in multidisciplinary and multi-scale processes. The creation of excellent greenspaces at different planning levels is hampered by a lack of standardized methods and guidelines. The idea of "green infrastructure," which originated in Western nations, has evolved into a sophisticated planning strategy that incorporates social and environmental factors [28].

7.3.11 Strategies and Best Practices for Integrating UGI into Climate Adaptation Planning

Integrating UGI into climate change adaptation and mitigation planning requires a strategic approach as success is vital. For UGI to be successful, it is important to assess the risks and vulnerabilities the city is facing so that UGI strategy could be planned accordingly as which would be more suitable. For instance, if the city is facing frequent challenge of urban heat island, then more vegetations on the road sides, parks and green roofs are the best strategy. Secondly, UGI development needs supporting policies and regulations that can guide and encourage more implementations of UGI projects. Reinwald et al. [18] states that goals are well-defined, and there is some social and political will but it is therefore necessary to swiftly modify the regulatory frameworks and policies in order to keep up with the fast changes that cities are seeing as a result of climate change. Updated policies and frameworks will show clear pathway the government is planning to follow, hence attracting investors to invest on UGI projects.

7.4 Conclusions

Like numerous other tiny island nations, Fiji is becoming more susceptible to the effects of climate change, such as cyclones, flooding, and rising sea levels. Urban Green Infrastructure (UGI) offers a comprehensive and sustainable approach to improving disaster resilience and addressing these concerns. This chapter has comprehensively analysed the various advantages of UGI (Urban Green Infrastructure) in urban regions of Fiji. It highlights the significant role of UGI in mitigating floods, protecting against cyclones, promoting biodiversity, reducing the effects of urban heat islands, providing economic benefits, improving public health, and facilitating climate adaptation. To effectively incorporate UGI into Fiji's urban environments, engaging in strategic planning and development is crucial. This includes the establishment of green spaces, parks, urban forests, and green roofs. These aspects offer environmental advantages, such as enhanced air quality and carbon sequestration, and promote social well-being by providing recreational spaces and improving mental health. From an economic standpoint, UGI can attract investments, lower infrastructure expenses, and generate employment opportunities, making it a financially efficient strategy for urban development. The chapter emphasises the importance of combining planning efforts and collaborating with stakeholders to properly implement UGI. This encompasses community engagement, regulatory measures, ongoing monitoring and maintenance, skill enhancement, and resilient landscaping methods. An all-encompassing strategy guarantees that UGI can make a substantial impact on disaster resilience, while simultaneously fostering sustainability and livability in urban areas. In addition, Fiji's dedication to addressing climate change at the international level, as seen by its leadership in global climate agreements and regulations, serves as a solid basis for promoting UGI efforts. By persistently giving importance to and enlarging its urban green infrastructure (UGI), Fiji can strengthen its ability to endure natural calamities and adjust to the evolving environment, ultimately guaranteeing the welfare and success of its urban populace. Urban Green Infrastructure is a progressive approach that reduces the effects of natural disasters and enhances the sustainability and quality of life in urban areas in Fiji. It is crucial to adopt and incorporate UGI concepts to establish urban settings that are resilient, dynamic, and sustainable and can effectively confront the difficulties posed by an increasingly unpredictable climate.

References:

1. Alam M, Ali MF (2022) Health and wellbeing and quality of life in the changing urban environment: challenges and government response In Fiji Island. J Positive School Psychol 6(9):2853–2863
2. Arifwidodo S, Abdulharis R, Kubota T (n.d.) Understanding urban heat island effect and its implications to climate change adaptation strategies in Major Southeast Asian cities. Asia-Pacific Network for Global Change Research. https://www.apn-gcr.org/project/understanding-urban-heat-island-effect-and-its-implications-to-climate-change-adaptation-strategies-in-major-southeast-asian-cities/
3. Brantley HL, Hagler GS, Deshmukh PJ, Baldauf RW (2014) Field assessment of the effects of roadside vegetation on near-road black carbon and particulate matter. Sci Total Environ 468:120–129
4. Chaudhury M (2015) A review of climate change impacts, adaptation practices and policies, and relevant institutions in Fiji. Fiji, Accessed August, 22, 2019
5. Demuzere M, Orru K, Heidrich O, Olazabal E, Geneletti D, Orru H, ... Faehnle M (2014) Mitigating and adapting to climate change: multi-functional and multi-scale assessment of green urban infrastructure. J Environ Manag 146:107–115
6. Fares S, Paoletti E, Calfapietra C, Mikkelsen TN, Samson R, Le Thiec D (2017) Carbon sequestration by urban trees. The urban forest: cultivating green infrastructure for people and the environment, 31–39
7. Farrugia S, Hudson MD, McCulloch L (2013) An evaluation of flood control and urban cooling ecosystem services delivered by urban green infrastructure. Int J Biodiversity Sci Ecosyst Serv Manag 9(2):136–145
8. Finucane M (2009) Why science alone won't solve the climate crisis: managing climate risks in the Pacific
9. Holland EA, Iese V, Nand M, Salili D, Wairiu M, Yuen L, Ocean cities of the pacific islands: re-naturing urbanization. https://www.unescap.org/sites/default/files/ESCAP_USP_Ocean%20Cities%20of%20the%20Pacific%20Islands_PB3.pdf
10. International Finance Corporation (2020) Fiji COVID-19 Business Survey: Tourism Focus. https://www.ifc.org/en/insights-reports/2020/fiji-covid-19-business-survey-tourism-focus. Accessed 17 May 2024
11. International Trade Administration (2024) Fiji—renewable energy. International Trade Administration | Trade.gov. https://www.trade.gov/country-commercial-guides/fiji-renewable-energy
12. Kumar R, Stephens M, Weir T (2013) Temperature trends in Fiji: a clear signal of climate change. South Pacific J Nat Appl Sci 31(1):27–38; Chaudhury M (2015) A review of climate change impacts, adaptation practices and policies, and relevant institutions in Fiji. Fiji, Accessed August, 22, 2019
13. Marquet O, Hirsch JA, Kerr J, Jankowska MM, Mitchell J, Hart JE, ... James P (2022) GPS-based activity space exposure to greenness and walkability is associated with increased accelerometer-based physical activity. Environ Int 165:107317
14. Mell IC (2009) Can green infrastructure promote urban sustainability?. In Proceedings of the institution of civil engineers-engineering sustainability (Vol 162, No 1, pp 23–34). Thomas Telford Ltd.
15. Ministry of Fisheries Fiji (2021) Annual report 2017–2018. Parliamentary Paper 37/2021. https://www.parliament.gov.fj/wp-content/uploads/2021/02/6-Ministry-of-Fisheries-2017-%E2%80%93-2018-Annual-Report.pdf. Accessed 17 May 202

16. Moreira TC, Polize JL, Brito M, da Silva Filho DF, Chiavegato Filho AD, Viana MC, ... Mauad T (2022) Assessing the impact of urban environment and green infrastructure on mental health: results from the São Paulo Megacity mental health survey. J of Exposure Sci Environ Epidemiol 32(2):205–212
17. Rauf HA, Wolff E, Hamel P (2023) Climate resilience in informal settlements: the role of natural infrastructure: a focus on climate adaptation. The Palgrave Encyclopedia of Urban and Regional Futures. Springer International Publishing, Cham, pp 269–277
18. Reinwald F, Weichselbaumer R, Schindelegger A, Damyanovic D (2024) From strategy to implementation: mainstreaming urban green infrastructure in Austria's spatial planning instruments for climate change adaptation. Urban Forestry & Urban Greening 94:128232
19. Scholze M, Knorr W, Arnell NW, Prentice IC (2006) A climate-change risk analysis for world ecosystems. Proc Natl Acad Sci 103(35):13116–13120
20. Shiiba N, Singh P, Charan D, Raj K, Stuart J, Pratap A, Maekawa M (2023) Climate change and coastal resiliency of Suva, Fiji: a holistic approach for measuring climate risk using the climate and ocean risk vulnerability index (CORVI). Mitig Adapt Strat Glob Change 28(2):9
21. Sia A, Tan PY, Er KB (2023) The contributions of urban horticulture to cities' liveability and resilience: insights from Singapore. Plants, People, Planet 5(6):828–841
22. Sia A, Tan PY, Wong JCM, Araib S, Ang WF, Er KBH (2022) The impact of gardening on mental resilience in times of stress: a case study during the COVID-19 pandemic in Singapore. Urban Forestry & Urban Greening 68:127448
23. Thompson BK, Escobedo FJ, Staudhammer CL, Matyas CJ, Qiu Y (2011) Modeling hurricane-caused urban forest debris in Houston. Texas. Landscape and Urban Planning 101(3):286–297
24. Turton SM (2012) Securing landscape resilience to tropical cyclones in Australia's Wet Tropics under a changing climate: lessons from Cyclones Larry (and Yasi). Geogr Res 50(1):15–30
25. United Nations Climate Change. (n.d.). What is the United Nations Framework Convention on Climate Change?. Unfccc.int. https://unfccc.int/process-and-meetings/what-is-the-united-nations-framework-convention-on-climate-change
26. Urry J (2015) Climate change and society. Why the social sciences matter. Palgrave Macmillan UK, London, pp 45–59
27. Van der Sommen FJ, Pearson DM, Boggs GS (2018) Analysis of the interrelationship between houses, trees and damage in a cyclone affected city: Can landscape design and planning utilising trees minimise cyclone impact?. Int J Disaster Risk Reduction 28:701–710
28. Vaňo S, Olafsson AS, Mederly P (2021) Advancing urban green infrastructure through participatory integrated planning: a case from Slovakia. Urban forestry & Urban Greening 58:126957
29. Velasco E, Roth M (2010) Cities as net sources of CO_2: review of atmospheric CO_2 exchange in urban environments measured by eddy covariance technique. Geogr Compass 4(9):1238–1259
30. Wycliffe J (2019) Green Infrastructure beats grey infrastructure economically, environmentally & socially. Ministry of Waterways & Environment, Fiji. https://www.mowe.gov.fj/wp-content/uploads/2019/08/Green-Infrastructure-Beats-Grey-Infrastructure-Economically-speech-F.pdf
31. Xing Y, Brimblecombe P (2019) Role of vegetation in deposition and dispersion of air pollution in urban parks. Atmos Environ 201:73–83
32. Yeo S (2013) A review of flood resilience in Fiji. In International conference on flood resilience,"Experiences in Asia and Europe," University of Exeter, UK, September, pp 5–7
33. Yu C, Hien WN (2006) Thermal benefits of city parks. Energy and buildings 38(2):105–120
34. Zari MP, Kiddle GL, Blaschke P, Gawler S, Loubser D (2019) Utilising nature-based solutions to increase resilience in Pacific Ocean Cities. Ecosyst Serv 38:100968
35. Zeadat ZF (2022) Urban green infrastructure in Jordan: a perceptive of hurdles and challenges. J Sustain Real Estate 14(1):21–41
36. Zheng S, Tang Y, Chan FKS, Cao L, Song R (2022) The demographic implication for promoting sponge city initiatives in the Chinese megacities: a case of Wuhan. Water 14(6):883

Chapter 8
The Impact of Green Spaces Characteristics on Enhancing Resilience of Urban Canyons Microclimates

Mohammadreza Baradaran Motie, Mohammadreza Bemanian, and Mansour Yeganeh

Abstract Following the Industrial Revolution and the rise of automobiles in cities, the shape and geometry of cities underwent significant changes, which affected urban morphology. This unsustainable development disrupted the balance between built and natural environments, leading to many problems for cities and their citizens over time. Environmental problems are the most significant. Damage to green infrastructure, natural ecosystems, water and air pollution, and extreme heat are just a few examples. Numerous Nature-based solutions have been proposed to achieve these goals. One primary strategy is the development of urban green spaces, which serve as a platform for restoring balance to cities and reducing environmental harm. Recent studies show that simply adding green spaces to cities is not enough. Their design and placement need to be carefully considered to maximize their benefits. Not considering these properties in the design and development of microclimates not only does not improve urban resilience but can also exacerbate these harms. Therefore, this chapter focuses on understanding the behavior of urban green spaces and their influential properties on urban microclimates. These influential properties include planting design and layout, tree proportions and geometric form, density and permeability, and other physiological characteristics.

Keywords Urban resilience · Greenery properties · Green spaces · Urban canyon · Urban Heat Island · Urban Pollution Island · Thermal comfort

M. Baradaran Motie · M. Bemanian (✉) · M. Yeganeh
Dept. Architecture, Tarbiat Modares University, Jalal AleAhmad, P.O. Box, 14115-111 NasrTehran, Iran
e-mail: bemanian@Modares.ac.ir

Digital Architecture and Artificial Intelligence Lab, Department of Architecture, Faculty of Art, Tarbiat Modares University, Tehran, Iran

Architectural Design, Modeling and Fabrication Lab, Department of Architecture, Faculty of Art, Tarbiat Modares University, Tehran, Iran

S. Ghosh et al. (eds.), *Cities of Tomorrow: Urban Resilience and Climate Change Preparedness*, Urban Sustainability, https://doi.org/10.1007/978-981-97-9658-8_8

Abbreviations

UHI	Urban Heat Island
UPI	Urban Pollution Island
H/W	Height-to-width
L/H	Length to height

8.1 Introduction

The Industrial Revolution, with the development of factories and the concentration of production in cities, led to an increase in urbanization and migration to cities, resulting in cities' development and megacities' emergence. The need to accommodate more people within city limits and close to workshops and industrial centers gradually increased the value and construction of land within cities.

Due to their potential for housing development, the increasing value of land and natural areas in cities led to a change in their land use. This development and further housing construction, with the introduction of cars and the development of streets and highways, decreased cities' connection to natural environments and damaged green infrastructure. As a result, cities took on a new form and morphology.

Over time, as cities expanded horizontally, a gradual shift towards vertical growth began. This transformative process fundamentally altered the urban form, creating built environments that increasingly disconnected human life from the city's surrounding natural environment. The central cores of cities, surrounded by high-rise buildings, had created a new environment, or rather a new "microclimate," that did not follow the natural environment around it. One of the most prevalent microclimates developed is the "urban canyon" or "urban street canyon."

8.2 Urban Canyon

Urban canyons, also known as street canyons, are spaces where the buildings around them surround urban streets and extend along the street. Since the cross-section of these streets creates a valley-like shape, they are called urban canyons [49].

Geometric properties are crucial to understanding urban canyons and classifying these unique urban zones. The three main geometric components of urban canyons are: Width (W), Height (H), and Length (L) of urban canyons, and aspect ratios such as H/W and L/H, are used to classify them and identify their impact on other components [71].

In addition to proportions, the orientation of canyons (East–West, North–South) and sky view factor (SVF) are other factors that affect the microclimate of urban canyons [16].

The confinement of urban canyons by buildings has a profound and far-reaching impact on the meteorological factors of the canyon, which in turn significantly influences other factors such as air pollution. In the following, we will delve into the crucial effect of urban canyons on several critical variables:

8.2.1 Temperature

City streets can trap heat, becoming significantly warmer than nearby areas. This intricate issue has been tackled by many researchers, each uncovering a different piece of the puzzle that explains this temperature disparity. The intensity and angle of solar radiation, orientation, albedo of canyon surfaces, thermal coefficients and heat transfer coefficients of canyon materials, sky view factor (SVF), wind speed and direction, etc., are among the factors that contribute to this temperature difference in urban canyons. The more sunlight the canyon surfaces receive during daylight hours, the more energy will enter the canyon [1, 10, 17, 58, 77]. An increase in the reflectivity coefficients of surfaces in urban canyons significantly impacts MRT (Mean Radiant Temperature) and disrupts thermal comfort [46, 59].

Changing the aspect ratios of urban canyons and ratios such as H/W or the orientation (N-S) or (E-W) of urban canyons causes noticeable temperature changes. This trapped heat disrupts how walls in the canyon exchange heat, causing the entire area to stay hotter, day and night. [2, 70].

8.2.2 Wind

Wind flow within urban canyons deviates from typical open-air patterns, leading to distinct microclimatic conditions. It can be generally categorized into two types:

8.2.2.1 Wind Parallel to the Canyon Orientation

When the wind direction aligns with the orientation of the urban canyon, a channeling effect occurs. Wind speed changes as the canyon cross-section varies, exhibiting a Venturi effect. Therefore, as the canyon cross-section narrows, we observe an increase in wind speed within the canyons [64].

8.2.2.2 Wind Perpendicular to the Canyon

When wind flows perpendicular to the orientation of urban canyons, it forms a vortex flow within the canyon. This vortex flows, and their behavior is highly influenced by

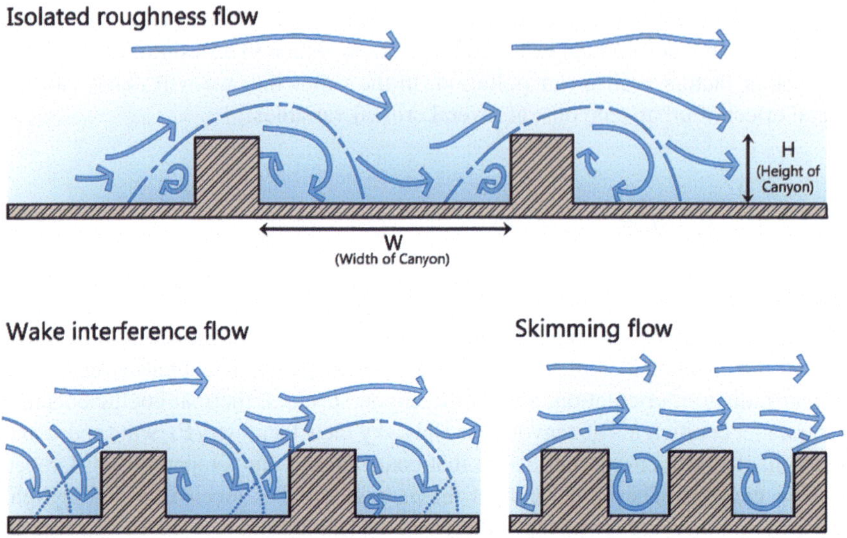

Fig. 8.1 The impact of canyon aspect ratio (H/W) on airflow patterns (Based on [52])

the canyon's aspect ratios. The main wind flow pattern changes in different aspect ratios [64].

The impact of H/W changes on perpendicular canyon flows is briefly illustrated in Fig. 8.1.

8.2.3 Air Quality

The temperature and wind flow variations within urban canyons also affect the air quality inside the canyons. With their channeling and Venturi effects, wind flows parallel to the canyon disperse air pollutants and cause them to move along the canyons. In contrast, vortex flow traps air pollutants when the wind blows perpendicular to the canyon (Fig. 8.2). These air pollutants can come from outside the canyon or linger from local sources, accumulating at street level due to the slower wind speeds [64].

8.3 Challenging Phenomena of Urban Canyons

According to the investigated cases, the main microclimate challenges of urban canyons can be called "urban heat islands (UHI)" and "urban pollution islands(UPI)."

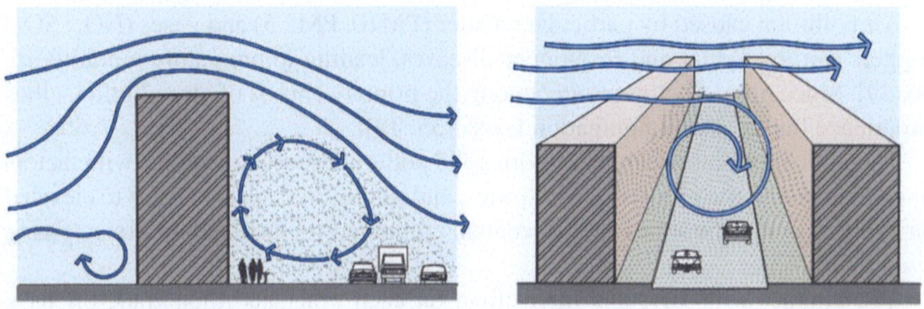

Fig. 8.2 Primary vortex flow in urban canyon

8.3.1 Urban Heat Islands (UHI)

Concrete, asphalt, and other building materials in cities have a high capacity for absorbing and re-emitting solar heat compared to natural elements like trees and soil. Additionally, urban spaces, particularly the central cores of cities with higher concentrations of buildings and structures, are more affected by this heat absorption. Dense urban areas further stifle natural air circulation, trapping heat and making them even hotter than other parts of the city. This effect is known as the "urban heat island." [20, 55]. Research results state a significant UHI effect in the US, with city temperatures exceeding surrounding areas by 7 °C during the day and 5 °C at night [31].

The transformation of cities plays a critical role in creating UHIs. Key factors include.

- The loss of natural environment and green spaces
- Properties of urban materials (such as color, thermal conductivity, specific heat, etc.) [59, 65, 66]
- Urban Geometry
- Anthropogenic Heat Emissions [55]

UHI pose a significant health risk to residents. Soaring temperatures are linked to a rise in heat-related illnesses, like heatstroke, exhaustion, fainting, and muscle cramps. [37].

8.3.2 Urban Pollution Islands

Another critical challenge in urban canyons is air pollution. Air pollution has emerged as a severe public health crisis in recent years. A report by UNEP reveals that air pollution affects over a billion people globally, with tragic consequences. More than a million deaths each year are linked to this pollution and the illnesses it causes. [69].

Air pollution caused by particulate matter (PM10, PM2.5) and gases (NO_x, SO_x) triggers cardiovascular and respiratory diseases, leading to premature mortality [6, 36, 39]. Since human activities are among the primary causes of air pollution, cities experience higher air contamination levels [35, 43].

This concentration of human activities and pollution sources, coupled with factors like urban geometry, reduced green spaces, and diminished airflow, leads to elevated air pollutant concentrations in city centers compared to surrounding areas, giving rise to UPI [13].

UPI interact with UHI and their effect on each other and their trade-off have been studied in many researches [9, 30, 33, 40, 41, 44, 60, 63, 68]. It is crucial to recognize that urban canyons are a primary platform for forming and concentrating these crisis-inducing phenomena in cities.

8.4 Urban Green Spaces

Numerous studies have acknowledged the significance of expanding green spaces in cities to combat climate change, global warming, and the detrimental effects of human activities. Green space development can restore balance in cities and alleviate harmful impacts by enhancing all aspects of ecosystem services within urban environments.

Green space development effectively addresses the heat-related and thermal discomfort effects of UHIs. During the hottest months, plants and green spaces mitigate heat through shading and heat absorption via evapotranspiration [5, 29, 45, 51].

Urban green spaces can be categorized into three groups:

- Extensive (e.g., parks)
- Linear (e.g., green corridors)
- Vertical (e.g., green wall)

Extensive green spaces are recognized as large natural patches within the city, encompassing parks and gardens.

High rise buildings and streets often confine extensive green spaces, reducing their effectiveness in central urban areas and limiting their influence to their immediate surroundings.

Hence, strengthening green corridors as linkages between extensive green spaces and maintaining connectivity between these patches is crucial [34].

Green corridors, also known as linear natural infrastructures, constitute a significant component of the urban ecological landscape. These pathways comprise trees, plants, and green elements that connect green spaces within cities and establish a green network [72].

Among their most notable advantages are mitigating UHI effects and reducing pollution islands in urban canyons. The shade provided by green corridors diminishes pedestrians' exposure to harmful UV radiation [25, 37, 62].

Additionally, trees and plants absorb gaseous pollutants through their pores and utilize them in their life cycles. Airborne particulate matter also settles on and is absorbed by plant leaves. Numerous studies have demonstrated the effectiveness of green spaces in reducing air pollution [4, 15, 23, 38, 50, 56].

8.5 The Duality of Green Spaces in Urban Canyon

While the development of green corridors in urban canyons offers numerous benefits for enhancing urban resilience, other studies have highlighted sensitivities in implementing trees and vegetation within these spaces. Research using computer models, simulations, and real-world observations shows that trees and plants in urban canyons can significantly influence their ability to combat problems like UHI and UPI. This happened due to their impact on critical meteorological factors within the canyons. [1, 14, 18, 27, 40, 42, 48, 75].

Trees in street canyons induce quantitative and qualitative changes in airflow, leading to an average increase in air pollutant depositions of up to 13.2% [28]. The foliage of trees acts as an obstacle and windbreak against airflow within the canyon, disrupting wind patterns and causing air pollution to settle beneath the tree canopy [32, 57, 74].

In urban canyons, trees play a dual role. On the one hand, they act as natural air filters, trapping air pollutants on their leaves and bark and facilitating their breakdown into harmless substances. On the other hand, their foliage can disrupt airflow patterns, potentially hindering the dispersion and removal of these pollutants, especially those originating from traffic emissions (canyon pedestrian level).

Studies have shown that trees' negative aerodynamic impact on air circulation in urban canyons often outweighs their air-purifying benefits (Fig. 8.3). This point highlights the need for careful consideration of tree placement and species selection when developing green corridors in urban settings [18, 73, 75].

While removing trees from urban canyons may offer certain benefits, it is not a viable solution due to the numerous advantages of green spaces discussed above. Therefore, resilience in urban canyons requires a thoughtful approach to green spaces.

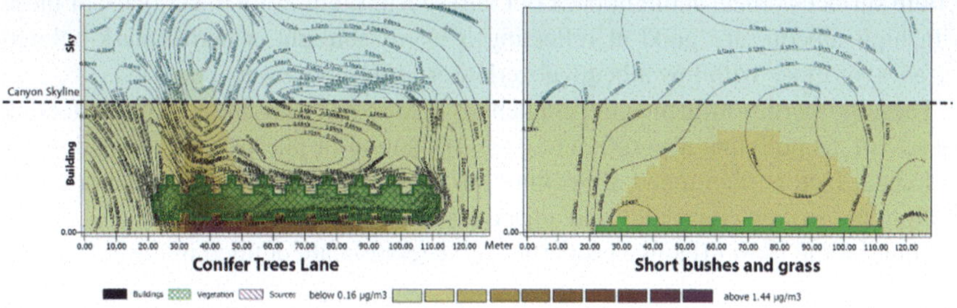

Fig. 8.3 Vegetation Effects on AirPollution deposition& wind dynamics in urban canyons (Simulated by ENVI-met software)

It needs to consider the specific characteristics of vegetation and how to utilize them effectively for a holistic solution.

Factors affecting the effectiveness of green spaces in enhancing resilience in urban canyons can be categorized into three groups:

1. *Meteorological Factors:*
   - Sunlight exposure and orientation of the urban canyon.
   - Relative humidity.
   - Air temperature.
   - Wind speed, wind direction.
2. *Spatial Factors:*
   - Geometry of the urban canyons.
   - Dimension and aspect ratios of the urban canyons (Length, Width, and Height, H/W,...).
   - Material properties of the urban canyons and their thermal and optical properties.
   - Land uses and population density in buildings adjacent to the urban canyon.
3. *Properties of Green Spaces and Their Elements.*

Trees in urban canyons play a crucial role by acting as porous barriers to wind flow. Various studies have identified their physical form and degree of porosity as key factors in the accumulation of air pollution.

The dimensions, proportions, and indices, such as LAD and LCR of each green element individually, as well as the spacing and density of their distribution in urban canyons, are among the influencing factors.

Additionally, specific physiological properties of plants, such as albedo and transmittance, influence the amount of sunlight absorbed and transmitted by plants and the intensity of their biological processes, such as photosynthesis.

In climate studies, a key factor is albedo, which refers to the amount of light a surface reflects [8].

The albedo effect, often known as planetary albedo, measures how efficiently certain surfaces reflect sunlight back out into space. Light-colored surfaces, or those with high albedo, are good at reflecting a lot of sunlight, whereas dark-colored surfaces, or those with low albedo, absorb more solar radiation.

These properties influence the challenges discussed (UHI and UPI) by affecting the rate of air pollutant absorption (e.g., CO_2 fixation or photosynthesis) and meteorological factors(mentioned above)through evaporative coolingand the amount of sunlight absorption (for plants and adjacent surfaces).

Some of the most effective properties of vegetation are defined in the Table 8.1:

Table 8.1 Green elements properties

Name	Definition
Height	The height of the mature plant from the highest point to the ground
Width	The maximum width of a plant species at maturity (Canopy width)
H/W	The plant's height-to-width ratio
LCR: (Live Crown ratio)	The Live Crown Ratio (LCR) is the percentage of that height that has healthy leaves and branches *No living branch = 0% *Top of tree to the ground covered with foliated branch = 100% [54]
Leaf type	Plants can be divided into different categories according to the type of leaves. However, the most appropriate classification is deciduous and conifer
LAD: (Leaf Area Density)	Leaf Area Density (LAD) is the total one-sided leaf area per unit volume of a plant canopy. It is typically expressed in square meters per cubic meter (m^2 m^{-3}). LAD is a critical parameter in various ecological studies, influencing factors such as light use efficiency, photosynthetic capacity, canopy gas exchange, forest canopy structure, radiation balance, and urban microclimate [24]
Vegetation silhouette	This concept can better show the distribution of green elements in urban canyons. Different sections along the length and width of the canyon can help us understand the distribution system of green elements in urban canyons (Fig. 8.4)
Foliage shortwave albedo	The percentage of shortwave sunlight reflected from plant foliage back into space
Foliage shortwave transmittance	The percentage of shortwave sunlight that penetrates through the plant's foliage
Emissivity of leaves	Leaf Emissivity quantifies a leaf's ability to emit thermal radiation, typically ranging from 0.8 to 1. A higher value indicates a more ability to release heat [26]

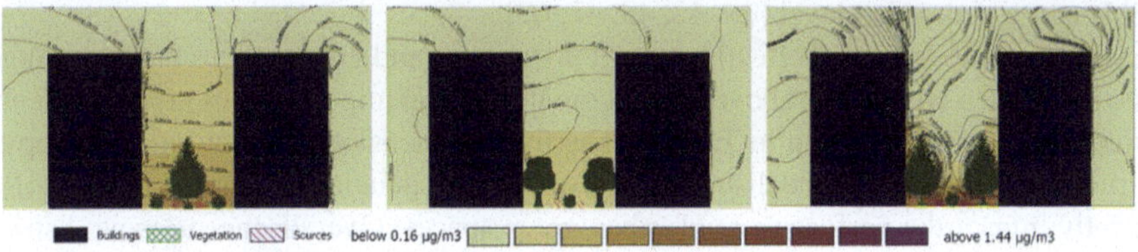

Fig. 8.4 Unveiling Urban Canyon vegetation effects through silhouette analysis

8.6 Towards an Integrated Understanding

Studies explore how urban canyon greenspace impacts resilience against UHI and UPI. A primary focus of this research has been the role of greenspace in reducing temperatures in urban canyons, attributed to the shading effect of plants and trees on other surfaces and evaporative cooling. The canopy of trees and plants acts as a sunshade, preventing the absorption of sunlight and solar energy by lower surfaces in urban canyons. The lack of solar energy reception during the day and its non-storage in the surfaces of the canyon leads to a reduction in the nocturnal UHI effect and an increase in nighttime temperatures in urban canyons [7, 22, 53].

Trees promote temperature reduction, enhance thermal comfort, and mitigate the UHI effect through evaporative cooling and by increasing the ambient relative humidity. (a process driven by their internal physiological mechanisms) [11, 12, 19, 22, 30].

Ettinger and colleagues investigated the role of trees in enhancing urban resilience by reducing extreme heat and lowering temperatures in a Washington, neighborhood during the summer. Their study found that areas without tree canopy cover were up to five times more likely to experience maximum temperatures than fully covered areas. Additionally, tree-covered spaces exhibited an average temperature reduction of up to 2.57 °C [21].

Another study, involving satellite-based high-resolution analysis of 293 European cities, investigated the impact of urban trees on reducing land surface temperature (LST). The findings revealed that tree-covered green spaces in southern Europe exhibited up to 4 degrees Celsius lower temperatures than non-treed areas. In comparison, central European cities showed a temperature difference of 8 to 12 °C [61].

Also, wong and colleagues identified the role of green infrastructure characteristics in influencing temperature cooling in urban areas. The results highlights the importance of scale, extent, type, and placement of vegetation in determining the effectiveness of temperature reduction [76].

Studies, both simulations and field experiments, have highlighted the sensitivity of using green elements and trees in urban canyons due to their potential to exacerbate UPI. Proper wind circulation and vortex flow are crucial for transporting and removing traffic-related air pollution in urban canyons. However, trees and hedges can disrupt this air circulation, accumulating and depositing gaseous pollutants and PMs at the pedestrian level [28, 42, 48, 47, 67].

The varying roles and effectiveness of urban trees and green spaces in addressing the mentioned challenges above underscore the critical importance of green element properties, as summarized in Table 1. The form, dimensions, aspect ratios, canopy structure, permeability, and distribution of trees and vegetation within canyons significantly impact the performance of urban green spaces. Overall, the strategic distribution of green spaces with varying densities and permeabilities throughout urban canyons is crucial for establishing a dynamic and holistic interplay that not only addresses the outlined challenges but also fulfills the broader spectrum of urban green space values. This is a crucial consideration for landscape designers, architects, and urban planners to incorporate into their designs and define appropriate scenarios.

Studies have explored and evaluated various scenarios to identify optimal configurations of these elements toward achieving more sustainable and holistic solutions. In examining 19 urban canyon greenspace design scenarios, trees were identified as the most significant contributors to air pollution exacerbation, followed by hedges. Additionally, the height of hedges and trees emerged as a critical parameter influencing the deposition of air pollution at lower elevations. This includes the height at which foliage begins and extends from the ground [73].

Furthermore, the permeability of green elements, measured by the Leaf Area Density (LAD) index, and the spacing and density of vegetation and trees also directly impacted air pollution exacerbation in urban canyons [73].

The interplay and relationship between greenspace properties, meteorological factors (mentioned above), and the canyon's spatial configuration must be considered holistically. Alterations to any of these factors significantly impacted the performance of various greenspace scenarios in enhancing the resilience of urban canyons [75].

A valuable approach to achieving an integrated solution for enhancing resilience in urban canyons is to bring the scenarios closer to real-world conditions. In a study conducted in Tehran's urban canyons, different greenspace scenarios were evaluated throughout the hottest summer days and coldest winter days, with hourly meteorological data defined in the simulation environment. The study found that moving broadleaf trees from sidewalks to the center of the canyon in north–south-oriented urban canyons significantly reduced air pollution. In summer, this relocation resulted in up to a 67% decrease, and in winter, a 54% decrease. This relocation reduced air pollution by up to 200% in summer and 90% in winter for coniferous trees. Furthermore, a more detailed examination of the results in the two seasons revealed correlations between meteorological factorsand the indicators for measuring the challenges discussed in different seasons [3]. These results indicate that the correlation coefficient of urban canyon meteorological factors in controlling UHI or UPI varies across seasons. Figure 8.5illustrates the operational framework of these components' relationship and intensity of the correlation coefficiency.

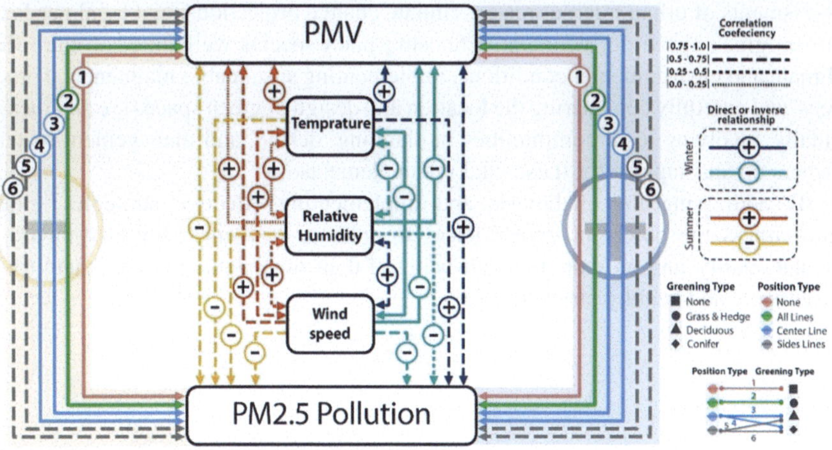

Fig. 8.5 Seasonal Impact of Canyon greening: a multi-factor correlation overview [3]

8.7 Discussion

Urban green spaces are crucial in enhancing urban resilience by providing ecosystem services and benefits to cities. These spaces contribute to the ecological balance of built and natural environments and strengthen various aspects of urban resilience.

Urban green spaces act as natural air filters, absorbing pollutants and improving air quality. Vegetation provides shade and evaporative cooling, reducing air and surface temperatures and mitigating the urban heat island effect. Plants release and absorb CO2, contributing to a healthier urban environment and mitigating climate change. Additionally, green spaces can absorb and retain rainwater, reducing runoff and the risk of flooding while filtering pollutants from stormwater improving water quality. They even support biodiversity by providing habitats for diverse plant and animal species.

Urban green spaces, beyond their environmental benefits, offer significant social and cultural advantages. They provide opportunities for recreation and relaxation, promoting mental health and well-being. Exposure to nature has been shown to enhance cognitive function and creativity, while green spaces facilitate social interaction and strengthen community bonds.

However, challenges exist in urban green space development. Unscientific and unplanned development without proper consideration of urban infrastructure, climate, and microclimates can lead to inefficiencies and even new problems. Failure to select species suited to local climate conditions and provide adequate maintenance can compromise the sustainability and effectiveness of green spaces. Establishing and maintaining these spaces can be costly, requiring careful budgeting and resource allocation. Improper placement and design can exacerbate existing environmental issues.

To overcome these challenges and enhance resilience, cities must conduct detailed assessments of urban microclimates, climate change projections, and local needs to inform green space development. Choosing plant species well-suited to the local climate and environmental conditions, implementing sustainable maintenance practices, and carefully considering the location and design of green spaces are all crucial. Finally, involving local communities in planning, design, and management fosters ownership and support for these vital urban elements.

By addressing the challenges and implementing effective strategies, cities can harness the power of green spaces to enhance resilience, improve environmental quality, and promote the well-being of their inhabitants, making them truly sustainable and healthy places to live.

8.8 Conclusion

Unbalanced urban development, driven by rapid population growth, has weakened green infrastructure and a growing disconnect between cities and their natural surroundings. This has resulted in a multitude of challenges and reduced urban resilience. Air pollution, global warming, climate change, and its consequences are among the most pressing issues. While Nature-based Solutions and green infrastructure development have been identified as the most effective strategies for overcoming these crises, their implementation at various scales requires careful consideration.

The microclimatic differences between central city areas and their peripheries necessitate localized planning approaches. Urban canyons are among the most critical microclimates. Urban canyons' environmental properties, such as geometry, morphology, aspect ratios, and orientation, along with meteorological factors like wind speed and wind direction, air temperature, surface temperature, radiation, and relative humidity, significantly impact the sustainability and resilience of these urban spaces.

Findings from field studies and software simulations suggest that utilizing urban green spaces and green elements like vegetation and trees in urban canyons requires a comprehensive and sustainable solution that considers the properties of green spaces, including the geometric proportions of vegetation, its distribution and placement within urban canyons, its permeability and density, and other physiological characteristics.

Further exploration of more realistic, practical, and compatible scenarios that consider social, psychological, and other aspects of the role of green spaces in urban-canyons in different climates is a topic that can be discussed and analyzed in further research.

Enhancing the role of urban green spaces in building resilient cities against climate change necessitates the implementation of robust policies. These policies should focus on integrating and prioritizing the development of green infrastructure in urban projects, conducting climate risk assessments for such projects, and formulating comprehensive plans at both macro and local levels to expand green spaces. Additionally, these initiatives' effectiveness and considering crucial elements such as site location, neighborhood morphology and characteristics, and design features are pivotal aspects that demand attention.

To ensure the successful execution of these policies, decision-makers and city managers should engage in strategic planning and target-setting. Adequate funding and investment in this sector are essential, along with fostering public participation. By involving local communities, the sustainability and effectiveness of green spaces can be guaranteed.

Furthermore, urban planners and specialists in related fields, including urban designers and landscape architects, play a crucial role in this process. Their expertise and dedication to prioritizing the qualities of urban green spaces and urban microclimates in their planning and architectural designs are vital to building resilient cities that effectively manage and mitigate climate change.

Overall, by implementing these measures and considering the social, economic, and environmental aspects, cities have the potential to develop into resilient urban environments. With the right strategies and investments, they can effectively address climate change challenges and pave the way for a more sustainable future.

References

1. Ahmadi S, Yeganeh M, Motie MB, Gilandoust A (2022) The role of neighborhood morphology in enhancing thermal comfort and resident's satisfaction. Energy Rep 8:9046–9056. https://doi.org/10.1016/J.EGYR.2022.07.042
2. Ali-Toudert F (2021) Exploration of the thermal behaviour and energy balance of urban canyons in relation to their geometrical and constructive properties. Build Environ 188:107466. https://doi.org/10.1016/j.buildenv.2020.107466
3. Baradaran Motie M, Yeganeh M, Bemanian M (2023) Assessment of greenery in urban canyons to enhance thermal comfort & air quality in an integrated seasonal model. Appl Geogr 151:102861. https://doi.org/10.1016/J.APGEOG.2022.102861
4. Beckett KP, Freer-Smith PH, Taylor G (1998) Urban woodlands: their role in reducing the effects of particulate pollution. Environ Pollut 99:347–360. https://doi.org/10.1016/S0269-7491(98)00016-5
5. Bolund P, Hunhammar S, Bolund P, Hunhammar S (1999) Ecosystem services in urban areas. Ecol Econ 29:293–301
6. Boningari T, Smirniotis PG (2016) Impact of nitrogen oxides on the environment and human health: Mn-based materials for the NOx abatement. Curr Opin Chem Eng 13:133–141. https://doi.org/10.1016/J.COCHE.2016.09.004
7. Bretz S, Akbari H, Rosenfeld A (1998) Practical issues for using solar-reflective materials to mitigate urban heat islands. Atmos Environ 32:95–101
8. Britannica TE of E (2024) albedo. Encyclopedia Britannica
9. Cao C, Lee X, Liu S et al (2016) Urban heat islands in China enhanced by haze pollution. Nat Commun 7:12509. https://doi.org/10.1038/ncomms12509
10. Chiang Y-C, Liu H-H, Li D, Ho L-C (2023) Quantification through deep learning of sky view factor and greenery on urban streets during hot and cool seasons. Landsc Urban Plan 232:104679. https://doi.org/10.1016/j.landurbplan.2022.104679
11. Cortes A, Rejuso AJ, Santos JA, Blanco A (2022) Evaluating mitigation strategies for urban heat island in Mandaue City using ENVI-met. Journal of Urban Management 11:97–106. https://doi.org/10.1016/J.JUM.2022.01.002
12. Crank PJ, Sailor DJ, Ban-Weiss G, Taleghani M (2018) Evaluating the ENVI-met microscale model for suitability in analysis of targeted urban heat mitigation strategies. Urban Clim 26:188–197. https://doi.org/10.1016/j.uclim.2018.09.002
13. Crutzen PJ (2004) New directions: the growing urban heat and pollution" island" effect-impact on chemistry and climate. Atmos Environ 38:3539–3540
14. Czader K, Balczó M, Eichhorn J (2009) Modelling of flow and dispersion in a street canyon with vegetation by means of numerical simulation. Proceedings of the XXIII MicroCAD international scientific conference
15. Diener A, Mudu P (2021) How can vegetation protect us from air pollution? A critical review on green spaces' mitigation abilities for air-borne particles from a public health perspective—with implications for urban planning. Sci Total Environ 796:148605. https://doi.org/10.1016/j.scitotenv.2021.148605
16. Dirksen M, Ronda RJ, Theeuwes NE, Pagani GA (2019) Sky view factor calculations and its application in urban heat island studies. Urban Clim 30:100498. https://doi.org/10.1016/j.uclim.2019.100498

17. Du Y, Mak CM, Li Y (2019) A multi-stage optimization of pedestrian level wind environment and thermal comfort with lift-up design in ideal urban canyons. Sustain Cities Soc 46:101424. https://doi.org/10.1016/j.scs.2019.101424
18. Eisenman TS, Churkina G, Jariwala SP et al (2019) Urban trees, air quality, and asthma: an interdisciplinary review. Landsc Urban Plan 187:47–59. https://doi.org/10.1016/j.landurbplan.2019.02.010
19. Elsayed IS (2012) Mitigation of the urban heat island of the city of Kuala Lumpur, Malaysia. Middle-East J Sci Res 11:1602–1613
20. EPA (2008) Reducing urban heat islands: compendium of strategies. Environmental Protection Agency, U.S
21. Ettinger AK, Bratman GN, Carey M et al (2024) Street trees provide an opportunity to mitigate urban heat and reduce risk of high heat exposure. Sci Rep 14:3266. https://doi.org/10.1038/s41598-024-51921-y
22. Faragallah RN, Ragheb RA (2022) Evaluation of thermal comfort and urban heat island through cool paving materials using ENVI-Met. Ain Shams Engi J 13:101609
23. Freer-Smith PH, Beckett KP, Taylor G (2005) Deposition velocities to Sorbus aria, Acer campestre, Populus deltoides × trichocarpa 'Beaupré', Pinus nigra and × Cupressocyparis leylandii for coarse, fine and ultra-fine particles in the urban environment. Environ Pollut 133:157–167. https://doi.org/10.1016/j.envpol.2004.03.031
24. Gao G, Qi J, Lin S et al (2023) Estimating plant area density of individual trees from discrete airborne laser scanning data using intensity information and path length distribution. Int J Appl Earth Obs Geoinf 118:103281. https://doi.org/10.1016/j.jag.2023.103281
25. Gehrels H, van der Meulen S, Schasfoort F et al (2016) Designing green and blue infrastructure to support healthy urban living. TO_2 federatie
26. Gerber F, Marion R, Olioso A et al (2011) Modeling directional–hemispherical reflectance and transmittance of fresh and dry leaves from 0.4μm to 5.7μm with the PROSPECT-VISIR model. Remote Sens Environ 115:404–414. https://doi.org/10.1016/j.rse.2010.09.011
27. Gromke C (2011) A vegetation modeling concept for building and environmental aerodynamics wind tunnel tests and its application in pollutant dispersion studies. Environ Pollut 159:2094–2099
28. Gromke C, Blocken B (2015) Influence of avenue-trees on air quality at the urban neighborhood scale. Part II: traffic pollutant concentrations at pedestrian level. Environ Pollut 196:176–184. https://doi.org/10.1016/j.envpol.2014.10.015
29. Hardin PJ, Jensen RR (2007) The effect of urban leaf area on summertime urban surface kinetic temperatures: a Terre Haute case study. Urban Urban Green 6:63–72
30. Henao JJ, Rendón AM, Salazar JF (2020) Trade-off between urban heat island mitigation and air quality in urban valleys. Urban Clim 31:100542. https://doi.org/10.1016/j.uclim.2019.100542
31. Hibbard KA, Hoffman FM, Huntzinger D, West TO (2017) Ch. 10: changes in land cover and terrestrial biogeochemistry. Climate science special report: fourth national climate assessment, Volume I. Washington, DC
32. Jin S, Guo J, Wheeler S et al (2014) Evaluation of impacts of trees on PM2.5 dispersion in urban streets. Atmos Environ 99:277–287. https://doi.org/10.1016/j.atmosenv.2014.10.002
33. Joachim Fallmann (2014) Numerical simulations to assess the effect of urban heat island mitigation strategies on regional air quality
34. Jongman B, Osmanoglou D, Van Zanten BT et al (2021) A catalogue of nature-based solutions for urban resilience (English). World Bank Group, Washington, D.C.
35. Ketzel M, Berkowicz R (2004) Modelling the fate of ultrafine particles from exhaust pipe to rural background: an analysis of time scales for dilution, coagulation and deposition. Atmos Environ 38:2639–2652. https://doi.org/10.1016/J.ATMOSENV.2004.02.020
36. Kim KH, Kabir E, Kabir S (2015) A review on the human health impact of airborne particulate matter. Environ Int 74:136–143. https://doi.org/10.1016/J.ENVINT.2014.10.005
37. Kovats RS, Hajat S (2008) Heat stress and public health: a critical review. Annu Rev Public Health 29:41–55. https://doi.org/10.1146/annurev.publhealth.29.020907.090843

38. Kumar P, Druckman A, Gallagher J et al (2019) The nexus between air pollution, green infrastructure and human health. Environ Int 133:105181. https://doi.org/10.1016/j.envint.2019.105181
39. Lelieveld J, Haines A, Burnett R, et al (2023) Air pollution deaths attributable to fossil fuels: observational and modelling study. BMJ e077784. https://doi.org/10.1136/bmj-2023-077784
40. Li H, Meier F, Lee X et al (2018) Interaction between urban heat island and urban pollution island during summer in Berlin. Sci Total Environ 636:818–828. https://doi.org/10.1016/j.scitotenv.2018.04.254
41. Li H, Sodoudi S, Liu J, Tao W (2020) Temporal variation of urban aerosol pollution island and its relationship with urban heat island. Atmos Res 241:104957. https://doi.org/10.1016/j.atmosres.2020.104957
42. Li X-B, Lu Q-C, Lu S-J et al (2016) The impacts of roadside vegetation barriers on the dispersion of gaseous traffic pollution in urban street canyons. Urban Urban Green 17:80–91. https://doi.org/10.1016/j.ufug.2016.03.006
43. Lutz M (2013) Mitigation strategies: Berlin, Germany. In: Particulate matter: environmental monitoring and mitigation. Future Science Ltd, pp 74–95
44. Meng X, Zhang Y, Zhao Z et al (2012) Temperature modifies the acute effect of particulate air pollution on mortality in eight Chinese cities. Sci Total Environ 435–436:215–221
45. Moreno-garcia MC (1994) Intensity and form of the urban heat island in Barcelona. Int J Climatol 14:705–710
46. Morini E, Castellani B, De Ciantis S et al (2018) Planning for cooler urban canyons: comparative analysis of the influence of façades reflective properties on urban canyon thermal behavior. Sol Energy 162:14–27. https://doi.org/10.1016/j.solener.2017.12.064
47. Motie MB, Yeganeh M, Bemanian M (2023) Assessment of greenery in urban canyons to enhance thermal comfort & air quality in an integrated seasonal model. Appl Geogr 151:102861. https://doi.org/10.1016/j.apgeog.2022.102861
48. Neofytou P, Venetsanos AG, Rafailidis S, Bartzis JG (2006) Numerical investigation of the pollution dispersion in an urban street canyon. Environ Model Softw 21:525–531. https://doi.org/10.1016/j.envsoft.2004.08.012
49. Nicholson SE (1967) (1975) A pollution model for street-level air. Atmos Environ 9:19–31. https://doi.org/10.1016/0004-6981(75)90051-7
50. Nowak DJ (1994) Atmospheric carbon dioxide reduction by Chicago's urban forest. Chicago's urban forest ecosystem: results of the Chicago urban forest climate project, 83–94
51. Nowak DJ, Crane DE (2000) The Urban Forest Effects (UFORE) Model: quantifying urban forest structure and functions. In: Hansen, Mark; Burk, Tom (eds) Integrated tools for natural resources inventories in the 21st century Gen Tech Rep NC-212 St Paul, MN: US Dept of Agriculture, Forest Service, North Central Forest Experiment Station 714–720 212
52. Oke TR (1988) Street design and urban canopy layer climate. Energy Build 11:103–113. https://doi.org/10.1016/0378-7788(88)90026-6
53. Oke TR (1981) Canyon geometry and the nocturnal urban heat island: comparison of scale model and field observations. J Climatol 1:237–254. https://doi.org/10.1002/joc.3370010304
54. Parsons A, Robichaud P, Lewis S, Napper C (2010) Field guide for mapping post-fire soil burn severity. General Technical Report, RMRS-GTR-243
55. Phelan PE, Kaloush K, Miner M et al (2015) Urban heat island: mechanisms, implications, and possible remedies. Annu Rev Environ Resour 40:285–307. https://doi.org/10.1146/annurev-environ-102014-021155
56. Pope CA, Dockery DW (2006) Health effects of fine particulate air pollution: lines that connect. J Air Waste Manage Assoc 56:709–742. https://doi.org/10.1080/10473289.2006.10464485
57. Ries K, Eichhorn J (2001) Simulation of effects of vegetation on the dispersion of pollutants in street canyons. Meteorol Z 10:229–233. https://doi.org/10.1127/0941-2948/2001/0010-0229
58. Rosso F, Golasi I, Castaldo VL et al (2018) On the impact of innovative materials on outdoor thermal comfort of pedestrians in historical urban canyons. Renew Energy 118:825–839. https://doi.org/10.1016/j.renene.2017.11.074

59. Salvati A, Kolokotroni M, Kotopouleas A et al (2022) Impact of reflective materials on urban canyon albedo, outdoor and indoor microclimates. Build Environ 207:108459. https://doi.org/10.1016/j.buildenv.2021.108459
60. Sarrat C, Lemonsu A, Masson V, Guedalia D (2006) Impact of urban heat island on regional atmospheric pollution. Atmos Environ 40:1743–1758
61. Schwaab J, Meier R, Mussetti G et al (2021) The role of urban trees in reducing land surface temperatures in European cities. Nat Commun 12:6763. https://doi.org/10.1038/s41467-021-26768-w
62. Sharbafian M, Yeganeh M, Baradaran Motie M (2024) Evaluation of shading of green facades on visual comfort and thermal load of the buildings. 317::114303. https://doi.org/10.1016/j.enbuild.2024.114303
63. Sokhi RS, Baklanov A, Piringer M (2017) Special issue of journal of urban climate: modelling of urban air pollution and climate interactions. Urban Clim 22:1. https://doi.org/10.1016/j.uclim.2017.11.002
64. Spirn AW (1986) Air quality at street-level: strategies for urban design. Boston Redevelopment Authority
65. Taleghani M (2018) The impact of increasing urban surface albedo on outdoor summer thermal comfort within a university campus. Urban Clim 24:175–184. https://doi.org/10.1016/j.uclim.2018.03.001
66. Taleghani M, Berardi U (2018) The effect of pavement characteristics on pedestrians' thermal comfort in Toronto. Urban Clim 24:449–459. https://doi.org/10.1016/j.uclim.2017.05.007
67. Taleghani M, Clark A, Swan W, Mohegh A (2020) Air pollution in a microclimate; the impact of different green barriers on the dispersion. Sci Total Environ 711:134649. https://doi.org/10.1016/j.scitotenv.2019.134649
68. Ulpiani G (2020) On the linkage between urban heat island and urban pollution island: three-decade literature review towards a conceptual framework. Science of the Total Environment 141727
69. UNEP (2005) 2005 Annual evaluation report
70. Vallati A, Mauri L, Colucci C, Ocłoń P (2017) Effects of radiative exchange in an urban canyon on building surfaces' loads and temperatures. Energy Build 149:260–271. https://doi.org/10.1016/j.enbuild.2017.05.072
71. Vardoulakis S, Fisher BEA, Pericleous K, Gonzalez-Flesca N (2003) Modelling air quality in street canyons: a review. Atmos Environ 37:155–182. https://doi.org/10.1016/S1352-2310(02)00857-9
72. Vink J, Vollaard P, Zwarte N de (2017) Stadsnatuur maken = Making urban nature. nai010 uitgevers/publishers, Rotterdam
73. Vos PEJ, Maiheu B, Vankerkom J, Janssen S (2013) Improving local air quality in cities: to tree or not to tree? Environ Pollut 183:113–122
74. Wang X, Zhou Z, Xiang Y et al (2024) Effects of street plants on atmospheric particulate dispersion in urban streets: a review. Environ Rev. https://doi.org/10.1139/er-2023-0103
75. Wania A, Bruse M, Blond N, Weber C (2012) Analysing the influence of different street vegetation on traffic-induced particle dispersion using microscale simulations. J Environ Manage 94:91–101
76. Wong NH, Tan CL, Kolokotsa DD, Takebayashi H (2021) Greenery as a mitigation and adaptation strategy to urban heat. Nat Rev Earth Environ 2:166–181. https://doi.org/10.1038/s43017-020-00129-5
77. Yuan J, Farnham C, Emura K (2021) Effect of different reflection directional characteristics of building facades on outdoor thermal environment and indoor heat loads by CFD analysis. Urban Clim 38:100875. https://doi.org/10.1016/j.uclim.2021.100875

Chapter 9
Leveraging Nature-Based Solutions for Urban Resilience and Climate Adaptation

Manish M. Nair, Shama Parween, and Rajan Chandra Sinha

Abstract In the last few decades, the escalation of pluvial flood occurrences has posed a significant concern for densely populated urban centres. This surge is attributed to amplified precipitation patterns driven by climate change, insufficient infiltration capacities due to land use and cover alterations, increased surface sealing, and inadequate stormwater infrastructure, all consequences of rapid urbanization. Over recent years, pluvial flood incidences have increased globally, particularly affecting Southeast Asian nations. Delhi, for instance, witnessed six urban floods between 2002 and 2016, indicative of the alarming pace of urbanization in the national capital, with 75.09% of its geographical area now classified as urban. According to the 2011 census, 97.5% of Delhi's populace resides in urban locales, with 60% of residents inhabiting unplanned settlements. These urban expansions, fuelled by migration, have encroached upon agricultural and green spaces, exacerbating the risk of pluvial floods. During the peak monsoon periods, numerous unplanned neighbourhoods and transport networks in Delhi are inundated, causing traffic disruptions and inconvenience for commuters. Addressing this challenge requires sustainable and resilient infrastructure solutions, prioritizing the implementation of nature-based approaches. This chapter proposes an integrated four-stage stormwater planning system employing Nature-Based Solutions (NBS) such as swales, rain gardens, permeable pavements, stormwater ponds, curb extensions, planters, and vegetative filter strips. The Soil Conservation Service Curve Number (SCS-CN) method was used to estimate runoff generated during the 10-year maximum rainfall event in Aug 2020. The runoff left after initial abstraction and the quantity conveyed through stormwater pipelines was estimated to be 29% of the runoff generated per hour. Through a four-stage planning approach, NBS features were strategically integrated within the study area, resulting in an 85.67% reduction in excess runoff, predominantly through stormwater collection (80.48%) and infiltration (5.19%). This study

M. M. Nair · S. Parween (✉) · R. C. Sinha
Department of Architecture and Planning, Birla Institute of Technology, Mesra, Ranchi, Jharkhand 835215, India
e-mail: ar_shama@bitmesra.ac.in

M. M. Nair
e-mail: manishnair3496@gmail.com

S. Ghosh et al. (eds.), *Cities of Tomorrow: Urban Resilience and Climate Change Preparedness*, Urban Sustainability, https://doi.org/10.1007/978-981-97-9658-8_9

provides valuable insights for stakeholders in devising strategies for the selection, placement, and execution of future nature-based planning strategies, aligning with the Sustainable Development Goals (SDGs) 11, 6, and 13, thus contributing to the 2030 agenda for sustainable development.

Keywords Pluvial floods · Nature-based solutions · Climate change · Resilient cities

9.1 Introduction

The world's ability to withstand extreme events has been impacted by climate change and climate-induced disasters, which have been scientifically confirmed [1]. Roughly 50% of South Asia's population was impacted by at least one kind of disaster, per an Asian Development Bank report [2]. The IPCC (2023) [3] predicts that the present rate of temperature increase would surpass the safe threshold of 1.5 °C, which might raise the likelihood of extreme weather events (heavy precipitation) that cause natural catastrophes, including floods (fluvial, coastal, and flash floods) in India [4]. If these problems are not addressed, they may serve as obstacles to the nation's socioeconomic advancement. This study focuses on pluvial floods caused by high surface runoff due to anthropogenic influences (such as high impervious surfaces) and natural factors (like high precipitation and climate change). Pluvial floods occur when increased precipitation coincides with decreased infiltration capacity [5, 6]. The risk of floods is attributed to climate change (increased precipitation), reduced pervious surfaces (increased runoff), and inadequate stormwater conveyance infrastructure.

Climate change exacerbates flood risks by altering precipitation patterns and increasing the frequency and intensity of heavy rainfall events. The main causes of pluvial flooding include retention, reduced infiltration, and changes in land use. India accounted for 36.6% of South Asian floods, with significant urban flood events in cities like Chennai (2005), Mumbai (2005), Surat (2006), Bhopal (2006), Kolkata (2007), Hyderabad (2008), Noida (2010), Gurgaon (2020), and Bengaluru (2022) [7, 8]. Between 2009 and 2020, India experienced approximately 316 floods, resulting in financial damages of 58 million USD, 17,540 fatalities, and affecting 100 million people (Emergency Event Database). Direct factors contributing to flooding include urbanization and global climate change, while indirect factors include inadequate drainage infrastructure and inefficient or non-existent solid waste management [8]. These factors significantly strain existing infrastructure, often unable to withstand the impacts of unforeseen natural disasters [9]. Addressing the increased flood risk due to climate change and urbanization requires robust mitigation and adaptation strategies. Cities need to adopt alternative storm water management plans and implement more ecologically and climate-adaptively designed interventions, such as Nature-Based Solutions (NBS). These strategies are referred to by various nomenclatures worldwide, including Best Management Practices in the US and Canada, Low-Impact Development in the US and New Zealand, Water-Sensitive Urban Design in Australia,

Sustainable Urban Drainage Systems in the UK, Green Infrastructure in the US and UK, Blue-Green Infrastructure in the UK, and Nature-Based Solutions in Europe [10, 11].

Nature-based solutions (NBS) have evolved significantly since their inception in the 1970s and 1990s, initially aimed at understanding the relationship between people and nature [12]. By the late 2000s, organizations such as the International Union for Conservation of Nature (IUCN), the World Bank, and the European Commission (EC) began to promote NBS as sustainable alternatives to address societal challenges, particularly those posed by climate change [12–14]. NBS first appeared in a report by the World Bank [15], with more structured definitions later provided by IUCN and EC. The IUCN defines NBS as "actions to protect, sustainably manage and restore natural or modified ecosystems that address societal challenges effectively and adaptively, providing human well-being and biodiversity benefits" [16].

NBS play a crucial role in mitigating climate change by enhancing natural carbon sinks, which absorb CO_2 and other greenhouse gases, thereby reducing atmospheric concentrations of these pollutants [17]. By restoring and managing ecosystems such as forests, wetlands, and urban green spaces, NBS contribute to carbon sequestration and help regulate local climates, reducing the overall impact of global warming. These solutions also improve urban resilience by enhancing the capacity of cities to adapt to changing climate conditions. Beyond climate mitigation, NBS offers a range of co-benefits that support biodiversity, create green jobs, and improve public health. These solutions also address water and food security by managing resources more sustainably and reducing the risk of natural disasters [18, 19]. The promotion of NBS by international organizations underscores their global relevance. The World Bank, for instance, views NBS as essential for strengthening urban resilience through infrastructure services inspired by natural processes [4]. Early research programs on NBS emphasize their potential to benefit both people and ecosystems, highlighting the symbiotic relationship between human well-being and environmental health [20, 21].

According to the European Commission (2015), there are three main pillars of NBS. Those are Social, Economic, and Environmental. The present study focuses on the environmental parameters, focusing on water regulation. The viability of NBS implementation at the neighbourhood level is investigated and assessed. To lessen urban runoff and the risk of pluvial floods for susceptible milli-watersheds with sizes between 500 and 1000 ha, which are defined using historical pluvial flood data and the proportion of built-up area, the study proposes a four-stage NBS planning strategy and demonstrates the effectiveness of specific urban NBS. Nature- based solutions can be categorized based on the type of issues and the scale of implementation [4]. A variety of NBS can be retrofitted or constructed into the existing urban. Out of the many varieties of NBS, swales, rain gardens, permeable pavement, storm water ponds, Rejuvenation of open green spaces, curb extensions, green filter strips, and storm water planters have been selected for the study, based on the application and the ability to reduce pluvial flood by effectively managing storm water runoff.

9.2 Study Area

According to the Delhi Development Authority (2010), the city of Delhi is home to 16.78 million people and occupies 47.31% of the state's land area of 148,300 Ha. Delhi's increased flood risk is caused by modifications to the catchment's hydrological characteristics, as stated by [22]. With Delhi's current population and rate of urbanization, 80% of the city will be submerged by 2041, predicts [23]. Delhi's situation has worsened, according to [24], because no effective urban planning strategies are in place. Floods in Delhi can be of two types: (1) fluvial flood, caused by the Yamuna River overflowing, (2) urban flood, caused by clogged drains and dense urbanization [22–25]. Delhi has seen multiple urban floods in the past that have caused extensive damage in the years 2002, 2003, 2009, 2010, 2013, and 2016 [9, 8].

The research is limited to a milli-watershed (1000–10000 ha) (Fig. 9.1) that has been delineated based on vulnerability which has been assessed using two indicators: (1) Percentage (%) of built-up, (2) Past pluvial flood locations. The watershed is part of the Barapullah basin, a part of Delhi's catchment system. The north and south of the study area are bounded by hilly terrain and protected forest, respectively. It reaches a maximum elevation of 279 meters and a minimum elevation of 224 meters above mean sea level (MSL). The elevation drops from the southeast to the northwest, and the runoff has been discharged into Pushpa Vihar Nallah, which drains into the Yamuna River. The terrain is primarily flat, with an average slope of 4.71%. Unauthorized colonies (30.84%), residential (21.37%), forest (21.47%), and recreational areas (10.34%) make up the study area. In recent years, there have been several instances of pluvial floods. During the monsoon months, unplanned areas and transport systems are prone to flooding, causing commuters discomfort and traffic delays.

9.3 Tools Used

This research has used the SCS-CN method over another method, like the rational method, because of the area of the watershed. Rational methods are more effective for watersheds having an area between 800 and 1000 ha [26]. With the watershed area being known, the runoff volume can be estimated using Equation 9.1

$$Q = (I - 0.2S)^2 \div (I + 0.8S) \tag{9.1}$$

where Q = Direct surface run off depth (in mm)

I = Storm rain fall depth (mm)

S = Maximum potential difference between rainfall and runoff (mm)

The curve number (CN) is a crucial component in estimating surface runoff and the retention parameter "S" in hydrologic modelling. CN is a whole number ranging from 0 to 100, directly related to surface runoff: higher CN values indicate higher

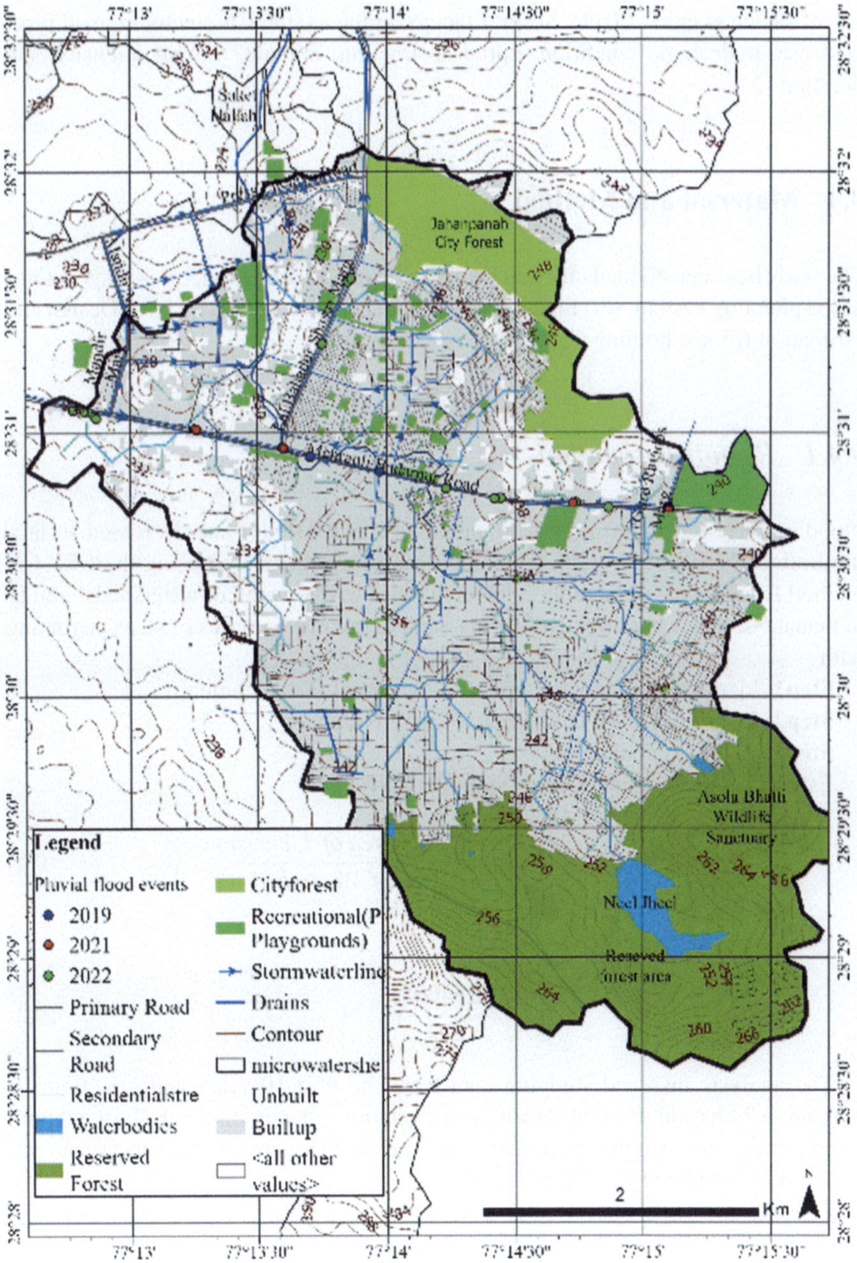

Fig. 9.1 Study area (Compiled by Author)

runoff (Rodney et al., 2013). Several factors influence the CN, including soil type, land use, hydrologic condition, initial abstraction, climate, rainfall intensity, and duration [27].

9.4 Material and Method

The study has been divided into four stages, (i) Run off Estimation, (ii) 4-Stagenature-based planning system, (iii) Identification of intervention areas, (iv) NBS Design and estimation (runoff holding Capacity)

9.4.1 *Runoff Estimation*

Flooding increases manifold due to increased surface runoff and decreased surface infiltration [5, 10]. Therefore, estimating the runoff generated using the SCS-CN method for the most vulnerable catchment helps to assess micro watersheds' ability to manage storm water effectively. This method involved a series of steps beginning with

Step1: Identifying the hydrological soil group of the catchment.
Step2: Identification of land use and Curve Number
Step3: Runoff Calculation.
Estimation of Weighted Curve Number (CN)

$$Weighted\ CN = \frac{(CN \times Area\ of\ Landuse)}{Area\ of\ Watershed} \tag{9.2}$$

Maximum Potential difference (S)

$$S = \frac{25400}{CN} - 254 \tag{9.3}$$

The analysis involved studying data from the past 10 years and selecting the maximum 24 h rainfall event for further calculations. A rainfall depth (I) of 318 mm (24 h storm event, August 2020) was used. The final Direct Surface runoff depth (Q) in mm was calculated using Eq. (9.1)

Then, Q (mm) was converted to Q (m^3) using Eq. (9.4)

$$\left(m^3\right) = \frac{(Q \times Total\ area\ of\ landuse \times 10000)}{1000} \tag{9.4}$$

9.4.2 Four-Stage Nature-Based Planning System

Urban hydrology is an integral part of storm water system planning. During the storm water movement from the

point when it rains to the point where the water reaches the sea, it undergoes several different naturally occurring hydrological processes like evaporation, infiltration, evapotranspiration, runoff, and detention (collection of storm waters in naturally occurring depressions) [28–30]. When the overall world water balance is split of the precipitation received around 61% reaches back to the atmosphere as evaporation, 38% as surface out flow, and 1% as ground water outflow which reaches back to sea. A considerable amount of storm water gets converted as overland flow. The changes in the catchment's topography due to urbanization have changed these water balance estimates. They are causing increased runoff (50% for low density, 65% for medium density, and 85% for High density), less evaporation and infiltration, increasing the risk of pluvial flooding [28, 31]. Following the concepts of the water cycle and urban hydrology, a new storm water planning approach (Nature-based solutions) has been developed using the application and concepts of NBS, where an integrated network follows the application of storm water collection, conveyance, and infiltration concepts. The NBSs have been located based on the topography of the catchment and the possible direction of overland flow (based on topography) to increase infiltration and collection and decrease surface runoff to reduce the risk of pluvial floods.

9.4.3 Identification of Intervention Areas

The study area is a milli-watershed with an area of 1764 ha. Several locations within the watershed have been selected to place these NBS features. The sites for NBS integration were identified during a reconnaissance survey conducted in January 2023. The selection of sites was further determined by knowing the landownership details. Placement of Nature Based Solutions (NBSs) like swales, rain gardens, storm water ponds, etc., was understood based on parameters like slope, contour, natural drainage, flood vulnerability, and NBS opportunity map developed using six indicators (slope, elevation, groundwater level, land use, soil infiltration, and built-up).

9.4.4 NBS Design and estimation (Runoff Holding Capacity)

This section deals with the identification of parameters and variables used to estimate the length (swale), area (curb extension, rain garden, green filter strips, and permeable pavement), or volume (retention and detention ponds) of the NBS features.

9.4.4.1 Design of Swale

Swales have been designed based on the outlined steps, with selected areas and their contributing drainage areas assessed. For designing these nature-based solutions (NBS) features, certain assumptions were made: the swale should be designed to hold the first inch of rainfall runoff, trapezoidal shapes are most suitable, the base width of the swale should be 2–10 feet wide, the side slope should be 25–33%, the design water depth should be 1 foot, and the freeboard should be 4–6 inches. To calculate the length of the swale, the total swale depth was first estimated by adding the design water depth and freeboard. Next, the slope along the length of the swale was calculated using Eq. 9.5.

$$Slope = \{Rise \div Run\} \times 100 \tag{9.5}$$

In designing the swales, the top width was calculated using Eq. 9.6, with a 25% side slope assumed.

$$Top\ width = (Side\ slope\ width + base\ width + Side\ slope\ width) \tag{9.6}$$

The length of the swale was then calculated using information such as the volume of runoff generated during rainfall events, the width of the water surface, the designed water depth, and the cross-sectional area of the swale. The volume of runoff was determined by multiplying the area of the contributing drainage area (impervious surface) by the rainfall depth (1 inch). The width of the water surface was calculated similarly to the top width of the swale but with the design water depth considered instead of the total swale depth. The slope was calculated using Eq. 9.5 where the slope was 25% and the rise was the design water depth. The width of the water surface was determined as (Basewidth + 2*Rise). Given that the swale has a cross-section similar to a trapezoid, the cross-sectional area was calculated using the trapezoid area formula (Eq. 9.7). Finally, the length of the swale was calculated using the Eq. 9.8. These calculations ensured that the swales were designed effectively to manage stormwater runoff.

$$Trapezoid\ area = \left\lfloor \frac{(Surface\ water\ width + base\ width)}{2} \right\rfloor \times (water\ depth) \tag{9.7}$$

$$Volume\ of\ swale = (Crosssectional\ area\ of\ swale \times Length\ of\ swale) \tag{9.8}$$

9.4.4.2 Design of Rain Garden

This section explains the calculation method adopted for determining the size and depth of the rain gardens. The first step involved estimating the drainage area (impervious surface, such as roads) that contributes to the overland flow of stormwater. For

Table 9.1 Slope/depth chart to find the rain garden depth [32]

Slope	Depth
< 4%	3–5 inches
5–7%	6–7 inches
8–12%	8 + inches

instance, if a linear rain garden is to be placed adjacent to a road network, the contributing drainage area would be the total surface area of the road from which the runoff is received. In the second step, the runoff volume generated from the drainage area was calculated by multiplying the drainage area by the 1 inch of storm received on the impervious surface. The third step involved calculating the slope using Eq. 9.5. Although the values for rise and run were initially calculated through an onsite experiment using the string method, due to constraints and prior knowledge of the land's dimensions, a slope of 8–10% was assumed. Finally, the fourth step determined the depth of the rain garden based on the slope, using the guidelines provided in Table 9.1. Further, based on depth and soil types, the size factors were decided, followed by the rain garden's area calculation.

9.4.4.3 Design of Storm Water Curb Extension

The following steps were used to calculate the dimensions of the curb extension. First, the drainage area contributing to the surface runoff was calculated. Next, the overall surface area of the stormwater curb extension was estimated to be 5–10%. A standard size (length by breadth) was then determined based on site constraints. Finally, the total number of stormwater curb extensions was estimated using Eq. 9.9.

$$Total\ number\ of\ curb\ extenions = Total\ Surface\ area/Area\ of\ one\ unit \tag{9.9}$$

9.4.4.4 Design of Storm Water Pond

The dimensions of stormwater ponds were calculated through the following steps. Initially, the catchment boundary and area from which the pond would receive surface runoff were identified and computed using GIS watershed analysis. Then, based on the assumption that the pond's area should be 2% of the catchment area (CWP, 2009), and with a height of 2 m, the total area of the pond and its capacity (in cubic meters) was determined using Eqs. 9.10 and 9.11

$$Area\ of\ the\ Pond\,(m2) = \frac{Area\ of\ catchment}{2} \times 100 \tag{9.10}$$

$$Volume\ of\ the\ pond\,(m3) = Area\ of\ the\ pond \times Depth\ of\ the\ pond \tag{9.11}$$

Finally, the design of permeable surfaces, stormwater planters, and vegetative filter strips were all decided based on space availability.

9.5 Analysis and Results

9.5.1 Run off Estimation

In the process of calculating runoff from a catchment, the first step involves identifying the Hydrological Soil Group (HSG) to which the catchment belongs. Soils are categorized into groups A, B, C, and D based on their infiltration, permeability, and runoff characteristics, collectively known as the Hydrological Soil Group. Group A soils exhibit low runoff potential, group B soils have moderately low runoff potential, group C soils have moderately high runoff potential, and group D soils have high runoff potential. Utilizing the HYSOGS 250 m gridded dataset, the HSG of the study area (Fig. 9.2) was determined, aiding in the runoff estimation. Upon analysis in a GIS environment, it was observed that most of the soil in the study area belongs to the HSG-C classification, characterized by moderately high runoff potential and an infiltration rate ranging from 1.3 to 3.8 mm/hr, with clay content between 20–40% and sand content less than 50%.

The subsequent step involves identifying land use and land cover, as they significantly influence the infiltration rate, directly impacting runoff. Land use categories such as high-density residential, medium-density residential, low-density residential, commercial & business, open spaces (lawns, parks, gardens), impervious areas (paved parking lots, roads), water bodies, and forest areas were identified using the draft master plan for Delhi (2041) and Google Earth imagery. Following this, a dimensionless Curve Number (CN) is determined, ranging from 0 to 100, reflecting surface runoff; a higher CN value indicates higher runoff. The CN corresponding to the HSG classification was selected for further assessment, as outlined in Table 9.2. This comprehensive approach enables a more accurate estimation of runoff from the catchment, which is essential for effective water resource management and urban planning.

HSG -C was selected, and the curve numbers belonging to the particular land use and HSG were selected. The sum of all the land use areas is 17983568 m^2.

Further, runoff estimation was conducted for the maximum rainfall event recorded over the past ten years. The calculation revealed that the watershed generated a surface runoff of 208,392 m^3/hr, as indicated in Table 9.3. Notably, 29% of this runoff persisted as a pluvial flood, as depicted in Table 9.4. This phenomenon could be attributed to the prevalence of impervious surfaces within the watershed. The table below presents comprehensive data and values derived from the procedures outlined in the method section.

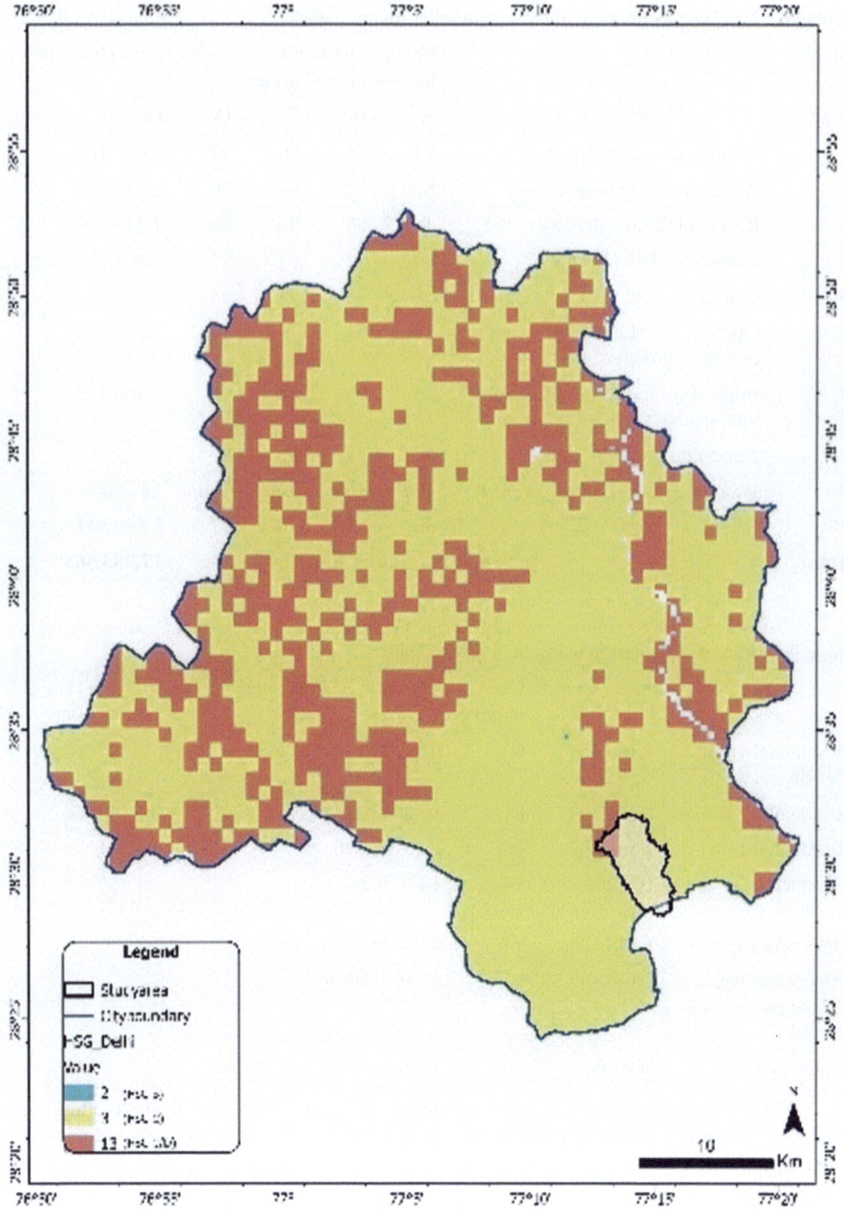

Fig. 9.2 HSG classification (Compiled by Author)

Table 9.2 Land use assessment and curve number identification

Sl. No	Landuse type	Curve number for hydrologic soil group				Area (In Sq.m.)
		A	B	C	D	LA1
1	Residential: High density	77	85	90	92	7,469,168
2	Residential: Medium density	57	72	81	86	1,259,794
3	Residential: Low density	48	66	78	83	1,539,792
4	Commercial and business	89	92	94	95	109,762
5	Industrial	81	88	91	93	
6	Open spaces (Lawns, Parks, golf courses, garden)	39	61	74	80	1,345,128
7	Impervious areas (paved parking lots, roads)	98	98	98	98	2,160,183
8	Agricultural	67	77	83	87	0
9	Waterbody	100	100	100	100	199,900
10	Forest	30	55	70	77	3,899,841
Total						**17,983,568**

Table 9.3 Run off calculation using SCS-CN method

Cover Type	Area (Sq m.)	Curve number	CN*Area	Weighted CN	S	I	Q (m^3)/ hour
Residential-High Density	7,469,168	90	67,222.512	86.688	39.00	318	208,391.7
Residential-Medium	1,259,794	81	10,204.3314				
Residential-Low	1,539,792	78	12,010.3776				
Commercial and Business	109,762	94	1031.7628				
Open Spaces	1,345,128	74	9953.9472				
Impervious Areas (Paved parking lots, Roads)	2,160,183.29	98	21,169.79624				
Waterbody	199,900	100	1999				

S = Maximum Potential Difference between Rainfall and Runoff (mm) I = Storm Rainfall Depth (Aug 2020) in mm, Q = Direct Surface Run off in Volume (m3) and HSG = Hydrological Soil Group.

9.5.2 Four-Stage Nature-Based Planning System

The selected nature-based solutions serve significant functions when integrated into an urban catchment. The point from where the storm water reached the earth's surface up to the end at the outfall location, the storm water was given a well-planned route to flow. Upper catchment areas were selected for storm water collection systems (Storm

Table 9.4 Existing gaps in storm water infrastructure

Location	Artificial (m)	Cross section (m^2)	Capacity (m^3)	Runoff (m^3)	Excess (m3)	Percentage (%)
Restof the areas	50,000	0.54	27,000	208,391.6555	60,716.66	29.13584
Sangam Vihar	2500	0.27	675			
Natural	2000	60	120,000			

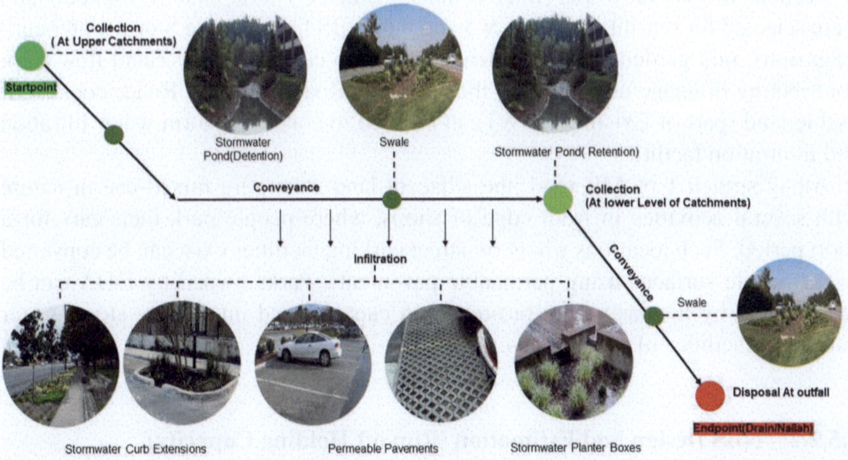

Fig. 9.3 Four stages NbS planning process (Compiled by Author)

water retention and detention ponds). The storm water ponds at the upper catchment were connected to a conveyance system like swales, which will carry the storm water to the collection system placed at areas in lower catchments. Between storm water conveyances, several bio-retention areas, permeable pavement, vegetative filter strips, and storm water planters were planned to increase storm water infiltration.

These urban nature-based solutions were selected based on their inherent applications. e.g., storm water ponds are meant for collection and swale because they convey storm water, and bio-retention spaces are suitable for infiltration, evaporation, and evapotranspiration. The result was an integrated storm water planning network based on applications like collection and infiltration (Fig. 9.3).

9.5.2.1 Identification of Intervention Area

The intervention area is shown in Fig. 9.4. During the site survey, many underutilized areas (Fig. 9.6) along the length of transport corridors, like MB Road, MV

Road, NBCC Road, and Pankaj Marg, were identified. The study areas also have many DDA-owned planned greens where nature-based solutions can be integrated, increasing their potential to hold storm water. The local action area was further delineated to watershed less than ha to identify where storm water ponds can be placed. Three watersheds (W1, W2, andW3) were selected within the study area. The recreational areas were identified to locate storm water ponds and government-owned parks were selected.

Road networks: MB Road, NBCC Road, Pankaj Marg, Mandir Marg, and MV Road were selected, and some DDA parks were also identified as potential sites where NBS can be developed.

Sections of road networks (Intersections, medians, Traffic Islands, roundabouts) were selected for retrofitting (Figs. 9.5 and 9.6) NBS features like Storm water curb extensions, rain gardens, and bioretention areas to capture the overland flow from contributing drainage areas. Along the length of Roads like MB Road, pockets of usable land (part of Existing ROW) can be used to construct storm water filtration and infiltration facilities.

Along stretch 1 of MB road, the adjacent land uses were mixed-use in nature with several activities in front edge of shops, where people park their cars for a short period. Such locations where on-street parking facilities exist can be converted to permeable surfaces using permeable pavements. Parks owned by DDA can be redeveloped by increasing the bio-retention capacity and integrating storm water collection facilities like wet ponds.

9.5.2.2 NBS Design and Estimation (Run off Holding Capacity)

According to runoff estimation results, a catchment with an area of 1764 ha generated 208,392 m^3 of storm water for a 1 h rainfall event in August 2020. The inefficiency of the current storm water infrastructure was also taken into account to comprehend the flooding that took place during that storm event, and it was found that approximately 29.13% of the runoff accumulated in the catchment as stagnation in an hour. An effort has been made to demonstrate how excess runoff can be managed through infiltration and storm water collection using natural solutions. (Table 9.5) lists the estimated individual design capacities for each Nature-based solution that was chosen.

All intervention areas are mentioned in column 1, and NBS integration has primarily integrated along transportation corridors. The total infiltration capacity achieved by using features is 2195 m^3 of water. The left and right lanes (LL and RL) were studied separately for the possibility of integrating NBS. According to the previous calculation, excess storm water was 60716 m^3 (30% approx.) for the first hour of the storm event (Aug 2020). Table (9.7) shows the amount of storm water managed by the collection and infiltration process and the amount that must be disposed of to the nearest nallah. Four water bodies were planned in recreational areas like neighborhood parks (Table 9.6). A total of 48867 m^3 of storm water collection capacity was achieved. Using infiltration, 5.19% of the storm water was managed, and 80% used collection facilities (Table 9.7).

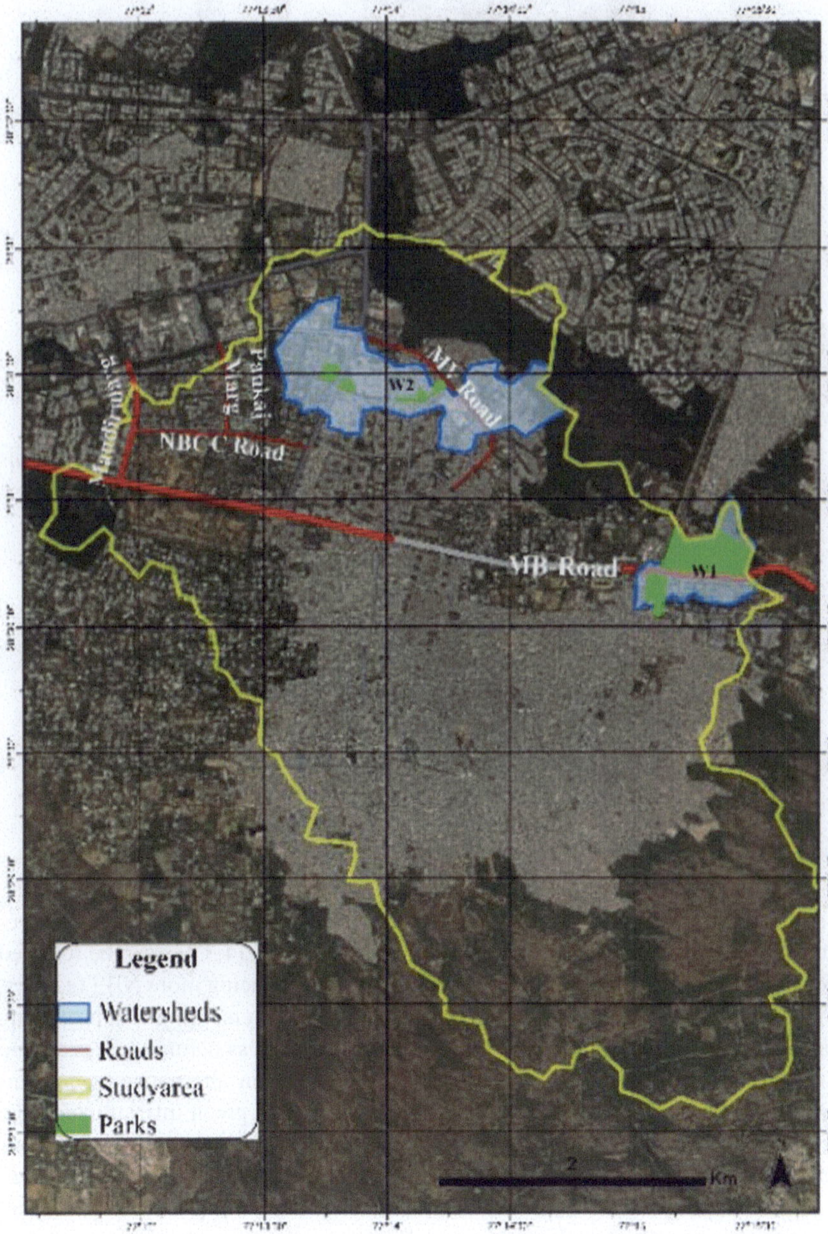

Fig. 9.4 Intervention areas (Compiled by Author)

Fig. 9.5 Traffic junctions where bio-retention areas can be integrated

Fig. 9.6 Underutilized areas and potential sites for NBS features

9.6 Discussion

The results show that 85.67% of excess storm water can be managed using collection and infiltration applied NBS features. The remaining, 14.33%, can be released through storm water drains to rivers or managed by introducing more NBS features.

Nature-based solutions (NBS) for managing floods are emerging as a promising strategy amid rising climate risks. These approaches harness natural processes and ecosystems to handle floodwaters sustainably, offering an alternative to conventional infrastructure. Key trends include integrating urban green infrastructure like green roofs and permeable surfaces, which mimic natural hydrological functions and reduce urban flooding. There is also a growing focus on ecosystem-based methods, utilizing wetlands, floodplains, and forests to mitigate flood peaks and enhance resilience. Beyond flood control, NBS provides multifaceted benefits such as biodiversity conservation and recreational opportunities, underscoring their broader value. Effective community engagement is essential for local ownership and knowledge exchange while evolving policy frameworks support the economic viability and long-term sustainability of NBS. As climate change intensifies, these solutions offer

Table 9.5 NBS design for intervention areas

Road Name	Infiltration Rate (m/hr)	Bio Swale (m)		Rain Garden (m2)		Permeable Pavements (m^2)		Stormwater curb extension (m^2)		Storm water Planters (m^2)		Green Filter Strips(m^2)	Total area (m^2)	IC (m^3/hr)
		LL	RL	LL	RL	LL	RL	LL	RL	LL	RL	0		
MBRoad-1	0.035	150	0	1043	0	4100	3425	235		0	0	0	8953	313.35
MBRoad-2	0.0455	0	0	6424	0	6285	1225	0	0	819	0	0	15,173	592.9
MBRoad-3	0.0455	0	0	0	0	0	0	0	0	0	0	1100	1100	50.05
Mandir Marg	0.035	0	0	0	640	0	425	0	62.5	0	17.5	0	1145	40.075
BirlaVidya Niketan Marg	0.0755	0	0	0	0	1750	1750	0	0	0	0	0	3500	264.25
ShaheedLT Pankaj Juyal Marg	0.0455	0	0	0	0	2800	2800	0	0	0	0	2763	8363	380.5165
RajaRam Marg	0.0755	0	0	0	0	1984	0	162.5	0	0	0	0	2146.3	162.0419
M. Valmiki Marg	0.0455	0	290	0	380	2880	2880	175	175	0	0	300	7080	322.14
Total		**410**	**290**	**8487**	**1020**	**20,049**	**12,505**	**572. 5**	**238**	**819**	**17.5**	**3063**	**0**	**0**

Table 9.6 Water body design area statement

	No water bodies	Area of Catchment (m^2)	Percent allocation	Area of one water body (m^2)	Volume (m^3)
W1	1.00	310,815.11	2.00	6216.30	12,432.60
W2	3.00	910,867.00		6072.45	36,434.68
Total capacity in m^3					48,867.28

Table 9.7 Storm water managed using collection and infiltration practices

	Excess storm water (m^3)	Total storm water managed (m^3)	Percentage (%)
Infiltration	60,716.66	3151.25	5.19
Collection		48,867.28	80.48
Disposal		8698.12	14.33

adaptive strategies to improve ecosystem health and reduce flood risks for vulnerable communities worldwide.

9.6.1 Planning and Design Interventions

- **Mehrauli-Badarpur Road**

The stretch was divided into three parts, and interventions like bioswale, rain garden, permeable surfaces, curb extensions, and stormwater planters, constitute an area of 25000 m^2 with an Infiltration capacity of 3800 m^3/hr.

Stretch 1 (Fig. 9.8) of the road section has been retrofitted using features like a rain garden, curb extension, and permeable pavements (Fig. 9.8). The higher impervious surface area was converted to rain gardens (Fig. 9.7), and the shop frontage was improved with permeable pavements and storm water planters. A total of 313.35 m^3/hr (Table 9.5) of infiltration capacity was achieved on this stretch. The impervious surface was replaced with 8953 m2 of permeable surface.

On Stretch 2 (Fig. 9.9) of MB road, using a feature like a rain garden, storm water planters, green filter strips, and curb extension, an infiltration capacity of 592.9m^3/hr was achieved, where an areaof15173 m2 of impervious surface was converted to pervious using nature-based solutions.

Only a few interventions are possible in stretch 4 as the adjacent land use premises are parts of an unauthorized colony (Sangam Vihar), leading to less road width. Medians were converted to bio-retention areas where 1100 m^2 of impervious surface has been converted to pervious with an infiltration capacity of 50.05 m^3 of storm water.

Using combination of Rain Garden, swales, permeable pavements, and storm water curb extension, a total of 7080m^2 of impervious surface has been converted

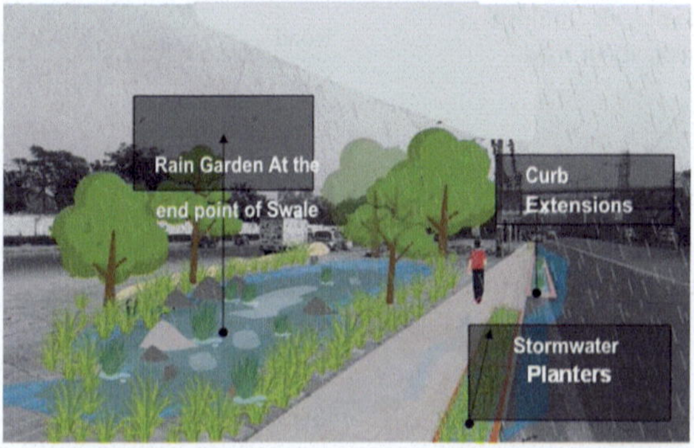

Fig. 9.7 MB road stretch1: Photo montage for NBS recommendations

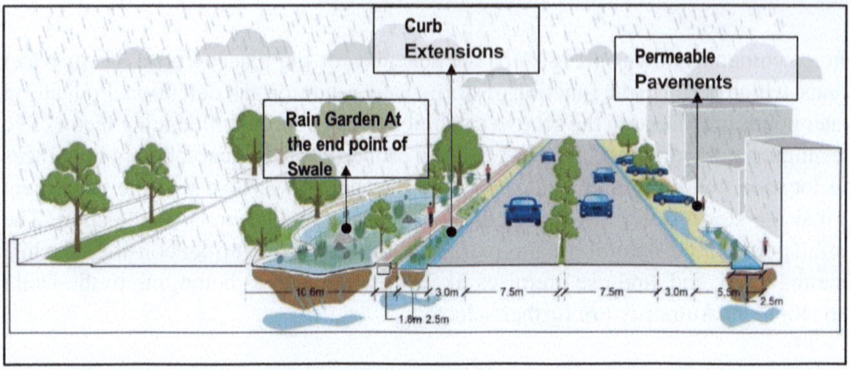

Fig. 9.8 MB road (Stretch1): Retrofitted road section with NBS features

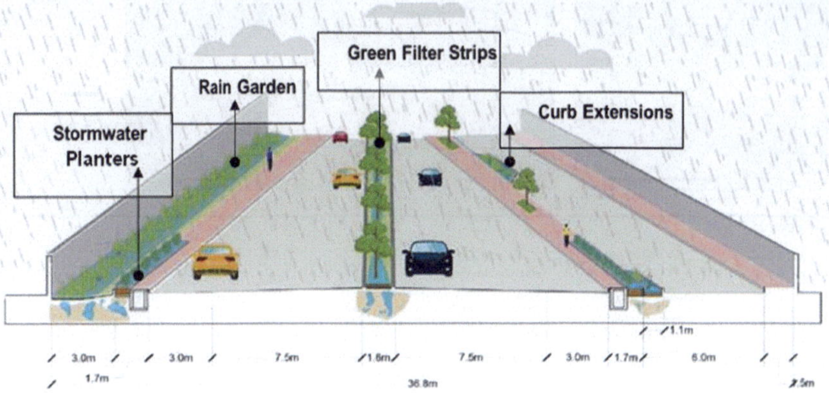

Fig. 9.9 Mb road (Stretch 2): Road section with NBS features

Table 9.8 Infiltration capacity of roads

Road	NBS area	Infiltration capacity
NBCC	1530	70at 0.0455 m3/hr
Mandir Marg	1145	40at 0.035 m3/hr
Pankaj Juyal Marg	8363	380at 0.0455 m3/hr

to pervious surfaces, achieving an infiltration capacity of 322 m^3/hour of infiltration capacity (Table 9.8).

9.6.2 Storm Water Ponds at Watersheds W1 and W2

The catchments shown in Fig. 9.10 are selected for the placement of storm water ponds, which perform the function of storm water retention and detention. Four storm water ponds are placed in the set two catchments, considering influencing factors like elevation, natural drainage, landuse, and soil properties. In this case, height influences the location condition and is located at higher and lower levels of the catchment to manage storm water effectively through primary and secondary collection. The outfall points of the natural streams extracted from DEM are further considered while locating them, and land use premises like recreational areas belonging to the Delhi Development Authority are further selected.

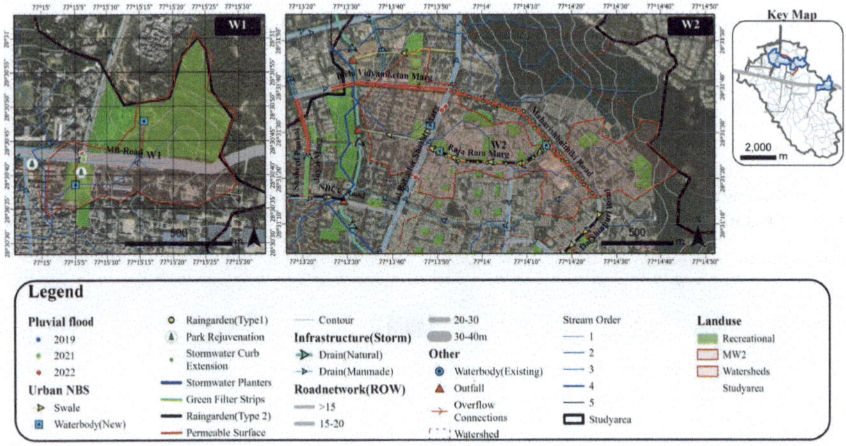

Fig. 9.10 The location of water bodies

9.7 Conclusion

Pluvial flooding, an outcome of highly impervious surfaces, inefficient storm water infrastructure, and climate change, has been causing significant and often irretrievable damage. Many countries are increasingly concerned about protecting their social-physical infrastructure and environmental resources from the impacts of floods. Urbanized cities have traditionally relied on grey infrastructure to manage flood risks. While these measures have been adequate to some extent, there is a growing need for more resilient approaches to adapt to the changing climate. Integrating conventional flood management strategies with sustainable, nature-inspired solutions, such as Nature-Based Solutions (NBS), is essential in India.

To address pluvial flooding effectively, this article proposes a four-stage NBS stormwater planning system comprising primary collection, conveyance, infiltration, and secondary collection. These stages aim to manage excess runoff that leads to pluvial floods. NBS features such as swales, rain gardens, permeable pavements, stormwater ponds, curb extensions, planters, and vegetative filter strips are recommended for implementation within a milli-watershed (1000–10,000 hectares). These features perform critical functions like water collection, conveyance, and infiltration. The Soil Conservation Service Curve Number (SCS-CN) method was applied using real-time rainfall data to understand the impact of land use and soil type on runoff. This method estimates the excess stormwater that remains as the pluvial flood. The generated data were then used to evaluate the efficiency of the installed NBS features. The results demonstrate that these NBS features can significantly reduce excess runoff through effective collection and infiltration practices, thereby mitigating the risk of pluvial floods.

This research provides a blueprint for developing new sustainable strategies to reduce pluvial flood risks. Other cities can adopt similar measures using this generalized four-stage stormwater planning system concept. Retrofitting existing road networks with NBS, such as swales, permeable pavements, curb extensions, and stormwater planters, can offer long-term benefits. This approach aligns with the broader goals of sustainable urban development and can be utilized by various stakeholders in planning and implementing future nature-based strategies. By adopting NBS, cities can make significant progress toward achieving the 2030 Agenda for Sustainable Development, specifically addressing Sustainable Development Goals (SDGs) 11 (Sustainable Cities and Communities), 6 (Clean Water and Sanitation), and 13 (Climate Action).

References

1. Cerutti E, Blagrave P, Agarwal R, Balasundharam V, Gudmundsson R, Mousa R (2021) Climate change in South Asia: Further need for mitigation and adaptation. IMF Working Papers (217):1. https://doi.org/10.5089/9781513590677.001

2. Asian Development Bank (ADB) (2010) Climate change in South Asia: Strong responses for building a sustainable future. Asian Development Bank. https://www.adb.org/sites/default/files/publication/27475/climate-change-sa.pdf
3. Intergovernmental Panel on Climate Change (IPCC) (2023) Summary for policymakers. In Climate Change 2023: Synthesis Report. J Cryst Growth 2:36
4. World Bank Group (2021) Climate risk country profile: India. The World Bank Group, 36. www.worldbank.org
5. Jha A, et al. (2012) Cities and flooding: A guide to integrated urban flood risk management for the 21st century. J Hydrol Eng 8(1). https://doi.org/10.1596/978-0-8213-8866-2
6. Yu Q et al (2022) Extreme flood disasters: Comprehensive impact and assessment. Water (Switzerland) 14(8):1–14. https://doi.org/10.3390/w14081211
7. Centre for Science and Environment (CSE) (2016) Why urban India floods: Indian cities grow at the cost of their wetlands. In Down to Earth
8. National Institute of Urban Affairs (NIUA) (2016) Urban flooding
9. De US, Singh GP, Rase DM (2013) Urban flooding in recent decades in four megacities of India. J Indian Geophys Union 17(2):153–165
10. Huang Y, Tian Z, Ke Q, Liu J, Irannezhad M, Fan D, Hou M, Sun L (2020) Nature-based solutions for urban pluvial flood risk management. WIREs Water 7(3). https://doi.org/10.1002/wat2.1421
11. Ruangpan L, Vojinovic Z, Di Sabatino S, Leo LS, Capobianco V, Oen AMP, McClain ME, Lopez-Gunn E (2020) Nature-based solutions for hydro-meteorological risk reduction: A state-of-the-art review of the research area. Nat Hazard 20(1):243–270. https://doi.org/10.5194/nhess-20-243-2020
12. Walters G, Janzen C, Maginnis S (2016) Nature-based solutions to address global societal challenges. Nat-Based Solut Address Glob Soc ChallS. https://doi.org/10.2305/iucn.ch.2016.13.en
13. Eggermont H, Balian E, Azevedo JMN, Beumer V, Brodin T, Claudet J, Fady B, Grube M, Keune H, Lamarque P, Reuter K, Smith M, Van Ham C, Weisser WW, Le Roux X (2015) Nature-based solutions: New influence for environmental management and research in Europe. GAIA-Ecol Perspect Sci Soc 24(4):243–248. https://doi.org/10.14512/gaia.24.4.9
14. McQuaid S, Kooijman E, Rizzi D, Andersson T, Schanté J (2022). The vital role of nature-based solutions in a nature positive economy. https://doi.org/10.2777/307761
15. Mackinnon K, Sobrevila C, Hickey V (2008) Biodiversity, climate change, and adaptation: Nature-based solutions from the World Bank portfolio. Clim Chang Adapt 1–381)
16. International Union for Conservation of Nature (IUCN) (2020) IUCN Global Standard for Nature-based Solutions: A user-friendly framework for the verification, design, and scaling up of NbS: First edition. https://doi.org/10.2305/iucn.ch.2020.08.en
17. Xiao Z, Ge H, Lacasse MA, Wang L, (Leon), & Zmeureanu, R. (2023) Nature-based solutions for carbon neutral climate resilient buildings and communities: A review of technical evidence, design guidelines, and policies. Buildings 13(6):1389. https://doi.org/10.3390/buildings13061389
18. Frantzeskaki N (2019) Seven lessons for planning nature-based solutions in cities. Environ Sci Policy 93:101–111. https://doi.org/10.1016/j.envsci.2018.12.033
19. UN-Habitat (2022) Envisaging the future of cities. In World City Report 2022
20. Sowińska-Świerkosz B, García J (2022) What are nature-based solutions (NBS)? Setting core ideas for concept clarification. Nat-Based Solut 2(January):100009. https://doi.org/10.1016/j.nbsj.2022.100009
21. Yuca H (2022) Capsicum annuum L. In Novel drug targets with traditional herbal medicines: Scientific and clinical evidence, pp 95–108. https://doi.org/10.1007/978-3-031-07753-1_7
22. Kumar M, Sharif M, Ahmed S (2019) Flood risk management strategies for National Capital Territory of Delhi India. ISH J Hydraul Eng 25(3):248–259. https://doi.org/10.1080/09715010.2017.1408434
23. Agrawal M (2020) Multisector exposure and vulnerability to climate change in India: Case of National Capital Territory of Delhi, India. Disaster Prev Manag: Int J 29(5):761–777. https://doi.org/10.1108/DPM-04-2020-0125

24. Delhi Disaster Management Authority (DDMA) (2015) Delhi disaster management plan (2016–27). Government of NCT of Delhi, Revenue Department. http://delhi.gov.in/wps/wcm/connect/DoIT_Planning/planning/economic+survey+of+delhi/economic+survey+of+delhi+2014+-+2015
25. Prashar S, Shaw R, Takeuchi Y (2012) Assessing the resilience of Delhi to climate-related disasters: A comprehensive approach. Nat Hazards 64(2):1609–1624. https://doi.org/10.1007/s11069-012-0320-4
26. Central Public Health and Environmental Engineering Organization (CPHEEO) (2019) Manual on storm water drainage system. In Volume 1 Part A: Engineering Design, vol 1, pp 1–370
27. Mishra SK, Singh VP (2003) Soil conservation service curve number (SCS-CN) methodology. Soil Conserv Serv Curve Number (SCS-CN) Methodol. https://doi.org/10.1007/978-94-017-0147-1_2
28. Chow VT, Maidment DR, Mays LW (1988) Applied hydrology (iv). McGraw-Hill. http://ponce.sdsu.edu/Applied_Hydrology_Chow_1988.pdf
29. IWP, ADRID, NIDM (2020) A report on urban flood management: Drainage and planning (Issue December)
30. Raghunath HM (2004) Hydrology: Principles, analysis and design (Revised ed., Issue 1). New Age International Publishers
31. Butler D, Davies J (2018) Urban drainage. In Urban Drainage, 4th ed. Spoon Press, Taylor & Francis Group. https://doi.org/10.1201/9781351174305
32. National Oceanic and Atmospheric Administration (NOAA) (2023) Stormwater solutions for homeowners fact sheet: Rain gardens
33. World Bank, WRI (2018) Nature-based solutions for disaster risk management

Chapter 10
Enhancing Urban Resilience with Green Infrastructure: The Impact of Genetic Variability

Saba Inamdar and Shaima Inamdar

Abstract Urban resilience remains imperative for sustainable development amidst rapid urbanization and climate change. Emphasizing the role of digital technologies, the United Nations Development Programme (UNDP) underscores their potential in shaping urban futures. Simultaneously, green infrastructure emerges as a vital tool in tackling urban challenges, bolstering air quality, biodiversity, and community well-being. This research studies the transformative capacity of green infrastructure, particularly in averting environmentally-linked cardiac conditions. Leveraging bioinformatics tools, the study forecasts variant pathogenicity, shedding light on genetic variation within urban populations. Further the study explore green infrastructure interventions that harness nature's therapeutic qualities, fortifying resilience against the environmental stressors. The study draws attention to the global health risks specifically the clinical significance of cardiac disorder posed by environmental pollution and pollutants. Emphasizing the urgency for interdisciplinary approaches and evidence-based strategies, the study highlights the imperative of creating healthier, more resilient cities worldwide. Nonetheless, challenges such as inter-agency coordination and sustained funding impede implementation. Addressing these challenges requires proactive strategies, effective leadership, and stakeholder collaboration to overcome barriers and ensure the successful implementation of initiatives to address various issues or achieve specific goals.

Keywords Bioinformatics · Climatic change · Environmental contaminants · Genetic variation · Green infrastructure

---

S. Inamdar (✉)
Department of Business Management, St. Joseph's Degree and Pg College, Gunfoundry, Hyderabad, India
e-mail: ms.sabainamdar@gmail.com; drsabainamdar@josephscollege.ac.in

S. Inamdar
Department of Genetics and Biotechnology, Shadan Degree College for Women, Khairtabad, Hyderabad, India

S. Ghosh et al. (eds.), *Cities of Tomorrow: Urban Resilience and Climate Change Preparedness*, Urban Sustainability, https://doi.org/10.1007/978-981-97-9658-8_10

10.1 Introduction

Urbanization is a widespread tendency, yet it may also result in health problems and a decline in wellbeing. Decision-makers, such as government officials and city planners, should prioritize people's everyday pleasure over data and statistics to improve city life [1, 2]. Accumulating more green belts can make cities pleasant by providing areas for rest, exercise, and the outdoors. Demonstrating that greenery increases well-being emphasizes the importance of making cities greener and emphasizing happiness, thereby enhancing everyone's quality of life in cities [3]. In the face of urgent environmental issues like climate change, sustainability depends on incorporating green infrastructure into corporate processes [4]. It delivers considerable cost savings and efficiency benefits thanks to features like lower maintenance costs and energy conservation. Through improved brand recognition and access to environmentally conscious customers, adopting green infrastructure gives businesses a competitive edge. In addition, adhering to changing environmental laws reduces risks and guarantees sustainability over the long run. Environmentally friendly infrastructure promotes social responsibility and community involvement, which helps to create a more sustainable future. By studying and implementing green infrastructure practices, management professionals may accomplish organizational goals and have a positive environmental impact [5].

This research studies the transformative capacity of green infrastructure, particularly in averting environmentally-linked cardiac conditions. The study forecasts variant pathogenicity, shedding light on genetic variation within urban populations. Further the study explores green infrastructure interventions that harness nature's therapeutic qualities, fortifying resilience against the environmental stressors. Exposure to environmental pollution is linked to various adverse health effects in both children and adults. While concerns often focus on neurological development, cancer, and pulmonary diseases, it's crucial to recognize the impact on cardiovascular health as well [6]. Cardiovascular disease stands as the leading cause of mortality globally, making it a key target for large-scale, multispectral interventions aimed at the population level [7, 8]. The study draws attention to the global health risks specifically the clinical significance of cardiac disorder posed by environmental pollution and pollutants. The PolyPhen2 (PolyPhen-2: Prediction of Functional Effects of Human nsSNPs, n.d.) software is utilized to predict the pathogenicity of variants reported in the ClinVar database for the genes.

10.2 Literature Review

Green infrastructure research has grown significantly recently, with the US and Europe leading the way. It is a crucial strategy for sustainable development that balances social, economic, and environmental goals. Studies cover various aspects, including ecosystems and human health, as well as its construction, assessment, management, and exploration [9, 10]. Green infrastructure is a multidisciplinary field that combines management, engineering, landscape architecture, urban planning, environmental science, life sciences and other disciplines to manage natural resources and improve urban settings [11]. The multidisciplinary approach assists in creating sustainable solutions for cities and improving air and water quality, lessening pollution, boosting biodiversity, lessening the effects of urban heat islands, reducing the repercussions of climate change and strengthening community health and well-being through green infrastructure solutions [12]. Green infrastructure also delivers financial advantages by boosting property prices, generating green jobs, and lowering the expenses related to conventional "grey" infrastructure [13]. Its interdisciplinary approach guarantees that the solutions satisfy the demands of society and the environment for a more resilient and habitable future societies [14, 15].

Contemporary research reveals multiple genetic variants associated with common ailments significantly contributing to physical wellness and mortality in affluent countries. These diseases encompass cardiac issues, cancer, diabetes, and psychiatric disorders such as schizophrenia and bipolar disorder (manic-depression). In contrast to conditions like cystic fibrosis or Huntington's disease, which result from mutations in single genes and manifest consistently across all environments, common diseases emerge from the complex interaction of multiple genes and environmental influences. Consequently, these diseases are termed polygenic and multifactorial. A confluence of several variables shapes most human characteristics, whether related to health or not. Rare single-gene disorders and common multifactorial diseases differ significantly in their genetic makeup. Single-gene disorders typically result from recent genetic variations that have a major, disruptive impact on homeostasis, often appearing at a young age, perhaps even in puerility. In contrast, the genetic differences that contribute to common multifactorial diseases are typically older and tend to influence body's appropriate balance. These diseases typically emerge in adulthood. Finding genetic variants linked to common diseases is relevant since individual shave time to modify their behavior, reducing the likelihood of disease development even in those with a genetic predisposition. It serves a purpose to apprehend human genetic variation to promote health and prevent unfavorable medical conditions [16–19].

Unless monozygotic (identical) twins, no two human individuals share identical genetic makeup. On an average, the DNA of any two persons varies at approximately one in every thousand nucleotide base pairs. The average variation in base pairs between any two individuals is approximately three million since each person's genome has about three billion base pairs. While this difference may seem substantial, it only accounts for 0.1% of the entire genetic composition. At an individual level, most human genetic variation doesn't have significant biological importance as it doesn't noticeably impact an individual's fitness. Only a small fraction of DNA variations occur in coding regions, which are responsible for protein production, or in regulatory regions that control gene expression. Variations in other DNA regions typically does not affect observable traits (phenotype). Even variations in coding regions may not always change how a protein functions; some may alter the protein's amino acid sequence without affecting its function, while others don't change the sequence at all [20]. Genetic and environmental variables can work together to produce regionally patterned deviations in phenotypic traits. The response of traits to environmental changes can vary between populations due to genetic drift or selection, leading to Genotype × Environment (G × E) interactions. Reaction norms, or genotype reactions to various circumstances, are generally believed to evolve through natural selection, even though the function of plasticity in adapting to novel settings is still up for discussion.

Cardiovascular diseases (CVDs), along with chronic respiratory and metabolic diseases, represent a significant portion of non-communicable diseases (NCDs). CVD encompasses a wide array of disorders, spanning diseases affecting the vasculature, myocardium, heart's electrical system, and congenital heart defects. This includes both acute and chronic conditions such as coronary artery disease, heart failure, arrhythmias, stroke, and arterial hypertension [21, 22]. CVDs accounts for the huge fatalities brought on by NCDs. According to [23], the contribution of CVDs to global mortality has steadily risen from 12.1 million in 1990 to 18.6 million in 2019. Notably, low- and middle-income countries bear the highest burden, with 86% of premature deaths attributed to NCDs. [23]. The majority of risk factors for NCDs originate from the environment, as evidenced by observations indicating that up to 25% of all cases of ischaemic heart disease (IHD) are linked to an unhealthy environment, particularly air pollution. However, environmental contributions to NCDs are often overlooked, as seen in the failure to acknowledge environmental risk factors in the 2013 WHO NCD Global Action Plan [24]. Comprehension of an extensive list of inherited cardiovascular diseases, such as cardiomyopathies, arrhythmic disorders, vascular problems, and lipid disorders, including familial hypercholesterolemia, has been facilitated by progressive research in human genetics [25].

Despite significant advancements in treatment and management, cardiovascular disease (CVD) remains the leading cause of mortality across all human populations. In developed countries, it surpasses all other diseases in terms of fatalities, while its prevalence is steadily rising in low and middle-income nations. Globally, deaths from ischaemic heart disease and stroke are on the rise. Although the United States has witnessed a decline in CVD mortality rates since the 1960s, this decline has slowed considerably since the 1990s [26]. In next six years, it is projected that 40.5% of the population will have some form of CVD [27]. While aging may contribute to this increase, the widespread occurrence of CVD underscores our ongoing challenges in preventing its escalation and understanding its underlying nature. The present consensus is that a complex interaction between genetic predisposition and environmental variables causes a range of chronic problems that eventually lead to progressive degradation of the structure and function of cardiovascular tissues, which is the cause of CVD. While genetic defects may underlie certain rare forms of heart disease, it is widely believed that the majority of CVD cases arise from interactions among multiple gene variants and lifestyle factors. Although the precise contribution of genetics and the environment remains poorly elucidated, there is a prevailing notion that external variables and way of living including personal health choices and preferences exert a more significant influence on CVD development. This belief is supported by numerous studies indicating that maintaining a healthy lifestyle can substantially reduce the risk of developing CVD [28].

10.3 Methodology

This research identified several obstructive CAD-related genes from a published paperby [29], which were subsequently subjected to in-silico analysis. To facilitate this analysis, the protein sequences required in FASTA format are retrieved from NCBI. PolyPhen2 was used for bioinformatics analysis to examine the sample.

10.4 Results

The SLC16A3 gene is expressed in tissues such as white skeletal muscle fibers, astrocytes, white blood cells, chondrocytes, testis, lung, placenta, heart and some mammalian cell lines (Halestrap & Meredith, 2004; Meredith & Christian, 2008).

The bioinformatics analysis using PolyPhen2 provided significant insights into the pathogenicity of variants as listed in Table 10.1 for the SLC16A3 and MORF4L1 genes, highlighting their potential interaction with environmental factors. Genome-wide association studies (GWAS) have identified these genes as being associated with coronary artery disease (CAD). Notably, the SLC16A3 gene is linked to CAD, with specific variants being influenced by exposure to traffic air pollution, as indicated in the GWAS catalog.

As displayed in Table 10.2 the SLC16A3 gene, out of the five analyzed variants, two were predicted to be benign, while three were classified as probably damaging and possibly damaging, respectively. This suggests a notable presence of potentially harmful mutations within this gene, which could contribute to the development of obstructive CAD, particularly in environments with high levels of air pollution. In the case of the MORF4L1 gene as shown in Table 10.3, the analysis predicted that two out of the three variants were benign, whereas one variant was deemed possibly

Table 10.1 Genes and variants

Sno	Gene	Variants
1	**SLC16A3**	1. NM_004207.4(SLC16A3):c.817A > G (p.Ile273Val) 2. NM_004207.4(SLC16A3):c.1187 T > G (p.Leu396Arg) 3. NM_004207.4(SLC16A3):c.1154 T > C (p.Met385Thr) 4. NM_004207.4(SLC16A3):c.571 T > G (p.Cys191Gly) 5. NM_004207.4(SLC16A3):c.875C > T (p.Pro292Leu)
2	**MORF4L1**	1. NM_006791.4(MORF4L1):c.379A > G (p.Ser127Gly) 2. NM_006791.4(MORF4L1):c.866A > G (p.Asn289Ser) 3. NM_006791.4(MORF4L1):c.368G > C (p.Gly123Ala)

damaging. These findings indicate a lower but still present risk of pathogenic variations in the MORF4L1 gene, which may be exacerbated by environmental conditions that promote cardiovascular stress.

The PolyPhen2 analysis, in conjunction with GWAS findings, affirms the importance of these gene variants in the context of CAD and their potential interaction with environmental factors. The identification of potentially damaging variants attests the need for further functional studies to understand their roles in disease pathogenesis, in particular, the impact of lifestyle choices and exposure to environmental factors like air pollution. Additionally, these results reinforce the utility of in-silico tools like PolyPhen2 in predicting the potential impact of genetic variations, aiding in identifying high-risk individuals and developing targeted therapeutic strategies that consider both genetic predispositions and environmental influences.

10.5 Discussion

The genes SLC16A3 and MORF4L1 have been identified as susceptible to damage from traffic-related air pollution. Individuals found to have variations in these genes should take necessary precautions to avoid potential cardiac damage [30]. Due to its synthesizing benefits, green infrastructure is crucial for reinforcing urban resilience to climate change. It mitigates the urban heat through shading and cooling mechanisms, enhancing the air quality by filtering pollutants. Additionally, green spaces harbour biodiversity, supporting vital ecosystem services for urban environments. Beyond environmental advantages, green infrastructure also enhances community well-being, reduces building energy consumption, and augments cities' aesthetic and economic value. Using green infrastructure, urban design and development may successfully alleviate climate change susceptibility, promote sustainability, and build an environmentally friendly infrastructure [31]. Green infrastructure profoundly impacts urban health, particularly in preventing environmentally-induced heart diseases [32]. This genetic diversity is vital for the population's health, providing resilience against pollution, pests, diseases, and environmental pressures. It also facilitates adaptation to various biological and ecological challenges, which are crucial for developing crossbred varieties with beneficial traits in green infrastructure and sustainable urban development [33]. The rich reservoir of diverse gene alleles in plant genetic resources is crucial for meeting global food demands. Unfortunately, the emphasis on cultivating high-yield crop varieties has led to a decrease in genetic diversity [34]. It is essential to protect, explore, collect, conserve, and utilize plant genetic resources judiciously to combat the risk of genetic erosion [35]. This strategy is critical to developing and enhancing crop varieties, essential to tackling present and future global challenges. The rapid extinction rate of plant species, driven by climate change and human activities, highlights the critical need for conservation to preserve ecological balance [36]. Ultimately, the sustainable use of plant genetic resources is fundamental for ensuring food security. It cushions the resilience of green infrastructure, enhancing urban health by reducing environmentally linked

Table 10.2 SLC6A3 result

NM_004207.4 (SLC16A3):c.817A > G (p.Ile273Val)	NM_004207.4 (SLC16A3):c.1187 T > G (p.Leu396Arg)
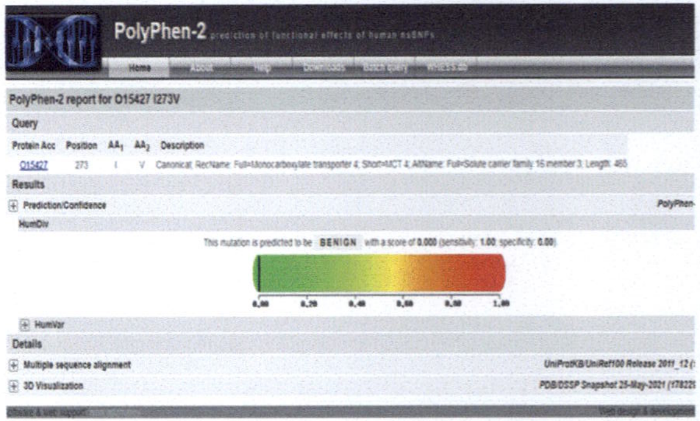	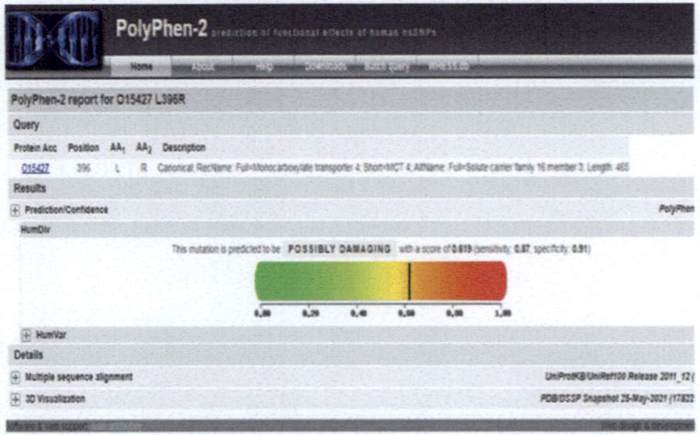
NM_004207.4 (SLC16A3):c.1154 T > C (p.Met385Thr)	**NM_004207.4 (SLC16A3):c.571 T > G (p.Cys191Gly)**
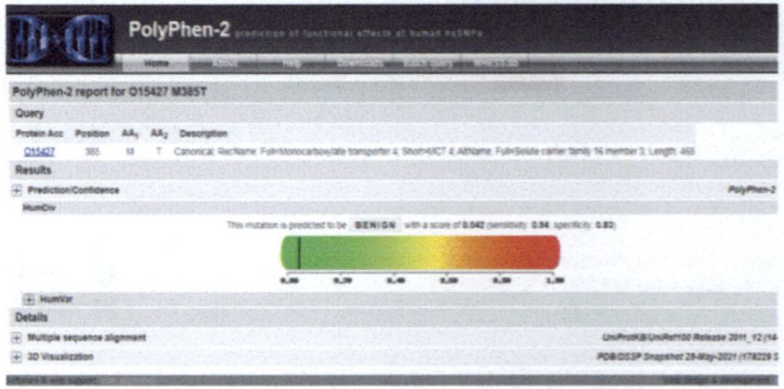	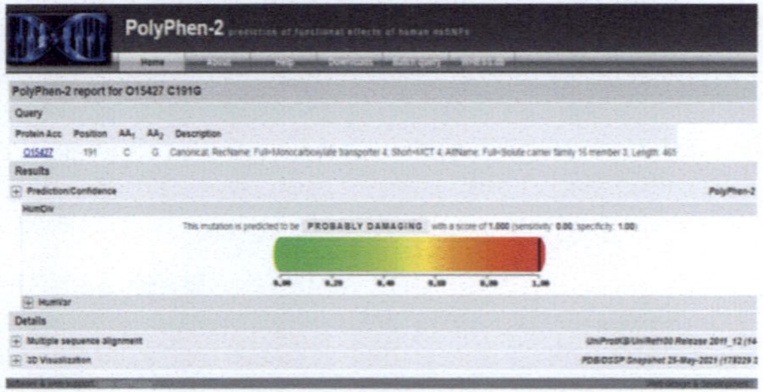

(continued)

Table 10.2 (continued)

NM_004207.4 (SLC16A3):c.817A > G (p.Ile273Val)	NM_004207.4 (SLC16A3):c.1187 T > G (p.Leu396Arg)
NM_004207.4 (SLC16A3):c.875C > T (p.Pro292Leu)	
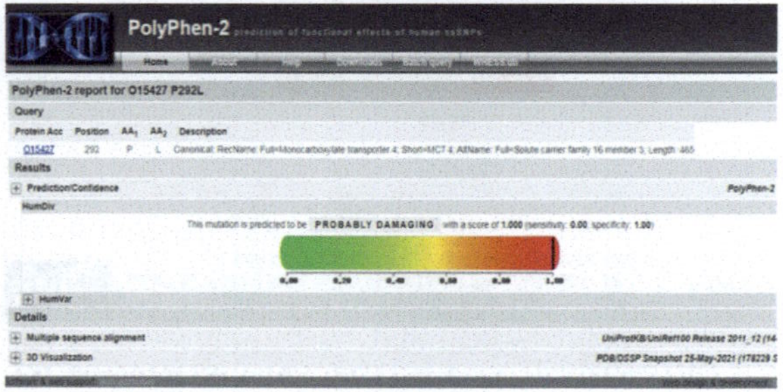	

Source PolyPhen2

Table 10.3 MORF4L1 result

NM_006791.4 (MORF4L1):c.379A > G (p.Ser127Gly)

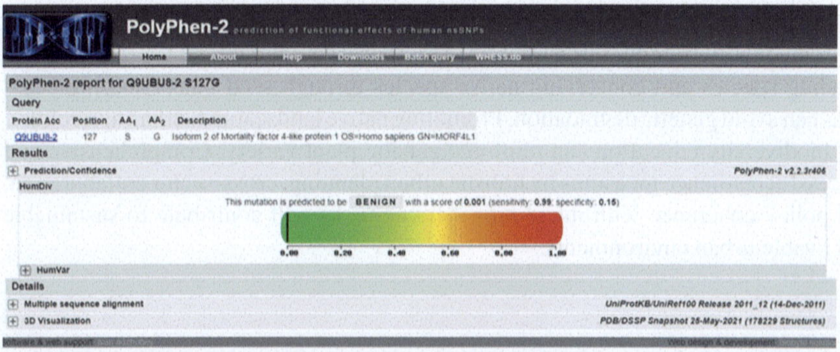

Variant NM_006791.4 (MORF4L1):c.866A > G (p.Asn289Ser)

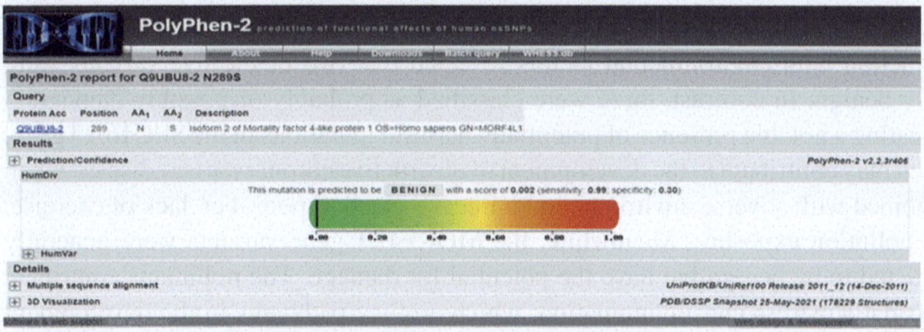

Variant NM_006791.4 (MORF4L1):c.368G > C (p.Gly123Ala)

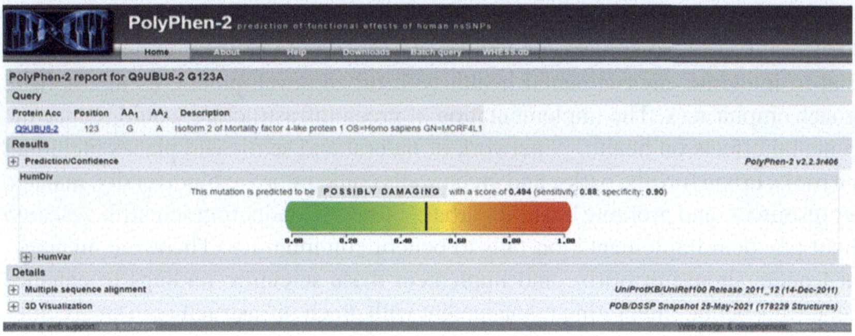

Source PolyPhen

health risks, such as cardiac conditions. Cities should use a multimodal strategy that addresses pollution management, genetic erosion prevention, and gene plant variety promotion to include climate resilience in urban design and infrastructure development [37]. Policymakers should implement stringent emission rules, fund environmentally friendly infrastructure, and improve public transit [38]. Regulating invading species and conserving native species through seed banks and protected areas can avoid genetic degradation. Promoting native landscaping, urban agriculture, and biodiversity education can reinforce genetic plant variety. Comprehensive and successful resilience measures by holistic urban planning, cross-sector collaboration, and policy coherence with international frameworks will contribute to sustainable and livable urban environments.

10.6 Conclusion

The study concludes after the analysis ofSLC16A3 and MORF4L1 genes and their interaction with environmental factors revealed that two of the five variants were likely benign. In contrast, three were classified as probably or possibly damaging, indicating a notable presence of potentially harmful mutations in the SLC16A3 gene, which may contribute to the development of obstructive coronary artery disease when combined with adverse environmental influences such as poor diet, lack of exercise, and pollution exposure. Meanwhile, the MORF4L1 gene variants were generally predicted to be benign but have the potential for damage. The pollutants contribute to oxidative stress and inflammation, which are key pathways in the development and progression of cardiac diseases. Considering the same the study suggests incorporating green infrastructure into urban development to improve the quality of the surrounding environment and acts as a proactive public health measure to lower the incidence of Cardiac disorders. To build sustainable and healthy communities, it is crucial to integrate environmental health with urban development, as this holistic approach emphasizes. The implementation of green infrastructure would reduce the detrimental effects on health. A network of natural and semi-wild places including green roofs, urban forests, parks, and green walls could improve biodiversity, manage water resources, and promote healthier urban settings. Conducting scientific research without practical implementation fails to benefit communities. Therefore, management leaders should prioritize and implement these scientific findings to enhance societal well-being. Integrating management with green infrastructure in smart cities can improve urban resilience by mitigating heat islands through sustainable practices like green roofs and permeable pavements. Pure Sciences should collaborate with these implementations to maximize their impact on urban resilience and create more sustainable, livable, and adaptive urban environments by fostering a harmonious relationship between technology and nature.

References

1. Hoover F-A, Meerow S, Coleman E, Grabowski Z, McPhearson T (2023) Why go green? Comparing rationales and planning criteria for green infrastructure in U.S. city plans. Landsc Urban Plan 237:104781. https://doi.org/10.1016/j.landurbplan.2023.104781
2. Navarrete-Hernandez P, Laffan K (2023) The impact of small-scale green infrastructure on the affective wellbeing associated with urban sites. Sci Rep 13(1):9687. https://doi.org/10.1038/s41598-023-35804-2
3. Shao H, Kim G, Li Q, Newman G (2021) Web of science-based green infrastructure: A bibliometric analysis in citespace. Land 10(7):711. https://doi.org/10.3390/land10070711
4. Kumar P, Mukul KD, Kaur A (2023) Green Infrastructure- A Roadmap towards sustainable development. IOP Conf Ser: Earth Environ Sci 1110(1):012060. https://doi.org/10.1088/1755-1315/1110/1/012060
5. Herath P, Bai X (2024) Benefits and co-benefits of urban green infrastructure for sustainable cities: Six current and emerging themes. Sustain Sci 19(3):1039–1063. https://doi.org/10.1007/s11625-024-01475-9
6. Bhatnagar A (2006) Environmental cardiology: Studying mechanistic links between pollution and heart disease. Circ Res 99(7):692–705. https://doi.org/10.1161/01.RES.0000243586.99701.cf
7. Beaglehole R, Bonita R, Alleyne G, Horton R, Li L, Lincoln P, Mbanya JC, McKee M, Moodie R, Nishtar S, Piot P, Reddy KS, Stuckler D (2011) Un high-level meeting on non-communicable diseases: Addressing four questions. The Lancet 378(9789):449–455. https://doi.org/10.1016/S0140-6736(11)60879-9
8. Lozano R, Naghavi M, Foreman K, Lim S, Shibuya K, Aboyans V, Abraham J, Adair T, Aggarwal R, Ahn SY, AlMazroa MA, Alvarado M, Anderson HR, Anderson LM, Andrews KG, Atkinson C, Baddour LM, Barker-Collo S, Bartels DH, Murray CJ (2012) Global and regional mortality from 235 causes of death for 20 age groups in 1990 and 2010: A systematic analysis for the global burden of disease study 2010. The Lancet 380(9859):2095–2128. https://doi.org/10.1016/S0140-6736(12)61728-0
9. Grabowski ZJ, McPhearson T, Matsler AM, Groffman P, Pickett ST (2022) What is green infrastructure? A study of definitions in US city planning. Front Ecol Environ 20(3):152–160. https://doi.org/10.1002/fee.2445
10. Ying J, Zhang X, Zhang Y, Bilan S (2022) Green infrastructure: Systematic literature review. Econ Res-EkonomskaIstraživanja 35(1):343–366. https://doi.org/10.1080/1331677X.2021.1893202
11. Adesoji T, Pearce A (2024) Interdisciplinary perspectives on green infrastructure: A systematic exploration of definitions and their origins. Environments 11(1):8. https://doi.org/10.3390/environments11010008
12. Korkou M, Tarigan AKM, Hanslin HM (2023) The multifunctionality concept in urban green infrastructure planning: A systematic literature review. Urban & Urban Green 85:127975. https://doi.org/10.1016/j.ufug.2023.127975
13. Janiszek M, Krzysztofik R (2023) Green infrastructure as an effective tool for urban adaptation—Solutions from a big city in a post-industrial region. Sustainability 15(11):8928. https://doi.org/10.3390/su15118928
14. Mell I (2022) Examining the role of green infrastructure as an advocate for regeneration. Front Sustain Cities 4:731975. https://doi.org/10.3389/frsc.2022.731975
15. Zuniga-Teran AA, Staddon C, De Vito L, Gerlak AK, Ward S, Schoeman Y, Hart A, Booth G (2020) Challenges of mainstreaming green infrastructure in built environment professions. J Environ Planning Manage 63(4):710–732. https://doi.org/10.1080/09640568.2019.1605890
16. De Jong G (2005) Evolution of phenotypic plasticity: Patterns of plasticity and the emergence of ecotypes. New Phytol 166(1):101–118. https://doi.org/10.1111/j.1469-8137.2005.01322.x
17. Pigliucci M, Murren CJ, Schlichting CD (2006) Phenotypic plasticity and evolution by genetic assimilation. J Exp Biol 209(12):2362–2367. https://doi.org/10.1242/jeb.02070

18. West-Eberhard MJ (2005) Developmental plasticity and the origin of species differences. Proc Natl Acad Sci 102(suppl_1):6543–6549. https://doi.org/10.1073/pnas.0501844102
19. Health (US), NI of, Study, BSC (2007) Understanding human genetic variation. In NIH Curriculum Supplement Series [Internet]. National Institutes of Health (US). https://www.ncbi.nlm.nih.gov/books/NBK20363/
20. Miller C (2020) 6. 2 Genetic Variation. https://humanbiology.pressbooks.tru.ca/chapter/8-2-genetic-variation/
21. Writing Group Members, Roger VL, Go AS, Lloyd-Jones DM, Benjamin EJ, Berry JD, Borden WB, Bravata DM, Dai S, Ford ES, Fox CS, Fullerton HJ, Gillespie C, Hailpern SM, Heit JA, Howard VJ, Kissela BM, Kittner SJ, Lackland DT, Turner MB (2012) Executive summary: heart disease and stroke statistics—2012 update: AReport from the American heart association. Circulation 125(1):188–197. https://doi.org/10.1161/CIR.0b013e3182456d46
22. Münzel T, Hahad O, Sørensen M, Lelieveld J, Duerr GD, Nieuwenhuijsen M, Daiber A (2022) Environmental risk factors and cardiovascular diseases: A comprehensive expert review. Cardiovasc Res 118(14):2880–2902. https://doi.org/10.1093/cvr/cvab316
23. Murray CJL, Aravkin AY, Zheng P, Abbafati C, Abbas KM, Abbasi-Kangevari M, Abd-Allah F, Abdelalim A, Abdollahi M, Abdollahpour I, Abegaz KH, Abolhassani H, Aboyans V, Abreu LG, Abrigo MRM, Abualhasan A, Abu-Raddad LJ, Abushouk AI, Adabi M, Lim SS (2020) Global burden of 87 risk factors in 204 countries and territories, 1990–2019: A systematic analysis for the global burden of disease study 2019. The Lancet 396(10258):1223–1249. https://doi.org/10.1016/S0140-6736(20)30752-2
24. Landrigan PJ, Fuller R, Acosta NJR, Adeyi O, Arnold R, Basu N (Nil), Baldé AB, Bertollini R, Bose-O'Reilly S, Boufford JI, Breysse PN, Chiles T, Mahidol C, Coll-Seck AM, Cropper ML, Fobil J, Fuster V, Greenstone M, Haines A, Zhong M (2018) The lancet commission on pollution and health. The Lancet 391(10119):462–512. https://doi.org/10.1016/S0140-6736(17)32345-0
25. Musunuru K, Hershberger RE, Day SM, Klinedinst NJ, Landstrom AP, Parikh VN, Prakash S, Semsarian C, Sturm AC, on behalf of the American Heart Association Council on Genomic and Precision Medicine; Council on Arteriosclerosis, Thrombosis and Vascular Biology; Council on Cardiovascular and Stroke Nursing; and Council on Clinical Cardiology (2020) Genetic testing for inherited cardiovascular diseases: A scientific statement from the American heart association. Circ: Genomic Precis Med 13(4):e000067. https://doi.org/10.1161/HCG.0000000000000067
26. Mensah GA, Wei GS, Sorlie PD, Fine LJ, Rosenberg Y, Kaufmann PG, Mussolino ME, Hsu LL, Addou E, Engelgau MM, Gordon D (2017) Decline in cardiovascular mortality: Possible causes and implications. Circ Res 120(2):366–380. https://doi.org/10.1161/CIRCRESAHA.116.309115
27. Heidenreich PA, Trogdon JG, Khavjou OA, Butler J, Dracup K, Ezekowitz MD, Finkelstein EA, Hong Y, Johnston SC, Khera A, Lloyd-Jones DM, Nelson SA, Nichol G, Orenstein D, Wilson PWF, Woo YJ (2011) Forecasting the future of cardiovascular disease in the United States: A policy statement from the American heart association. Circulation 123(8):933–944. https://doi.org/10.1161/CIR.0b013e31820a55f5
28. Bhatnagar A (2017) Environmental determinants of cardiovascular disease. Circ Res 121(2):162–180. https://doi.org/10.1161/CIRCRESAHA.117.306458
29. Lee T, Hwang S, Seo DM, Shin HC, Kim HS, Kim J-Y, Uh Y (2022) Identification of cardiovascular disease-related genes based on the co-expression network analysis of genome-wide blood transcriptome. Cells 11(18):2867. https://doi.org/10.3390/cells11182867
30. Ward-Caviness CK (2019) A review of gene-by-air pollution interactions for cardiovascular disease, risk factors, and biomarkers. Hum Genet 138(6):547–561. https://doi.org/10.1007/s00439-019-02004-w
31. Pamukcu-Albers P, Ugolini F, La Rosa D, Grădinaru SR, Azevedo JC, Wu J (2021) Building green infrastructure to enhance urban resilience to climate change and pandemics. Landscape Ecol 36(3):665–673. https://doi.org/10.1007/s10980-021-01212-y

32. Seo S, Choi S, Kim K, Kim SM, Park SM (2019) Association between urban green space and the risk of cardiovascular disease: A longitudinal study in seven korean metropolitan areas. Environ Int 125:51–57. https://doi.org/10.1016/j.envint.2019.01.038
33. Swarup S, Cargill EJ, Crosby K, Flagel L, Kniskern J, Glenn KC (2021) Genetic diversity is indispensable for plant breeding to improve crops. Crop Sci 61(2):839–852. https://doi.org/10.1002/csc2.20377
34. Rasheed A, Ilyas M, Khan TN, Mahmood A, Riaz U, Chattha MB, Al Kashgry NAT, Binothman N, Hassan MU, Wu Z, Qari SH (2023) Study of genetic variability, heritability, and genetic advance for yield-related traits in tomato (Solanum lycopersicon MILL.). Front Genet 13:1030309. https://doi.org/10.3389/fgene.2022.1030309
35. Khoury CK, Brush S, Costich DE, Curry HA, De Haan S, Engels JMM, Guarino L, Hoban S, Mercer KL, Miller AJ, Nabhan GP, Perales HR, Richards C, Riggins C, Thormann I (2022) Crop genetic erosion: Understanding and responding to loss of crop diversity. New Phytol 233(1):84–118. https://doi.org/10.1111/nph.17733
36. Le Roux JJ, Hui C, Castillo ML, Iriondo JM, Keet J-H, Khapugin AA, Médail F, Rejmánek M, Theron G, Yannelli FA, Hirsch H (2019) Recent anthropogenic plant extinctions differ in biodiversity hotspots and coldspots. Curr Biol 29(17):2912-2918.e2. https://doi.org/10.1016/j.cub.2019.07.063
37. Salgotra RK, Chauhan BS (2023) Genetic diversity, conservation, and utilization of plant genetic resources. Genes 14(1):174. https://doi.org/10.3390/genes14010174
38. Tol RSJ (2023) The fiscal implications of stringent climate policy. Economic Analysis and Policy 80:495–504. https://doi.org/10.1016/j.eap.2023.09.004

Part IV
The Role of Community Engagement in Resilience Building

Chapter 11
Forgotten Urban Waters and Green Areas of Dhaka: Analyzing the Scope of Urban Resilience by Revisiting Historical Maps and Policy Documents

Tahmina Rahman and M. N. Fatemi

Abstract Dhaka City, serving as Bangladesh's capital, has seen a gradual encroachment and destruction of its green and blue areas, a condition that has progressively worsened since its independence. Without considering the environmental consequences, the public and private sectors are aggressively developing natural areas inside and around the city through the encroachment of the water bodies and the destruction of green spaces. Recently, a growing body of research among scholars and policymakers is considering incorporating green and blue infrastructure (GBI) inside urban settings to augment urban resilience. This study, therefore, is initiated to conduct a macro-level analysis of the potential of GBI in urban resilience, ecological conservation, and regeneration of Dhaka city. To provide a guideline for preserving and regenerating the existing urban waters and green areas, the spatiotemporal changes in GBI of Dhaka city were analysed through the historical map (1924, 1960, 1987, 1995, 2001 and 2015) regression and policy content analysis. Using the historical map regression, the connections between urban waters and green areas, as well as their historical dynamics, were identified, which helped limit their encroachment and advance their resuscitation. This study revealed that Dhaka's urban waters and green areas have been drastically decreasing, especially after 1995. Consequently, the research findings led to the formulation of proposals aimed at including the preservation and regeneration of GBI in the urban design, policy and planning framework of Dhaka to make the city resilient.

Keywords Urban waters · Urban green areas · Green and blue infrastructure · Urban resilience · Historical maps · Policy documents

T. Rahman (✉) · M. N. Fatemi
Department of Architecture, University of Asia Pacific, Dhaka, Bangladesh
e-mail: tahminarahman@uap-bd.edu

M. N. Fatemi
e-mail: nawrose@uap-bd.edu

S. Ghosh et al. (eds.), *Cities of Tomorrow: Urban Resilience and Climate Change Preparedness*, Urban Sustainability, https://doi.org/10.1007/978-981-97-9658-8_11

11.1 Introduction

Presently, the majority of individuals are inclined towards embracing an urban lifestyle due to its well-being benefits and access to amenities. Cities currently account for 56.5% of the global population [1], with the Industrial Revolution accelerating mass migration from rural to urban areas [2]. By 2050, 68% of the global population is expected to be urbanites [1]. Bangladesh, with a population of 64.81 million, has one of the highest populations in the world, with 1,342 people per square kilometer [3]. Approximately 60 per cent of the whole population resides in the four main cities of Bangladesh (Dhaka, Chittagong, Khulna, and Rajshahi), while the capital, Dhaka, alone represents 35 per cent of the population [4]. However, Dhaka's rapid urbanization and unpredictable growth patterns of the infrastructure have led to the depletion of green spaces and urban waters in Dhaka, being infringed upon and destroying the valuable open spaces [5]. Likewise, ponds, lakes, and canals are at risk of the same issue. In Dhaka, a shortage of land, inadequate zoning regulations, insufficient land use planning and policy all contribute to an already dire situation, as does the increasing demand for space caused by cities' burgeoning populations [6].

On the contrary, historical evidence suggests that Dhaka's urban area was designed with parks, gardens, and canals as early as 1960 [7]. However, the relentless expansion in natural areas inside and outside of cities by public and private sectors has led to ecological sustainability threats and a decline in environmental quality [8]. This has affected the urban ecosystem and climate of Dhaka, causing serious consequences for its citizens [5]. The idea of green and blue infrastructure (GBI) is increasingly being considered as a way to make cities more resilient [9, 10]. This chapter aims to evaluate the potential of GBI in enhancing urban resilience and regeneration of the Dhaka metropolitan area. The spatiotemporal variations in GBI were examined using policy content analysis and regression analysis using historical maps (1924, 1960, 1987, 1995, 2001, and 2015). The historical map regression method was employed to identify the relationships between urban waterways and green spaces, as well as their historical changes for preventing their invasion and promoting their restoration.

11.2 Literature Review

11.2.1 Urban Resilience Through Green and Blue Infrastructure (GBI)

Urban resilience is crucial for addressing global challenges like climate change and rapid urbanization [11]. GBI plays a vital role in maintaining this resilience by offering ecosystem services such as mitigating flooding risks, controlling extreme temperatures, cleaning the air, and protecting biodiversity [12]. Integrating GBI into planning and design processes can improve resilience to environmental stressors and shock [13]. Climate change makes cities more vulnerable, and GBI acts as a natural barrier against its impacts [14, 15]. Permeable pavements and green roofs may reduce urban heat islands' impact, while wetlands and urban green areas can lessen flood severity [16]. For example, to better manage stormwater, increase biodiversity, and encourage active commuting, Copenhagen, Denmark, has established a thorough GBI plan [17]. Singapore's "City in a Garden" project improves living standards by minimizing flood risk and incorporating urban green areas into city design [18]. China's national policy for sponge city development integrates GBI into urban design to restore downstream ecosystems by reducing flood risks, surface runoff, and pollution [19]. These case studies demonstrate how GBI initiatives make cities more sustainable and livable by solving local and regional urban challenges.

When applied to Dhaka, GBI holds immense potential to deal with critical socioeconomic and environmental challenges while simultaneously making Dhaka more resilient [20]. Dhaka's rapid urbanization has led to environmental degradation, including loss of green areas, deterioration of wetlands, and increased susceptibility to natural disasters [21]. To adapt to climate change, increase biodiversity, and promote community well-being, Dhaka should revitalize its forgotten urban waters and green areas. To build more resilient and livable cities in Bangladesh, it is crucial to include GBI in urban planning and policy frameworks, alleviating environmental and socioeconomic challenges faced by the cities.

11.2.2 Provision of GBI in Urban Planning and Policies for Resilience

It has long been argued that to achieve sustainable and resilient urban development, it is crucial to mainstream environmental considerations into all government policy instruments, including budgets, plans, and strategies [22]. Urban planning and policy frameworks must prioritize urban resilience by integrating urban waters, green areas, and natural elements into the built environment [9, 10]. Recently, GBI has been recognized as a benchmark for local governments to make cities more resilient and sustainable [23]. The transformation of GBI concepts into practical initiatives for urban and regional areas requires effective policy instruments [10]. The adoption of GBI is influenced by policies concerning land-use planning, zoning ordinances, environmental protection, and urban development regulations [24]. Land-use planning ensures high-quality ecosystem services [25], while zoning ordinances provide incentives for developers and planners to adopt GBI [26]. Environmental regulations that mandate the protection of wetlands or riparian buffers contribute to the provision of GBI and attempts to enhance resilience in urban areas [27]. Integrating GBI into infrastructure projects enhances resilience to climate change impacts and is encouraged by policies promoting sustainable urban design and low-impact development strategies [28].

To adopt GBI initiatives in urban development plans, four areas of intervention proposed by Nesshöver et al. [29] include- (a) Preservation—preserving biodiversity and ecosystem services (b) Protection-repairing ecosystems damaged by human interventions, (c) Design- developing solutions to issues related to climate change through design; and (d) development of effective use of nature to mitigate the impact of problems like sociocultural and economic inequality, and vulnerability on live ability. Correspondingly, Kopp et al. [30] identified that GBI may be effectively integrated into urban development plans by focusing on geographical patterns such as water bio-corridors (blue networks with inland and peripheral waters), green networks, and land cover (greens and open areas, wetlands and riparian buffers).

For instance, all European Union policy sectors have actively advocated GBI, which was designed as a city planning as well as a city design strategy [31]. Swedish cities are encouraged to incorporate GBI plans in their urban development plans [50], while the UK has successfully integrated GBI initiatives into its national policy [32]. South Asian cities like Kochi and Chennai in India, have also incorporated GBI into their city planning and reconstruction efforts to enhance resilience [33]. Drawing these case studies and the conceptual framework shown here (Fig. 11.1), this research aims to revisit historical maps and policy documents of Dhaka, Bangladesh, to gain insights into GBI provision and resilience-building initiatives which can help shape future urban planning and policy choices.

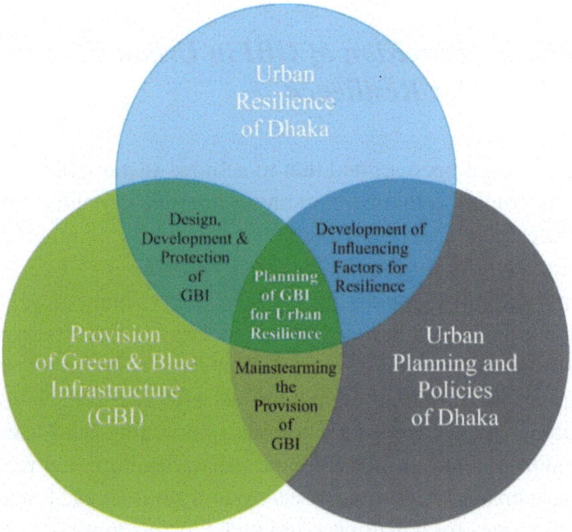

Fig. 11.1 Conceptual framework of the planning of GBI for urban resilience of Dhaka

11.3 Methodology

The term ‘Green–Blue Infrastructures’ (GBI) is used interchangeably with ‘urban waters and green areas’ throughout this research. The study utilized historical map regression and policy content analysis to track potential GBI linkages at the local level, as suggested by Tache et al. (2023), highlighting the importance of understanding the fragmentation of green–blue zones over time.

11.3.1 Historical Map Regression

As cited by Aston [34], the idea of historical map regression is to identify the alterations that have occurred in areas’ borders and alignments, roads and railroads, and construction sites. According to Glover [35], map regression can be used for three main purposes: (i) analyzing and determining changes in features, (ii) detecting features that were present in earlier maps but are no longer present in recent maps, and (iii) identifying boundary areas and their features, whether they are natural (such as fields, parks, and water bodies) or man-made (such as railroads and roads). In the same direction, Iojă et al. [36] examined the transformations in Bucharest’s urban waters and green areas over the last 160 years, while Rahman et al. [37] analyzed Dhaka’s physical evolution during the Mughal era in the seventeenth century.

The study examines changes in urban green spaces and water bodies over time in the Dhaka Metropolitan Area using historical maps from 1924 to 2015 (Fig. 11.2). The study aimed to examine land cover transformations over the past 91 years (from 1924 to 2015). The data was analyzed using data from six time-steps, covering the entire administrative jurisdiction (Table 11.1).

Identifying generic ground points, especially crossroads and important public buildings, between the maps is important to performing map georeferencing [36]. As one of the selected historical maps, T6: Dhaka City Map by RAJUK was considered

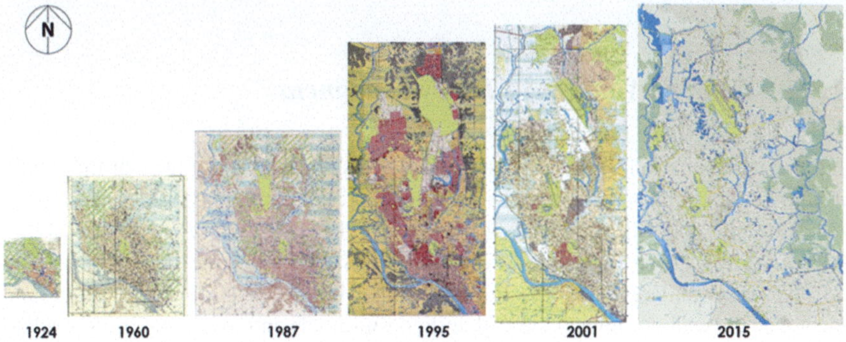

Fig. 11.2 Map of Dhaka from 1924 to 2015

Table 11.1 Maps that supported data collection

Time-steps	Map	Period of surveying
T1	Dacca City Map during the British Period by John Murray	1924
T2	Dacca City Map during the Pakistani Period by A. R. Quraishi	1960
T3	Dhaka Guide Map by M. Mahbubul Karim	1987
T4	Dhaka Guide Map by Surveyor General of Bangladesh	1995
T5	Dhaka Guide Map by Surveyor General of Bangladesh	2001
T6	Dhaka City Map by RAJUK	2015

Sources T1: [38], T2 and T5: Digital Maps collected from Department of Architecture, BUET; T3 and T4: Digital Maps collected from Department of Urban and Regional Planning, BUET; T6: Digital Maps collected from RAJUK Office, Dhaka

a reference for this study because of its accuracy. To reliably extract spatial data for parks and urban waterways, it has been verified that mapping from 1924 to 2015 is compatible. Whether the maps are colored or have legends, it is easy to see the parks and urban waterways. Each map was digitized using AutoCAD 2017 to extract data on urban green spaces (such as parks, playgrounds, informal green areas, restricted playgrounds, and restricted green areas), urban waters (such as rivers, canals, ponds, and wetlands), and road networks. Following the methodology [36], morphometric indicators were computed and selected from the collected spatial data to evaluate the changes in urban waterways and parks. These indicators include the area and percentage of GBI (Urban waterways and Greens), as well as their locations relative to built-up areas. Iojă et al. [36] recognized three components of urban green–blue dynamics: (i) GBI that remained in the same place, (ii) GBI that became part of built-up regions, and (iii) new GBI, which are man-made and result from changing other land uses. Several sections of Dhaka city's urban waters have their lost link pinpointed. Most of the time, these connections are turned into built-up areas or used as box culverts to build roads. Additionally, the Map Regression Analysis has confirmed the return of the previously vanished green network adjacent to one of the railway lines.

11.3.2 Content Analysis of Policy Documents

The provision of urban green areas and urban waters was examined in this study through Dhaka development policy instruments. Six plans were considered from the first planning proposal in 1917 to the most recent Dhaka Area Plan (DAP) 2022–2035 (Table 11.2). To investigate the possibilities of preserving the existing green and blue areas and restoring the previously lost interconnection of the green and blue areas, these documents review identified four interventions: water bio-corridors (preservation of river and river-front areas, protection of inland waterbodies and interlinkages), green networks (provision of green areas and interlinkages), and land

Table 11.2 Development policy instruments that supported data collection

Development policy instruments	Year	Interventions in policy instruments
Dacca City Planning (DCP)	1917	1. Water Bio-Corridors (i) Preservation of river and river-front areas (PRRA) (ii) Protection of inland waterbodies and interlinkages (PIWI) 2. Green Networks (i) Provision of green areas and interlinkages (PGAI) 3. Land Cover (i) Strategies for wetlands and riparian buffers (SWRB)
Dacca Master Plan (DMP)	1959	
Dhaka Metropolitan Area Integrated Urban Development Project (DMAIUDP)	1981	
Dhaka Metropolitan Development Plan (DMDP)	1995–2015	
Dhaka Structure Plan (DSP)	2016–2035	
Dhaka Area Plan (DAP)	2022–2035	

cover (strategies for wetlands, and riparian buffers); as mentioned by Kopp et al. [30] and Nesshöver et al. [29]; later modified according to the Dhaka context.

11.4 Findings and Analysis

11.4.1 GBI-Related Policy in Dhaka

Sir Patrick Geddes developed the Dacca City Planning (DCP 1917) report, focusing on Dhaka's open spaces and water channels. The report proposed interconnected green spaces along the river, protecting 25 miles of water channels, promoting landscape, water parks, and public amenities [39]. Geddes also recommended cleaning and dredging water bodies for trade and communication. He advised developing land above floodplains, protecting natural greenery, and establishing Dhaka's first continuous ecological network [40].

In 1958, the Dacca Improvement Trust (DIT) commissioned British architects Minoprio, Spencely, and Macfarlane to draft the initial comprehensive master plan (DMP 1959) for Dhaka city. The plan included the Buckland Bund, a 17-acre riverfront amenity park, riverfront urban renewal, and a water network with canals [41]. It also emphasized the importance of high land for residential areas and recommended structural measures in flood-prone areas. The plan suggested 3–4 acres of open spaces per 1,000 people for Dhaka [41].

The Dhaka Metropolitan Area Integrated Urban Development Project (DMAIUDP 1981) was Bangladesh's first strategic master plan, developed by the Shankl and Cox Partnership. It aimed to protect the city and development areas from floods through embankments but did not include the river-front area as an urban green space, as suggested in DCP 1917 and DMP 1959. The plan proposed a primary drainage network with natural canals interconnected to integrate flood plains. It did not specify green and open areas but recommended agricultural land in peri-urban areas and forested land in the north.

The Dhaka Metropolitan Development Plan (DMDP 1995), developed by Mott MacDonald Ltd. and Culp in Planning Ltd., was a Bangladeshi government plan designed to expand the city over 20 years to 2015. It included the Structure Plan, Urban Area Plan, and Detailed Area Plan, which aimed to consolidate urban resources and accelerate development. The plan proposed circular waterways to integrate riverfront development and beautify the city while considering the Buckland Embankment for tourism and recreation. DMDP also standardized 0.16 acres as green space per 1,000 populations, securing around 1590.6 acres of open spaces for recreational purposes. However, Dhaka's growth outpaced the plans, forcing government and semi-government bodies to occupy large-scale open and green spaces.

The Dhaka Structure Plan (DSP) was developed in 2015 by the Bangladesh Government and the Asian Development Bank. DSP aims to safeguard natural areas like rivers, canals, forests, and parks as preservation areas [42]. It recommends creating a 'Special Conservation Zone' with stricter guidelines to safeguard these areas. DSP also suggests developing an urban riverfront at Buckland Bund as a recreational corridor. It also recommends conserving water bodies to enhance local aesthetics and provide recreation. The plan also offers protection against the encroachment of flood flow zones and water retention areas, policies for a green network, and urban and peri-urban greenery and forestry. The plan also aims to protect open spaces in the RAJUK area.

The Detailed Area Plan (DAP) (2016–2035) was developed in 2022 by RAJUK, the Capital Development Authority. The plan includes eco-friendly initiatives such as a water-based eco-park, a GBI network, and rain gardening for buildings and urban areas. Key strategies include preserving river and river-front areas, inland waterbodies, and developing green spaces. The plan proposes progressive transformations in the city, including the design of a 574 km canal network as the 'Urban Lifelines'. To achieve hydrologically sensitive and sustainable land development, it is recommended to connect canals and waterbodies with rivers to establish a blue network and eliminate water obstruction. The plan also outlines specific land cover categories for different locations within the Dhaka Metropolitan Area.

The Dhaka City Development Plans and their GBI-related Policies can be analyzed quantitively through a scoring system (Table 11.3). The scoring themes have been identified from the study of Nesshöver et al. [29] and Kopp et al. [30]. In this analytical framework, the GBI Related Policies are scored between zero (0) and three (3). The score criteria are:

- Design and Development (DD-3 points): Three pointsareallotted if the plan addresses the new design and/or development of GBI options.
- Protection and Preservation (PP-2 points): Two pointsareallotted if the plan indicates the protection and/or preservation of existing GBI options.
- Interconnection and Network (IN-1 point): One point is allotted if the plan mentions interconnection and/or network between the existing GBIs.

Table 11.3 The score matrix for GBI-related policies of Dhaka city development plans

GBI Topic		Policy					
		DCP1917	DMP1959	DMAIUDP 1981	DMDP1995–2015	DSP2016–2035	DAP2022–2035
Preservation of river and river-front areas	DD (3)	-	3	-	3	3	3
	PP (2)	-	-	-	-	2	2
	IN (1)	1	1	-	1	-	1
	NI (0)	0	0	0	0	0	0
	Sub-Total	**1**	**4**	**0**	**4**	**5**	**6**
Protection of inland waterbodies and interlinkages	DD (3)	-	-	-	-	3	3
	PP (2)	2	-	-	2	2	2
	IN (1)	1	1	1	1	1	1
	NI (0)	0	0	0	0	0	0
	Sub-Total	**3**	**1**	**1**	**3**	**6**	**6**
Provision of green areas and interlinkages	DD (3)	-	3	3	3	3	3
	PP (2)	2	2	-	2	2	2
	IN (1)	-	-	-	-	1	1
	NI (0)	0	0	0	0	0	0
	Sub-Total	**2**	**5**	**3**	**5**	**6**	**6**
Strategies for wetlands and riparian buffers	DD (3)	-	-	-	-	-	-
	PP (2)	2	2	2	2	2	2
	IN (1)	-	-	-	-	-	1
	NI (0)	0	0	0	0	0	0
	Sub-Total	**2**	**2**	**2**	**2**	**2**	**3**
Total		**8**	**12**	**6**	**14**	**19**	**21**

- No indication (NI-0 points): Allotted in situations where the plan does not refer to or include this strategy.

Here, Design and Development (DD) -3 points; Protection and Preservation (PP)-2 points; Interconnection and Network (IN)-1 point; and No Indication (NI)-0 points.

The maximum achievable score is 24 in quantitative content analysis. With just basic GBI integration, DCP 1917 scores an 8 (33.3%). Preservation of river and river-front areas (4) and provision of green areas and interlinkages (5) were the primary areas of emphasis when the 1959 DMP was initiated, with a score of 12 (or 50%). However, the focus in DMAIUDP 1981 shifted away from policies on the GBI (score 6–25%). In DMDP 1995–2015, a little higher-level integration (score 14–58%), has been observed and a similar increased trend is shown in the DSP 2016–2035, with a score of 19 (79%) for GBI options. Meanwhile, the most recent DAP (2022–2035) has a score of 21 (88%), meaning that the majority of the GBI choices have been achieved. Upon reviewing all the plans, it has been observed that the 'strategies for wetlands and riparian buffers' is the most deficient sector.

11.4.2 Historical Evolution of Urban Waters and Green Areas

In 1924, approximately 48% of Dhaka's administrative boundary was comprised of urban greens, which equates to 3% of the city's administrative territory now (Fig. 11.3). Of this total, 37% was informal green space and 11% was parks and playgrounds (Table 11.4). Covering around 8 per cent of the administrative boundary, the 163 hectares included urban waterways (exclusively inland bodies of water) (Table 11.4). In 1960, the urban green spaces in Dhaka spanned 1,520 hectares, which accounted for almost 18% of the administrative limit at that time. However, it only covered 5% of the current administrative territory (Fig. 11.3). Of this, 11% was informal green space, 3% was restricted green area, and 4% was designated as parks and playgrounds (Table 11.4). This constitutes a decline in urban greens from 1924. Specifically, there were reductions in all types of green spaces, with the most significant reductions in informal green spaces and Parks and playgrounds. Similarly, the urban waterways covered an area of 2,912 hectares, consisting of 5% inland water bodies and 29% wetlands, which accounted for about 34% of the administrative boundary at that time (Table 11.4). This constitutes an expansion of urban waters from 1924.29% of wetlands were included inside the administrative boundaries of Dhaka in 1960. From Fig. 11.4, these urban waters and green spaces can be easily identified.

In 1987, the urban green spaces in Dhaka comprised an area of 3,039 hectares, which accounted for about 22% of the administrative limit at that time (Table 11.4). However, it only represented 9.9% of the current administrative area. These green spaces consisted of 19% informal green areas, 2% restricted green areas, and 1%

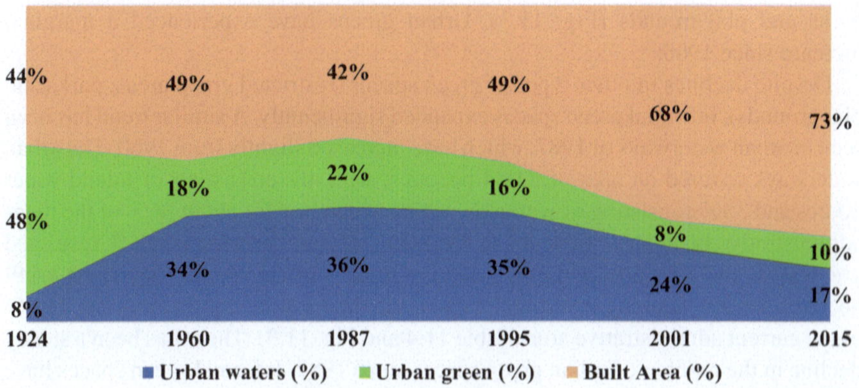

Figure 11.3 Evolution of Dhaka urban waters and green space between 1924 and 2015

Table 11.4 Evolution of urban waters and green areas in 1924–2016

Indicator	Time period					
	1924	1960	1987	1995	2001	2015
The entire area of the urban Greens (ha)	930	1520	3039	3449	1805	3038
Parks and playground (ha)	208	341	116	345	377	502
Restricted green area (ha)	0	213	279	1256	631	653
Informal green (ha)	723	965	2644	1848	797	1884
Share of urban green concerning the administrative area (%)	48	18	22	16	8	10
The entire area of the urban waters (ha):	163	2912	5084	7467	5297	5290
Inland waterbodies (ha)	163	405	508	611	791	1721
Wetlands (ha)	0	2507	4576	6856	4506	3568
Share of urban water concerning the administrative area (%)	8	34	36	34	24	17

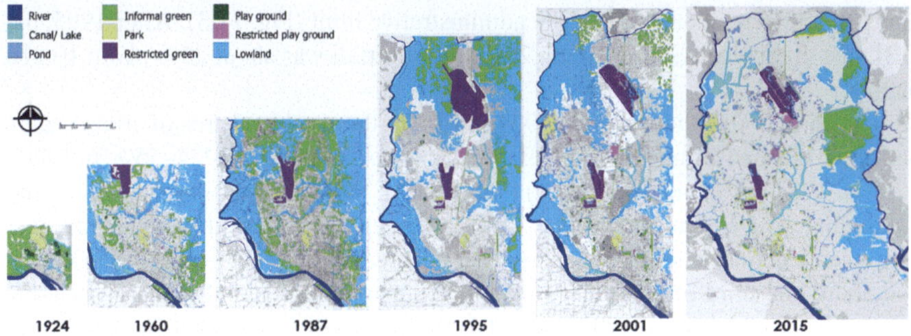

Fig. 11.4 Evolution of urban waters and green spaces of Dhaka

parks and playgrounds (Fig. 11.3). Urban greens have experienced a marginal increase since 1960.

Despite declines in other types of green spaces (restricted green areas, parks and playgrounds), informal green spaces expanded significantly. A similar trend has been seen in urban waterways of 1987, which have increased slightly from 1960. The urban waterways covered an area of 5,084 hectares, with 4% consisting of inland water bodies and 33% consisting of wetlands, which accounted for about 36% of the then-administrative boundary (Table 11.4). With 8% informal green space, 6% restricted green area, and 2% parks and playgrounds, Dhaka's urban greens comprised 3,449 hectares in 1995, making up around 16% of the then administrative limit but 11.3% of the current administrative area (Table 11.4 and Fig. 11.3). There has been a steady decline in the number of urban green spaces since 1987. Informal green spaces have shrunk dramatically, in contrast to other types of green space, such as restricted green areas, parks and playgrounds, which have expanded than those of 1987. However, as compared to 1987, the urban waterways in 1995 remain steady and exhibit no significant change in proportion. The urban waters presented 7,467 ha, with 3% inland water bodies and 31% wetlands which were about 34% of then administrative limit (Table 11.4). From Fig. 11.4, these urban waters and green spaces can be easily identified.

Since 1995, both the urban greens and urban waters have witnessed an abrupt drop in 2001. Urban green space in Dhaka included 1,805 hectares in 2001 (equivalent to 5.9% of the present-day administrative area, or about 8% of the then administrative limit), with 4% being informal green, 3% being restricted green, and 2% being made up of parks and playgrounds (Table 11.4 and Fig. 11.3). About 24% of the area under the then administrative control was comprised of urban waterways, which extended over 5,297 ha, with 4% being inland bodies of water and 20% being wetlands (Table 11.4 and Fig. 11.3). Moreover, the decline of urban waterways and natural spaces has persisted from 2001 to 2015. In 2015, urban greens in Dhaka presented 3038 ha (around 10% of the administrative limit) with 6% informal green, 2% Restricted green area and 2% Parks and playgrounds (Table 11.4 and Fig. 11.3). The urban waters covered 5,290 ha, with 5.5% inland water bodies and 11.5% low lands which were about 17% of the administrative limit (Fig. 11.3). After 1995, the GBI area dramatically dropped, by 2015, the shortfall was about 25% (from 10,916 ha to 8,328 ha).

It is worth noting that between 1924 and 1995, the combined area of urban waters and green reached around 50–60% of the total area of Dhaka (in 1924–56%, in 1960–52%, in 1987–58% and 1995–50%) (Table 11.4). However, starting from 1995 and continuing over 6 years until 2001, both the green spaces and urban waters saw a significant decline. The city witnessed a drastic reduction of half of its urban greens and about one-third of its urban waters. Upon examining the individual indicators, it is evident that between 1995 and 2001, the restricted green space declined from 1,256 hectares to 631 hectares, the informal green area decreased from 1,848 hectares to 797 hectares, and the wetlands decreased from 6,856 hectares to 4,506 hectares (Table 11.4). The downward trend persisted up to 2015 when urban waters made up only 17% of the total area and urban greens barely remained at 10%.

11.4.3 Correlation Between Built Area and Urban Green–Blue Spaces

The correlation between the overall size of Dhaka city and the total built area, urban waters, and green areas has also been determined. Statistical analysis revealed a 'large positive linear association' (r = 0.966, R-squared = 0.934) between Dhaka city's total area and its total built area between 1924 and 2015. It is evident from Fig. 11.5 that both the trend line of the area of Dhaka city and that of the built area exhibit a significant upward trend, indicating the above-mentioned 'large positive linear association' between these two. On the other hand, a "small positive linear association" (r = 0.723, R-squared = 0.525) was identified between the entire area and the urban greens of Dhaka city (Fig. 11.5). A comparable "small positive linear association" between the entire area and urban waters of Dhaka city was identified, with the correlation value r = 0.811 and R-squared = 0.658 (Fig. 11.5).

In addition, it was observed that while the built area trend line experienced a significant incremental increase, there has been a substantial decline in green spaces and urban water since 1995. Figure 11.5 shows that though it remained relatively constant from 1960 to 1995, the share of urban waters and green areas began to decline sharply in the years after that year. Here, three factors that contributed to the decline are as follows: first, the land being turned into urban areas in the 1990s as a result of real estate [43]; second, the flood defense at Dhaka West [44]; and third, the building of roadways on the canal using 'box-culvert' [45].

The study findings indicate that the proportion of GBI to the built-up area was 51:49 prior to Bangladesh's independence in 1971, and this ratio remained unchanged until 1995. Nevertheless, after the significant decline indicated earlier after 1995, the loss by 2015 amounted to around 25%, with the area decreasing from 10,916 ha to 8,328 ha. At present, the proportion of GBI in stock is a mere 27% (approximately

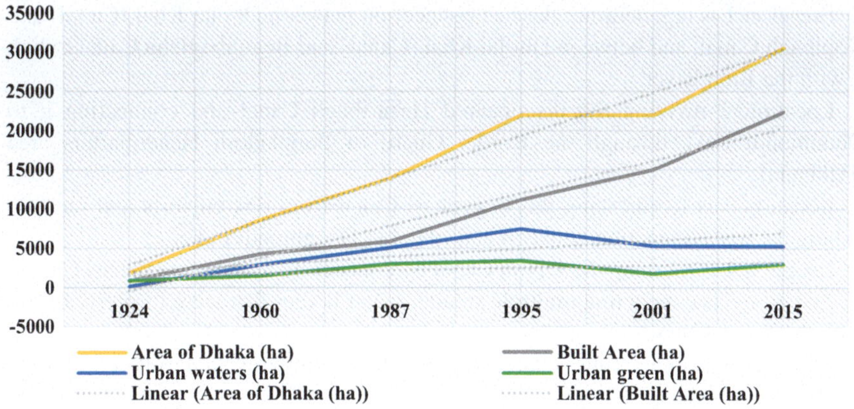

Fig. 11.5 Dhaka city area vs built area, urban waters and urban greens

8,328 hectares), which comprises privately owned informal greens and restricted greens that are mostly inaccessible to the general public.

11.4.4 Traces of the Forgotten Linkages of Urban Waters

From the historical evidence from 1924 to 2015, it can be confirmed that Dhaka has been developed with its urban blue network which is acting and will continuously act as its vascular system. However, flood defense measures at Dhaka West; roads and streets construction, and encroachment by public and private sectors in several areas of Dhaka, have led to the loss of canals and discontinuation of this canal network. In the last 20 years, hundreds of hectares of wetlands have been lost to the city [46]. Datta and Mandal [47] stressed that there are still opportunities to establish a blue network. Peeters and Shannon [48] suggested new connections and reservoirs to safeguard Dhaka's remaining water bodies. Therefore, by revisiting the historical map from 1924 to 2015 to trace Dhaka's water network and studying geomorphology (especially, the water flow direction and streamlines of Dhaka studied by Mandal [49], seven potential locations have been identified for new connections or reopening the existing water network by removing box culverts with minimal intervention to reestablish a GBI network.

In those seven potential locations, the study recommends four interconnected networks connecting east to west of Dhaka. The canal network of Dhaka East still now remains in good order, but it needs to be protected from further negative intervention. Through this study, a new proposal suggests resurfacing these canals without hampering existing roadside commercial and other facilities. The seven identified locations and their details have been given below to introduce new connections with the existing water network or resurface the existing water network by removing the box culverts through minimum intervention to reestablish a GBI network (Fig. 11.6):

Location 1 is regenerating the lost connection between Dholai Khal (Canal) and Gopibagh Canal, and between Dholai Khal (Canal) and Begunbagicha Canal to link it with the Balu River.

Location 2 reviving the disappeared Hazaribagh Canal and connecting it to Dhanmondi Lake through the Border Guard of Bangladesh Headquarters area (Pilkhana).

Location 3 is Resurfacing the Begunbari Canal from box culverts and reestablishing the lost connection between Hatirjheel to Dhanmondi Lake.

Location 4 is reviving the Digun Canal by connecting it with the Diabari Canal.

Location 5 is connecting multiple smaller water bodies that were formerly linked with the Tongi Canal.

Location 6 is connecting Gulshan Lake with Ibrahimpur Canal, linking multiple smaller waterbodies.

Location 7 is connecting Gulshan Lake with the Shahjadpur Canal to link it with the Balu River.

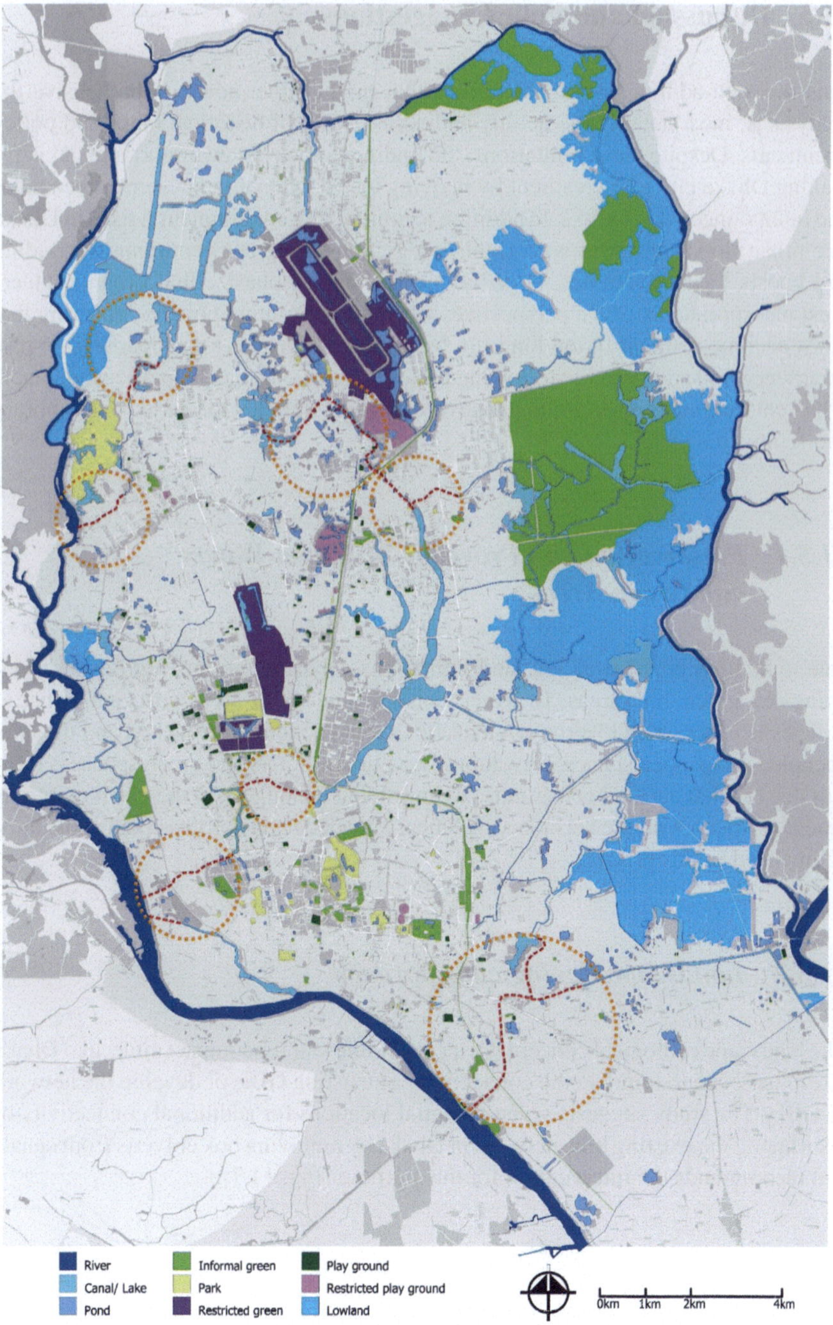

Fig. 11.6 The proposed interconnection to reestablish the urban water network

11.5 Discussion and Policy Suggestions

This research admits several limitations that might affect how comprehensive the analysis is, most notably the quality and accessibility of historical maps and policy documents. Despite these limitations, the findings showed how important GBI is for making Dhaka city more resilient by reviving its forgotten urban waters, green areas and interconnected networks. In addition to minimizing environmental risks in Dhaka like urban floods and heat waves, this integration strengthens community bonding and boosts local economies. Broadly, the findings emphasize that urban planners must incorporate historical perspectives when crafting future policies, ensuring that cities become resilient in the long run by leveraging these lessons learnt from past experiences. In addition, the following policy suggestions may help direct adaptable urban environments and sustainable growth, better equipped to tackle contemporary and future challenges.

11.5.1 Preservation and Provision of Urban Waters and Green Areas

Due to its high population and land limitations, Dhaka confronts hurdles in developing new green–blue areas. To protect existing green–blue areas, strict adherence to policies is necessary. Government agencies should acquire encroached and illegally occupied urban open spaces. Developing new urban waters and green spaces is difficult due to Dhaka's land constraints, so improving the quality of the existing ones is crucial. Experts with the necessary technical knowledge and professional skills must be involved in this process.

11.5.2 Landscape Ecological Approach

The landscape ecological concept is especially relevant for compact cities like Dhaka due to its scope to improve the connectivity of existing GBIs or develop the network of GBIs. The study suggests seven potential locations for additional connectivity or resurfacing the existing blue network in Dhaka by removing box culverts from canals and recommends the intervention for this purpose (Fig. 11.7).

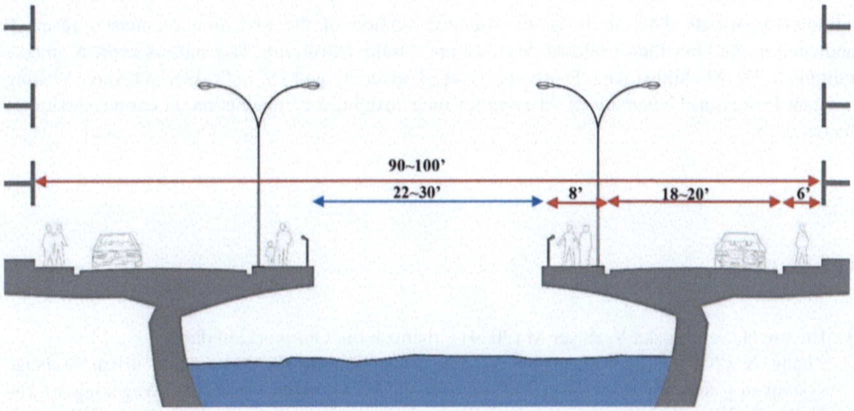

Fig. 11.7 Schematic sectional diagram illustrating the process of de-culverting the canal

11.5.3 Integrating Urban Waters and Green Areas in the Planning Process

A comprehensive and long-term strategy is required to actualize the green–blue network, as suggested by the DSP 2016–2035 and the DAP 2022–2035 into action. Evidence from China and Singapore [18, 19] indicates that green policies can significantly increase urban water and green spaces, even in relatively small, compact areas. The proposal should involve development along riversides, and plans for river-canal confluence while maintaining wetlands' protection and preservation. Additionally, new plans for wetland design and development should be formulated.

11.6 Conclusion

The city development plans of Dhaka (from DCP 1917 to DAP 2022–2035) have shaped its resilience throughout time, integrating historical perspectives into contemporary urban planning approaches. By reviving forgotten urban waters and green areas through GBI provision and integration, Dhaka can improve its urban resilience, meet its socioeconomic needs, and combat environmental challenges. Findings from this research stress the need to develop novel approaches for historical analysis to glean useful information from previous planning decisions that may guide present and future policymaking. In conclusion, the quantitative evaluation of GBI-related policies revealed in this study demonstrates urban planners and policymakers to get new insights regarding how to promote resilience through the provision and integration of GBI in planning documents.

Acknowledgements This study is an extended version of the first author's master research conducted in the Division of Global Architecture, Osaka University. The authors express sincere gratitude to Dr. Michihiro Kita, Professor, Osaka University and Dr. Seth Asare Okyere, Visiting Assistant Professor, University of Arizona for their insightful comments on an earlier version of this research.

References

1. Ritchie H, Samborska V, Roser M (2024) Urbanization. Our world in data
2. Zhang X (2017) Adapt green-blue space: Implementing the sustainable urban drainage system in rotterdam city context. Master Thesis, Wageningen University, Wageningen, The Netherlands
3. Ritchie H, Mathieu E (2019) Which countries are most densely populated? Our world in data
4. UN (2018) The world's cities in 2018–Data booklet. department of economic social affairs, Population Division, United Nations (UN), New York, USA
5. Byomkesh T, Nakagoshi N, Dewan AM (2012) Urbanization and green space dynamics in greater Dhaka, Bangladesh. Landscape Ecol Eng 8:45–58. https://doi.org/10.1007/s11355-010-0147-7
6. Fatemi N (2014) Urban Green Space in a High-Density City: User Expectations, Accessibility and Experience in Context of Dhaka. In: Dayaratne R, Wijesundara J (eds) Proceedings of International Conference on Cities, People and Places -ICCPP, 2014 Colombo, Sri Lanka, 2014. Department of Architecture, University of Moratuwa, pp 234–244
7. Nawar N, Sorker R, Chowdhury FJ, Rahman MM (2022) Present status and historical changes of urban green space in Dhaka city, Bangladesh: A remote sensing driven approach. Environ Chall 6:100425. https://doi.org/10.1016/j.envc.2021.100425
8. Islam N (2005) Dhaka now: Contemporary urban development. Bangladesh Geographical Society, Dhaka, Bangladesh
9. Bush J, Doyon A (2019) Building urban resilience with nature-based solutions: How can urban planning contribute? Cities 95:102483. https://doi.org/10.1016/j.cities.2019.102483
10. Wellmann T, Lausch A, Andersson E, Knapp S, Cortinovis C, Jache J, Scheuer S, Kremer P, Mascarenhas A, Kraemer R (2020) Remote sensing in urban planning: Contributions towards ecologically sound policies. Landsc Urban Plan 204:103921. https://doi.org/10.1016/j.landurbplan.2020.103921
11. Leichenko R (2011) Climate change and urban resilience. Curr Opin Environ Sustain 3(3):164–168. https://doi.org/10.1016/j.cosust.2010.12.014
12. Pinto LV, Inácio M, Pereira P (2023) Green and blue infrastructure (GBI) and urban nature-based solutions (NbS) contribution to human and ecological well-being and health. Oxf Open Infrastruct Health 1:1–14. https://doi.org/10.1093/ooih/ouad004
13. Seddon N, Smith A, Smith P, Key I, Chausson A, Girardin C, House J, Srivastava S, Turner B (2021) Getting the message right on nature-based solutions to climate change. Glob Change Biol 27(8):1518–1546. https://doi.org/10.1111/gcb.15513
14. Hobbie SE, Grimm NB (2020) Nature-based approaches to managing climate change impacts in cities. Philos Trans R Soc B 375(1794):20190124. https://doi.org/10.1098/rstb.2019.0124
15. Mersha E, Degefa S (2022) Ecosystem service of urban green and blue spaces towards sustainable urban ecosystem: A meta-analysis. J Ecosyst Ecography 12(11):1000363. https://doi.org/10.4172/2157-7625.1000363
16. Junqueira JR, Serrao-Neumann S, White I (2021) Managing Urban Climate Change Risks: Prospects for Using Green Infrastructure to Increase Urban Resilience to Floods. In: Letcher TM (ed) The Impacts of Climate Change. Elsevier, UK, pp 379–396. https://doi.org/10.1016/B978-0-12-822373-4.00013-6

17. Pauleit S, Fryd O, Backhaus A, Jensen MB (2020) Green infrastructures to face climate change in an urbanizing world. In: Loftness V (ed) Sustainable Built Environments. Springer, New York, USA, pp 207–234. https://doi.org/10.1007/978-1-0716-0684-1_212
18. Yi C (2020) Re-imagining nature in dense, high rise urban environment: The present and future of green building infrastructure in Singapore. Bachelor Thesis, Pomona College, Claremont, California, USA
19. Ma Y, Jiang Y, Swallow S (2020) China's sponge city development for urban water resilience and sustainability: A policy discussion. Sci Total Environ 729:139078. https://doi.org/10.1016/j.scitotenv.2020.139078
20. Rahman T (2020) A study on the potentials of green blue infrastructure (GBI) Network for Dhaka City in Bangladesh. Master Thesis, Osaka University, Osaka, Japan
21. Parvin GA, Esraz-Ul-Zannat M, Ahsan R, Islam I (2022) Disaster risks and resilience of urban Bangladesh: Role of blue-green infrastructure. In: Dhyani S, Basu M, Santhanam H, Dasgupta R (eds) Blue-Green Infrastructure Across Asian Countries. Springer, Singapore, pp 419–440. https://doi.org/10.1007/978-981-16-7128-9_19
22. Afionis S, Mkwambisi DD, Dallimer M (2020) Lack of cross-sector and cross-level policy coherence and consistency limits urban green infrastructure implementation in Malawi. Front Environ Sci 8:558619. https://doi.org/10.3389/fenvs.2020.558619
23. Roostaie S, Nawari N, Kibert CJ (2019) Sustainability and resilience: A review of definitions, relationships, and their integration into a combined building assessment framework. Build Environ 154:132–144. https://doi.org/10.1016/j.buildenv.2019.02.042
24. Barreto Rosa DW (2023) Assessing the benefits of green and blue infrastructure in peri-urban areas. Doctoral Thesis, University of Technology Sydney, Sydney, Australia
25. BenDor TK, Spurlock D, Woodruff SC, Olander L (2017) A research agenda for ecosystem services in American environmental and land use planning. Cities 60:260–271. https://doi.org/10.1016/j.cities.2016.09.006
26. Kronenberg J, Andersson E, Barton DN, Borgström S, Langemeyer J, Björklund T, Haase D, Kennedy C, Koprowska K, Łaszkiewicz E (2021) The Thorny Path Toward Greening: Unintended Consequences, Trade-Offs, and Constraints in Green and Blue Infrastructure Planning, Implementation, and Management. Ecology and Society 26 (2Toward):36. https://doi.org/10.5751/ES-12445-260236
27. Bel Fekih S, Cohen M, Khebour F (2022) Green and blue infrastructure design in a semi-arid region. Front Environ Sci 10:1061256. https://doi.org/10.3389/fenvs.2022.1061256
28. Nascimento N, Moura P, Reboita M, Rosa D, Drumond RA, Silva A, Almeida RP, Silva T, Fernandes W, Patrício PA (2022) Using green and blue infrastructure for urban flood mitigation: Simulating scenarios for climate change, GBI Technologies, and Land Policy. Lincoln Institute of Land Policy, Brazil
29. Nesshöver C, Assmuth T, Irvine KN, Rusch GM, Waylen KA, Delbaere B, Haase D, Jones-Walters L, Keune H, Kovacs E (2017) The science, policy and practice of nature-based solutions: An interdisciplinary perspective. Sci Total Environ 579:1215–1227. https://doi.org/10.1016/j.scitotenv.2016.11.106
30. Kopp J, Frajer J, Lehnert M, Kohout M, Ježek J (2021) Integrating Concepts of Blue-Green Infrastructure to Support Multidisciplinary Planning of Sustainable Cities. Problemy Ekorozwoju 16 (2):137–146. https://doi.org/10.35784/pe.2021.2.14
31. Rolf W, Diehl K, Zasada I, Wiggering H (2020) Integrating farmland in urban green infrastructure planning. An evidence synthesis for informed policymaking. Land Use Policy 99:104823. https://doi.org/10.1016/j.landusepol.2020.104823
32. Hrdalo I, Pereković P, Reljić DT (2021) Historical development of urban green infrastructure and possibilities of its implementation in the republic of croatia. Prostor 29:(61):56–71. https://doi.org/10.31522/p.29.1(61).5
33. Sanchez FG, Govindarajulu D (2023) Integrating blue-green infrastructure in urban planning for climate adaptation: Lessons from Chennai and Kochi India. Land Use Policy 124:106455. https://doi.org/10.1016/j.landusepol.2022.106455
34. Aston M (2013) Using maps: Current archaeology. Current Publishing, London, UK

35. Glover S (2013) Historic map regression. Skillspassport. Archaeology Skills, Online
36. Iojă I-C, Osaci-Costache G, Breuste J, Hossu CA, Grădinaru SR, Onose DA, Nită MR, Skokanová H (2018) Integrating urban blue and green areas based on historical evidence. Urban For Urban Green 34:217–225. https://doi.org/10.1016/j.ufug.2018.07.001
37. Rahman M, Zaman KB, Hafiz R (2016) Translating text into space for mapping the past territory of a city: A study on spatial development of Dhaka during Mughal period city. Territ Arch 3:1–16. https://doi.org/10.1186/s40410-016-0036-y
38. Murray J (1924) A handbook for travellers in India Burma and Ceylon including all British India, the Portuguese and French Possessions, and the Indian States. Thacker, Spink, & Co. Calcutta, London, UK
39. Geddes P (1917) Report on Dacca City Planning. Secretariate Press, Calcutta, India
40. Ahmed F (2017) To Flow, Or to Fortify?: Water, development, and urbanism in building a deltaic metropolis. The University of Adelaide, Adelaide, Australia
41. Minoprio CA, Spencely HGC, Macfarlane P (1959) Report on the Master Plan for Dacca 1958 Dacca Improvement Trust, Dacca
42. Fatemi MN, Okyere SA, Diko SK, Kita M (2020) Multi-level climate governance in Bangladesh via climate change mainstreaming: Lessons for local climate action in Dhaka City. Urban Science 4(2):24. https://doi.org/10.3390/urbansci4020024
43. Alam MJ (2018) Rapid urbanization and changing land values in Mega cities: Implications for housing development projects in Dhaka, Bangladesh. Bandung: J Glob South 5 (2):1–19. https://doi.org/10.1186/s40728-018-0046-0
44. JICA (1992) Master plan for greater Dhaka protection project- flood action plan: FAP 8A, (Main Report and Supporting Reports I and II). Japan International Cooperation Agency (JICA), Dhaka, Bangladesh
45. Ferdous F, Nilufar F (2007) Morphological transformation and evolution of Panthapath as a commercial belt of Dhaka City. Protibesh 11(2):55–66
46. Islam I (2009) Wetlands of Dhaka metro area: A study from social, economic, and institutional perspectives. AH Development Publishing House, Dhaka, Bangladesh
47. Datta U, Mandal D Creating Blue Network: A Solution of Present Water Logging and Transportation Problem for The Dhaka Metropolitan Area. In: Upendra R (ed) "Design that cares"—multi-disciplinary approaches to creating sustainable and meaningful built environment: Proceedings of the 10th International Conference of Faculty of Architecture Research Unit (FARU), Colombo, Sri Lanka, 2017. University of Moratuwa, pp 247–259
48. Peeters S, Shannon K (2011) Readdressing Dhaka's public water bodies: A design research. Nakhara: J Environ Des Plan 7:25–46. https://ph01.tci-thaijo.org/index.php/nakhara/article/view/104970
49. Mandal D (2019) Developing a strategic process to reintroduce a lost water way as a possible solution for water logging: Case study, an area of Dhaka Metropolitan City. Master Thesis, West Virginia University, West Virginia, USA
50. Sandstrom UG (2002) Green infrastructure planning in urban Sweden. Plan Pract Res 17(4):373–385. https://doi.org/10.1080/02697450216356

Chapter 12
Policymaking for Urban Climate Resilience in a Multi-Level Government System

Kirti Kusum Joshi

Abstract Many developing countries are grappling with rapid urbanization and climate change amidst low economic growth, regional disparities, and poor governance. Urbanization in these countries, driven mostly by rural-to-urban migration, strains cities already struggling with basic infrastructure and services, making urban populations highly vulnerable to climate hazards. Moreover, within the same regions or cities, climate vulnerability varies due to geographical and socioeconomic disparities. How effectively a country can address its climate risks is closely tied to its political system, which shapes the formulation and sustainability of climate policies and actions. This chapter explores opportunities and challenges presented by Nepal's federal political system in bolstering resilience against climate change in its urban and urbanizing areas, considering its complex topography, pace of urbanization, and prevailing regional disparities amidst limited experience with federalism. On one hand, the federal system offers a platform for tailored, context-specific climate measures as climate change concerns transcend the political boundaries of any single province or city. The system also holds promise for fostering community engagement and enhancing local resilience capacities. On the other hand, several challenges persist, including the need for effective and efficient collaboration among the levels of government in the formulation and implementation of climate change policies and plans across national, sub-national, and local levels.

Keywords Climate change · Disaster · Federal system · Land use · Local government · Urban resilience

K. K. Joshi (✉)
Institute of Engineering and Information Technology, Lumbini Technological University, Nepalgunj, Nepal
e-mail: kirti.joshi@ltu.edu.np

S. Ghosh et al. (eds.), *Cities of Tomorrow: Urban Resilience and Climate Change Preparedness*, Urban Sustainability, https://doi.org/10.1007/978-981-97-9658-8_12

12.1 Introduction

Global temperatures are now projected to surpass pre-industrial levels by 2 °C by around 2050, which will lead to more frequent, prolonged, and severe extreme weather events [1]. These effects will be particularly exacerbated in cities and urban areas where many people live. In many cities, urban flooding and urban heat island effects have already emerged as significant climate threats [2–6]. Urban climate change adaptation is particularly challenging in developing countries where urbanization mostly stems from migration-driven population growth rather than economic expansion, placing strain on cities already struggling to provide basic infrastructure and services [7–10]. Moreover, within the same region or city, communities often face uneven or differential levels of climate risks due to geographical locations or disparities in accessing socioeconomic and political resources [11].

The effects of climate change are experienced at local levels. Because of this, local authorities are generally seen as best positioned to build climate resilience. Cities and city-networks are now increasingly taking part in making and implementing climate change decisions [12–17]. However, the extensive reach of climate impacts across regions and the multi-sectoral nature of policies often demand coordination on broader scales including, forest, river basin, water, flood management, and spatial planning, but the connection between city climate resilience strategies and the related national and sub-national policies is not often clear, affecting the quality of coordination and collaboration between different government levels [18–23].

Numerous prior studies have explored the role of cities or local governments regarding urban climate change policymaking from the perspective of climate governance, focusing on the enablers and barriers to the formulation of such policies [24], 25–29. These studies show that how effectively a country can address its urban climate risks is closely tied to its government system, which shapes the formulation and sustainability of policies and actions regarding urban development and climate change.

This chapter examines the case of Nepal which ranks among the world's most climate-vulnerable countries. In 2015, Nepal transitioned to a federal, multi-government political system, delineating constitutional powers and functions and powers among federal, provincial, and local governments. This paper explores opportunities and challenges presented by Nepal's federal system in bolstering resilience against climate change in its urban and urbanizing areas, considering its complex topography, pace of urbanization, and prevailing regional disparities amidst limited experience with federalism. On one hand, the federal system offers a platform for tailored, context-specific climate measures as climate change concerns transcend the political boundaries of any single province or city. The system also holds promise for fostering community engagement and enhancing local resilience capacities. On the other hand, several challenges persist, including the need for effective and efficient collaboration among the levels of government in the formulation and implementation of climate change policies and plans across national, sub-national, and local levels.

12.1.1 Government System in Nepal

Nepal underwent a significant constitutional transformation with the adoption of the Constitution of Nepal in 2015. This marked a shift towards federalism, fostering sharing of constitutional powers among federal, provincial and local governments, guided by the principles of "coordination, cooperation, and coexistence" as envisioned in the Constitution. The Constitution delineates exclusive powers to the federal government (Schedule 5), provincial governments (Schedule 6), and local governments (Schedule 7) as well as and concurrent powers to the federal and provincial governments (Schedule 8) and to all government levels (Schedule 9). The lists of powers have further been unbundled or detailed for smooth functioning of all levels of government [30].

The country is divided into seven provinces, each governed by its provincial government. At the local level, there are 753 local government entities, comprising 293 municipalities and 460 rural municipalities. Additionally, for coordination purposes, local government entities are clustered into districts, with each province comprising several districts. Nepal has a total of 77 districts. Figure 12.1 shows the political division of the country.

There are also constitutional provisions for District Coordination Committees (DCCs) to facilitate coordination among municipalities and rural municipalities within the respective districts, and between the federal and provincial governments. The District Assembly consists of mayors/chairpersons and deputy mayors/vice-chairpersons of municipalities/rural municipalities within the district.

12.1.2 Climate Change and Urbanization in Nepal

The Germanwatch's Climate Risk Index ranked Nepal as the 12th most affected country in 2019 and the 10th most affected country from 2000 to 2019 in terms of the "impacts of weather-related loss events" [31]. The direct cost from climate variability and extreme events, such as floods and landslides, was estimated at 1.5–2% of the national gross domestic product (GDP) in 2014—the economic impact of climate change in the country is projected to reach an extra 2%–3% of GDP annually by 2050 [32, 33].

Nepal's vulnerability to climate change is attributed to its fragile and complicated topography—within 145 km to 241 km of its width, the altitude in Nepal varies from about 60 m in the Tarai plains in the south to 8,848 m at the peak of Mt. Everest in the north (Fig. 12.2). Rising temperatures are causing glaciers in the mountains to melt, heightening the likelihood of glacial lake outburst floods (GLOFs). Meanwhile, springs in the hills are drying up, and river discharge in the Tarai plains is decreasing along with the groundwater table. Likewise, shifts in precipitation patterns have resulted into more rainfall and less snowfall in the mountains, reducing snow deposition. Additionally, extreme weather events are becoming more frequent.

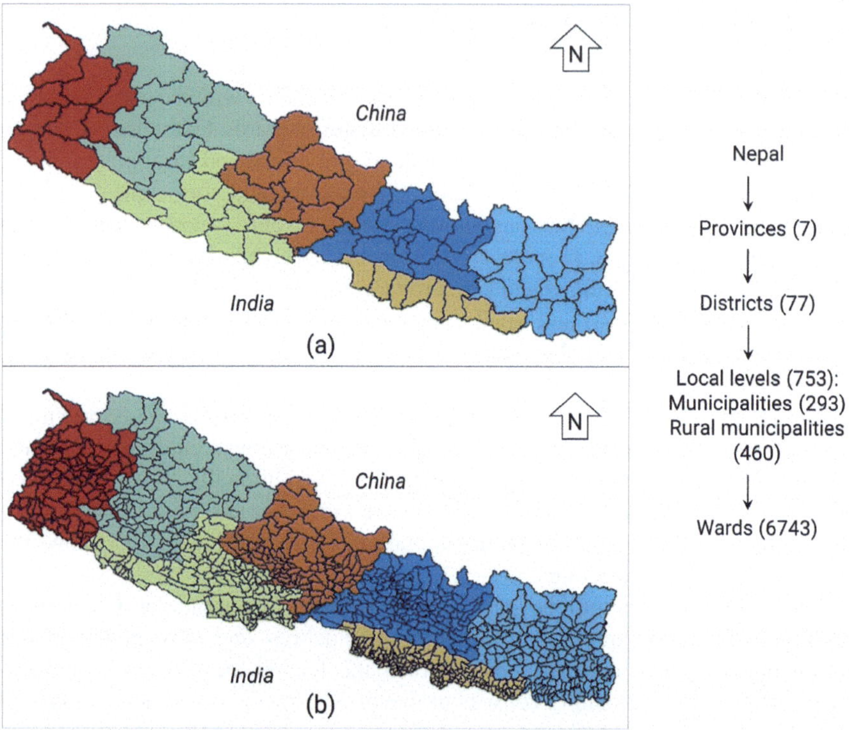

Fig. 12.1 Political division of Nepal into provinces and districts (Panel (**a**)) and municipalities/ rural municipalities (Panel (**b**))

Heavy and untimely rainfalls have escalated the threat of landslides in the mountain regions, floods and landslides in the hill- and Chure (mid-hill) regions, and floods in the Tarai plains. Poor economy and climate-sensitive, subsistence livelihoods of the majority population have furthered increased Nepal's climate vulnerability.

Climate change projectionssuggest that Nepal is likely to experience increase in temperature by 0.92–1.07° Celsius in the medium term (2016–2045) and by 1.3°–1.8° C in the long term (2036–2065) from the baseline period of 1981–2010 [35]. Likewise, annual precipitation is likely to rise by 2%–6% in the medium term and 8%–12% in the long term, indicating drier winters and wetter monsoon summers, with the possibility of up to three times more rainfall. Because of these changes in climate, extreme weather events (heavy rainfall, snowstorms, hailstorms, and high winds), along with heat and cold waves, landslides, floods, and forest fires, are likely to become more frequent and severe [36]. Additionally, slow-onset hazards like drought, shifting precipitation patterns, alterations in snow cover, glacier retreat, and GLOFs are predicted to increase in severity.

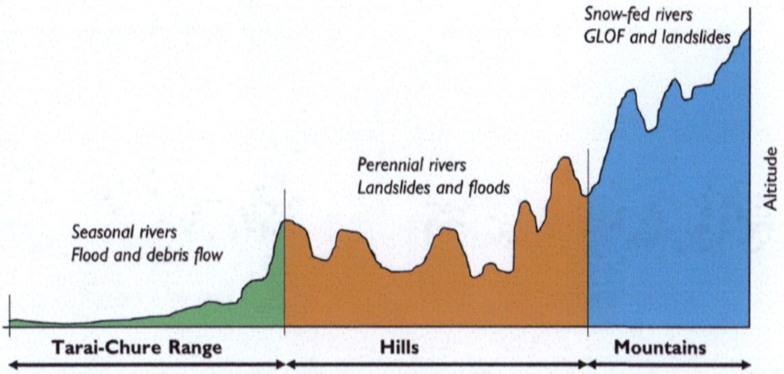

Fig. 12.2 Schematic representation of Nepal's topography and hazard risks (not to scale). *Source* Author's illustration, adapted from MoPE [34]

Along with escalating climate risks, Nepal has also experienced rapid urbanization, with annual growth rates reaching up to 5% [37]. In Nepal, there is no official definition of urban areas—although municipalities are envisioned to be developed as urban areas in near future, many municipalities possess rural characteristics. Likewise, some rural municipalities are more urbanized than many municipalities. Recently, NSO [38] has proposed a methodology for determining degree of urbanization in the country based on ward-wise population density. Accordingly, as of 2021, the urban population—population living in urban wards—accounts for 27.1% of the total population [38], Table 12.1). About two-thirds of Nepal's population lives in urban and peri-urban areas. It turns out that out of 293 municipalities and 460 rural municipalities, 193 municipalities and 73 rural municipalities have at least one urban ward. According to MoFE [36], out of 293 municipalities and 73 rural municipalities that have at least one urban ward, 38 municipalities and 15 rural municipalities, respectively, are prone to high to very high climate risk as of 2021 (Fig. 12.3).

The rapid urban growth, coupled with poor urban management, has increased climate risks in urban areas, affecting both people and infrastructure. Insufficient consideration of climate resilience in urban planning has exacerbated climate risks in

Table 12.1 Urban population in Nepal in 2021

Local government entities	Total	Entities with at least one urban ward	Entities with no urban ward	Population in million (2021)		
				Urban ward	Non-urban ward	Total
Municipalities	293	193	100	7.2 (38.0%)	11.8 (62.0%)	19.1
Rural municipalities	460	73	387	0.6 (5.9%)	9.2 (94.1%)	9.8
Total	**753**	**266**	**487**	**7.8 (27.1%)**	**21.1 (72.9)**	**28.9**

Source Author's calculation based on the data from NSO [38]

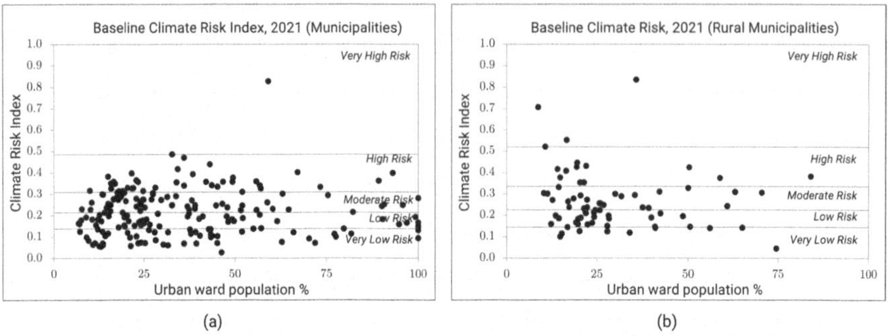

Fig. 12.3 Baseline climate risk index, 2021 for municipalities (Panel (**a**)) and rural municipalities (Panel (**b**)) with at least one urban ward. *Source* Author's illustration based on climate risk index data from MoFE [36]

Nepal's urban areas. Nepal's urban resilience to climate change significantly depends on its management of urbanization through the lens of climate resilience. However, the country faces a policy conundrum in addressing both climate change and urbanization in an integrated manner. The challenge stems from the involvement of multiple government agencies overseeing urban affairs with a history of poor coordination and collaboration among these agencies. The establishment of provincial and local governments as new tiers following the adoption of a federal political system in 2015 has added new challenges, but there could also be opportunities to be harnessed.

12.1.3 Materials and Methods

This chapter analyses Nepal's national urban climate change policy architecture and the content of other relevant policies and strategies at the national level including policies on urban development, climate change adaptation, disaster risk reduction and management (DRRM), land use, and municipal governance and management (Fig. 12.4). The analysis includes DRRM policies since many disaster risks stem from climate variability and extremes, with both DRRM and climate resilience targeting the reduction of community and infrastructure vulnerability [39]. Likewise, land use policies are included in the analysis because urban growth-related land use changes can reduce adaptive capacity and increase climate change vulnerability, while land use planning can facilitate urban climate resilience [40, 41].

The analysis focuses on the organizational and institutional structure of the policy domain and mandates of and provisions for engaging provincial and local governments for the adoption of these policies for urban climate resilience at the corresponding government levels. Only national policies and strategies formulated after the promulgation of Nepal's 2015 constitution are considered to focus on policy-making under the multi-level government system (Fig. 12.5). Policies and strategies at the national (federal) level are selected for analysis because even the exclusive

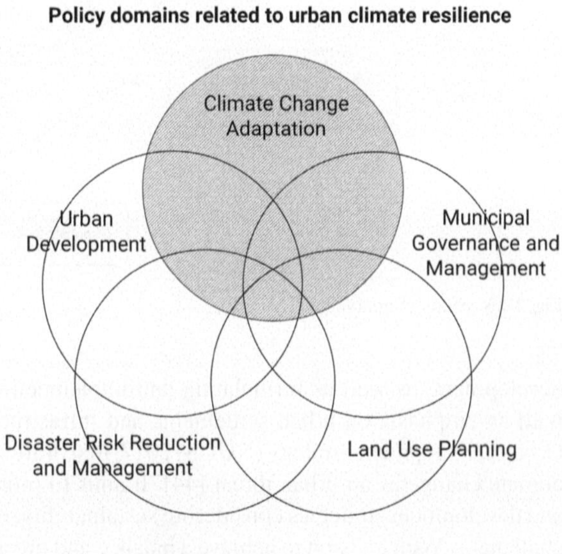

Fig. 12.4 Policy domains related to urban climate resilience

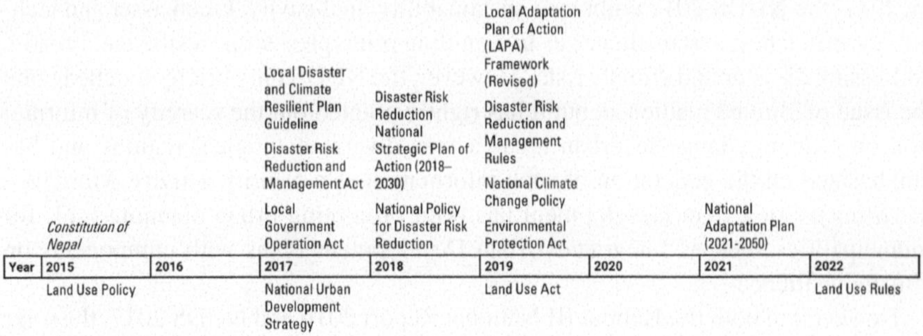

Fig. 12.5 Nepal's major post-2015 policies related to urban climate resilience

powers of provincial and local governments are mostly conditional on federal laws [42]. The policy analysis follows simple steps as shown in Fig. 12.6.

12.2 Results and Discussion

12.2.1 Addressing Climate Resilience in Urban Policies

As one of the first attempts to integrate climate change into urban policy in Nepal, the National Report for the Habitat III, which was prepared in 2016, included a plan of action for addressing climate change [43]. The plan of action notably emphasizes integrating climate resilience (particularly adaptation) into urban planning and

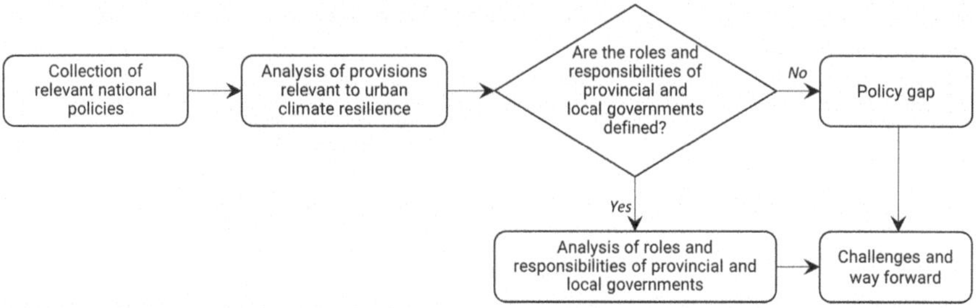

Fig. 12.6 Steps of analysis

development, as well as formulating and implementing a national adaptation plan with an emphasis on urban settlements and infrastructure. Likewise, the National Urban Development Strategy (NUDS) 2017, first drafted in 2015, has also recognized climate change as an urban threat [44]. It aims to promote innovative urbanization and development strategies considering sustainability, resilience, and climate change challenges. With a vision to achieve a balance and prosperous national urban system by 2031, the NUDS 2017 embraced sustainability, inclusivity, green asset and technology, efficiency, and resilience as five guiding principles, with 'resilience' broadly addressing disaster and climate risks. However, the NUDS only briefly touched upon the issue of climate resilience, but it has rightly pointed out the scarcity of information on climate change in urban areas across diverse ecological regions and has emphasized on the generation of such information as a priority activity. Moreover, the Ministry of Urban Development has been promoting urban planning tools for municipalities such as Integrated Urban Development Plans with components on climate resilience.

However, in both the Habitat III National Report 2016 and NUDS 2017, the roles of provincial and local governments have not been clearly defined. This could be attributed to the absence of specific legal frameworks detailing the constitutional roles of the provincial and local governments at the time of preparing these policy documents—the Local Government Operation Act was formulated only in 2017, and the detailed delineation of exclusive and concurrent powers of the federal, provincial, and local governments also took place in 2017 [30].

Urban development is a cross-cutting issue with numerous stakeholder agencies involved in urban planning and management. Municipalities and rural municipalities are in various stages of urbanization. Integrating climate concerns within the domain of urban development and planning can provide an opportunity for local governments to design targeted interventions that address climate change while aligning with urban development goals. However, such opportunity has not been utilized adequately in Nepal due to the lack of clarity in vertical and horizontal relationships between different agencies at the three levels of government. Even at the national level, the scope of work of the Ministry of Urban Development overlaps considerably with other ministries. Moreover, there is no direct working relationship between the urban ministry and local (municipal) governments. Because of this, interpreting urban policies (responsibility of the urban ministry) within the context of local development (responsibility of the municipal governments) remains a daunting challenge.

12.2.2 Addressing Urban Resilience in Climate Change Policies

Article 232 of the Constitution of Nepal 2015 exclusively emphasizes cooperation, coexistence, and coordination as the key principles for maintaining the relationship between/among the three levels of government. These principles are also crucial for addressing climate change across different levels of government. Accordingly, with the Prime Minister as its chair, a high-level national council for environmental protection and climate change managementhas been established for integrating environmental and climate change concerns into national policies, plans, and programs, as provisioned in the Environment Protection Act 2019 [32]. This council, which includes the chief ministers of all provinces as members, also provides guidance to provincial and local governments on environmental protection and climate change issues. Likewise, provincial-level climate change coordination committees have been formed in each province, tasked with integrating and mainstreaming climate change adaptation into sub-national policies, plans, and programs, while also ensuring vertical linkages with the federal government and coordinating with local governments within the concerned province [32].

Nepal's National Climate Change Policy (NCCP) 2019 emphasizes on the integration of climate actions into policies across all government levels. It also highlights the need to integrate disaster risk reduction (DRR) activities into climate change adaptation plans, embed adaptation components in socio-economic development, alternative income generation, and poverty alleviation initiatives, and conduct periodic vulnerability and risk assessments to inform adaptation planning. Regarding urban resilience, the NCCP 2019 envisions building climate-friendly cities by developing safe, sustainable, and resilient habitats and infrastructure. Relevant strategies include risk sensitive land use planning at local level, incorporation of adaptation programs in settlement development plans, and development of appropriate standards for settlements, infrastructure, and buildings for climate resilience.

In terms of institutional framework, the NCCP 2019 envisions formation of a national-level council under the functional coordination of the Ministry of Forests and Environment (MoFE) for policy coordination on the matters of climate change. The institutional setup also includes an inter-ministerial coordination committee under the coordination of MoFE, thematic ministry at each provincial level for the coordination of province-level climate actions, and climate change section or unit at each local level and in the concerned thematic ministries at national and provincial levels. The three levels of governments, within their jurisdictions, are responsible for formulating and implementing relevant laws, strategies, working policies, directives, procedures, guidelines, standards, and plans in conformity with the NCCP 2019.

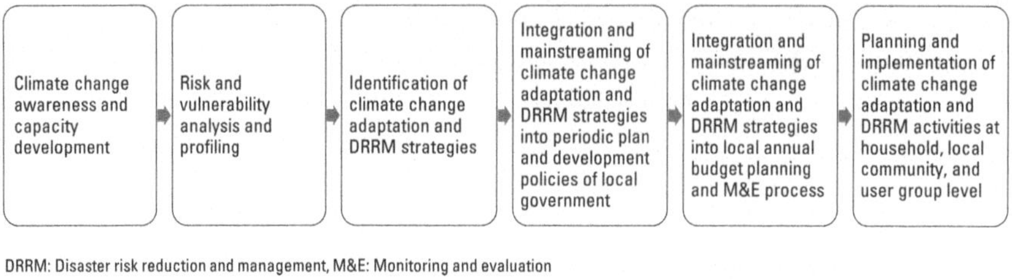

Fig. 12.7 The LAPA process. [45] *Source* Author's illustration based on the LAPA Framework 2019

The NCCP 2019 has been instrumental in the formulation of National Framework on Local Adaptation Plans for Action (LAPA) 2019, which aims to mainstream climate change adaptation and DRRM into the overall process of development and resource management at the local level [45] (Fig. 12.7). The LAPA Framework positions local governments at the core of local adaptation planning decisions and actions, with support from the federal and respective provincial government (Fig. 12.8). It is mandatory for all local governments to prepare medium-term (five-year) periodic plans, and the LAPA Framework 2019 rightly identifies periodic plans as a vehicle for integrating and mainstreaming climate change adaptation.

More recently, Nepal finalized its National Adaptation Plan (NAP) 2021 with a timeframe of 2021–2050 [32]. Although the NAP 2021 focuses on the national-level adaptation priorities, it includes predictions on climate extreme events and also provides climate risk scenarios at local (municipal) levels for the first time in the country [36]. Such information is crucial for local-level adaptation planning as envisioned in the LAPA Framework. However, the NAP was unable to provide settlement-level analysis due to insufficient data, missing a significant opportunity to inform local-level adaptation planning.

12.2.3 Integrating Climate Change Adaptation and DRRM

In Nepal, climate change and DRRM are often addressed separately in national policies and practices with separate legal provisions and institutional mechanism. Given the considerable overlap between climate change adaptation and DRRM such seen in the instances like floods and rainfall-induced landslides, local governments often face challenges in terms of harmonizing climate change adaptation and DRRM efforts due to the lack of concrete policies for an integrated approach. The DRRM policies are more inclined towards relief and response to disasters, focusing more on the functions of emergency operation centers (EOCs) at national, provincial, district, and local levels (Fig. 12.9).

As provisioned in the DRRM Act 2017 and Rules 2019, vertical integration between different levels of government in DRRM efforts is maintained through (i)

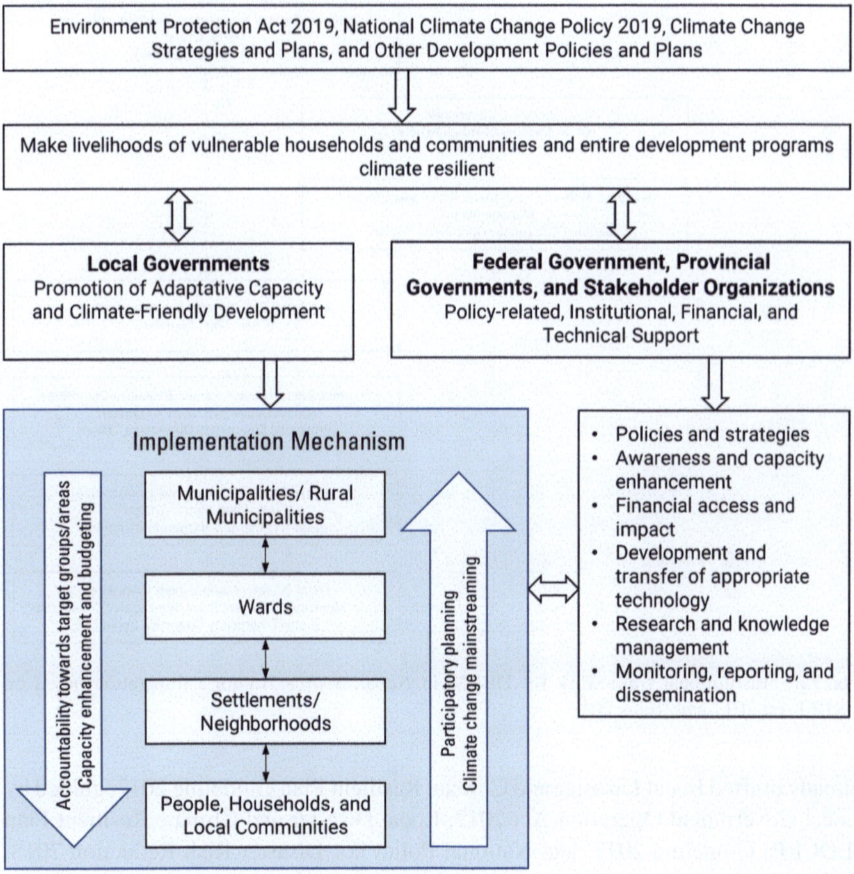

Fig. 12.8 Roles of different levels of government in the LAPA process. [45] *Source* Author's illustration, adapted from the LAPA Framework, 2019

a national council with the Prime Minister as its chair and chief ministers of all provinces as members, (ii) provincial councils in all provinces, chaired by respective chief ministers, (iii) disaster management committees in all districts, chaired by respective Chief District Officer, in which all chiefs of municipalities and rural municipalities within the respective districts are members, and (iv) local disaster management committees, chaired by the chiefs of respective municipalities. This elaborative institutional structure can be leveraged to address urban climate resilience at local levels.

The National Policy for DRR 2018 also emphasizes integrating DRR with climate change adaptation activities into all development processes. Likewise, the National DRR Strategic Action Plan (2018–2030) prioritizes preparing integrated guidelines for local-level disaster risk management (DRM) and climate change adaptation. Interestingly, the Ministry of Federal Affairs and General Administration has

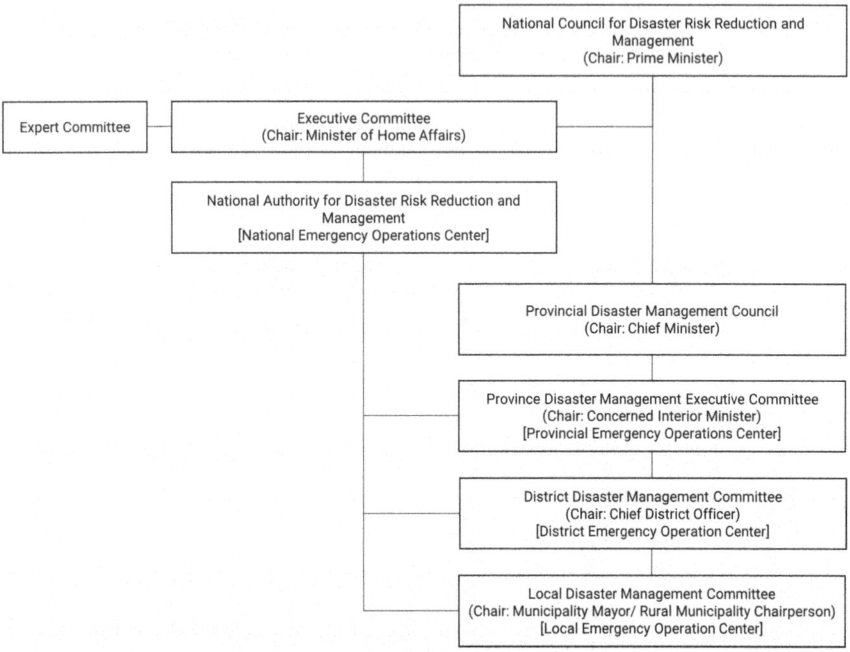

Fig. 12.9 Institutional framework for DRRM in Nepal. *Source* Author's illustration based on DRRM Act 2017 and Rules 2019

already drafted Local Disaster and Climate Resilient Plan Guideline 2017, guided by Local Government Operation Act 2017, Local Disaster and Climate Resilient Plan (LDCRP) Guideline 2017, and National Policy for Disaster Risk Reduction 2018. In fact, the federal government has been incentivizing local governments to prepare LDCRP to mainstream disaster and climate resilience in local development process [46]. However, the LDCRP's efforts to promote ownership and facilitate implementation have faltered due to its reliance on establishing new institutions, allocating financial resources, and navigating complex coordination with municipal, provincial, and federal departments [47]. On the other hand, with assistance from development partners, some local governments have begun developing climate action plans, focusing on the most relevant climate hazards in their areas (for example, [48–50].

12.2.4 Land Use Policies for Climate Risk Resilience

Haphazard urbanization, driven by the premature conversion of farmlands to urban use, stands as a significant factor in exacerbating urban vulnerability. Such unplanned urban expansion has contributed to various challenges, including floods resulting from residential development in flood plains and urban flooding due to the loss of farmlands that would otherwise absorb excess water [51]. The Land Use Act 2019, which is built on the Land Use Policy 2015, aims to discourage the plotting of farmlands and promote more sustainable land use practices to mitigate the adverse impacts of haphazard urbanization including increased climate vulnerability.

The Land Use Act 2019 classifies land use into agricultural, residential, commercial, industrial, mining, forest, waterbody, public use, cultural and archaeological, and other areas designated by the federal government. According to the Act, the federal ministry responsible for land management must prepare a land use map for each municipality and rural municipality and provide it within one year of the Act coming into effect. The land use map should also delineate areas prone to natural hazards. The Local Land Use Council, established in each municipality and rural municipality, can update the corresponding land use map based on the necessity assessment.

The Land Use Act 2019 mandates the local governments, within their jurisdiction, to update landowners' records and certificates of land ownership in alignment with the land use map [52]. The Act prohibits land use change; however, if change in land use is deemed necessary, the concerned landowner must apply to the Local Land Use Council with justification. This application is forwarded to the Provincial Land Use Council, which convenes a technical panel to study the matter. If the panel approves the change in land use, the Provincial Land Use Council forwards its recommendation to the Federal Land Use Council. However, in cases of settlements facing disaster risks, the concerned local government has legal mandates to change land use and relocate the affected households to other suitable locations. The Act and the corresponding rules, the Land Use Rules 2022, also set necessary criteria and standards for restricting land fragmentation and regulating land plotting for housing development, emphasizing the importance of preserving farmlands amidst urban development pressures.

As a key feature of the Land Use Act 2019, the federal, provincial, and local governments are responsible for formulating land use plans in their respective jurisdictions. This process begins with the preparation of a long-term approach paper on land use, which involves a comprehensive study of the land conditions, population growth, food and housing needs, and the demand for land for economic development and infrastructure [52]. Such land use plans at the national, provincial, and local levels have to be approved by the respective land use councils (Fig. 12.10). According to the Act, land use plans must clearly delineate, among others, disaster-prone areas. Despite the legal mandates, the federal and provincial governments are yet to develop land use plans. However, some municipalities have started preparing risk-sensitive land use planning, which also incorporates measures for planned urbanization and climate resilience besides DRR [53].

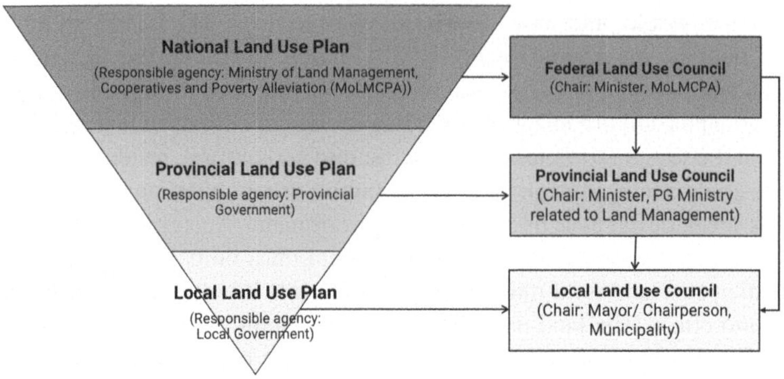

Fig. 12.10 Role of different levels of government in preparing land use plans. *Source* Author's illustration based on Land Use Act 2019

12.2.5 Challenges and Way Forward

The existing legal frameworks require municipalities and rural municipalities to prepare multiple policy documents of similar nature such as LAPA, LDCRP, and RSLUP in addition to the regular urban development plans and coordinate with multiple agencies for the same (Fig. 12.11). Despite possessing constitutional powers, local governments face the challenge of prioritizing one policy document over another, and there is risk of non-optimal use of financial resources. Moreover, the provincial governments are still struggling to function as a powerful tier of government, which has hampered the vertical linkages among the three levels of government. Most municipalities are poor in terms of financial, technical, and institutional capacity to prepare and implement climate resilience policies and strategies together with local development plans [54, 55]. Some municipalities have benefitted from the support of development agencies and some have hired external consultants to prepare resilience plan, raising question over the sustainability of such efforts, effectiveness of knowledge transfer, and quality of ownership by local governments.

The way forward for policymakers involves the preparation and implementation of an integrated municipal action plan that simultaneously addresses climate resilience, DRRM, land use, and urban development (Fig. 12.11). Instead of scattering investments across separate plans with overlapping objectives, this integrated approach ensures a more cohesive and efficient use of resources. For this, local governments must be strengthened both technically and financially. Technical capacity building includes training local officials in the field of climate science, urban planning, and disaster risk management, as well as providing them with the necessary tools and technology. Financial capacity building involves securing adequate funding and developing financial management skills to effectively allocate and utilize resources. The NCCP 2019 requires that a minimum of 80% of climate finance from international mechanisms be allocated to implementing programs at the local level, which can be leveraged to fund local climate projects and programs.

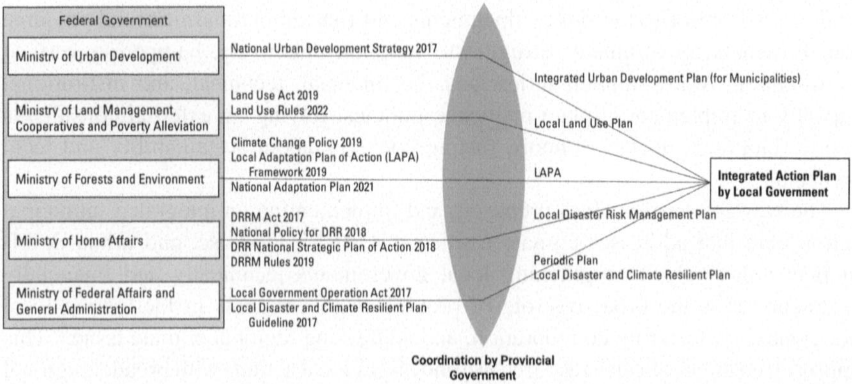

Fig. 12.11 Integrating national policies and municipal plans (recommended measures are highlighted in bold)

In addition, provincial governments need to enhance their capabilities to engage more proactively in facilitating policymaking at the local level. This includes providing guidance, resources, and support to local governments as they develop and implement their action plans. Provincial governments can also play a crucial role in fostering collaboration among local governments, promoting cross-learning, and addressing climate issues that span regional boundaries. By acting as intermediaries, provincial governments can ensure that local actions are aligned with broader regional and national strategies, creating a more unified and effective response to climate change and urban development challenges.

12.3 Conclusion

This paper explores the challenges of Nepal's federal political system with multi-level government structure in enhancing climate change resilience in urban and urbanizing areas, considering its complex topography, rapid urbanization, and regional disparities amidst limited federalism experience. The federal system allows for tailored, context-specific climate measures and promises greater community engagement and local resilience. However, effective coordination among federal, provincial, and local governments is essential for successful policy formulation and implementation.

The analysis includes Nepal's national climate change policy architecture and other relevant policies on urban development, climate adaptation, DRRM, land use, and municipal governance. DRRM policies are crucial as many disaster risks stem from climate variability, and both DRRM and climate resilience aim to reduce vulnerability. Land use policies are also important as urban growth-related changes can impact adaptive capacity and vulnerability. The analysis, focusing on policies and strategies since 2015, reveals that municipalities and rural municipalities face challenges in prioritizing policy documents and risk suboptimal financial resource use. Provincial governments struggle to function effectively, hampering vertical coordination. Many municipalities lack the financial, technical, and institutional capacity to implement climate resilience policies, relying on external consultants and development agency support, raising concerns about sustainability and local ownership.

The way forward involves preparing and implementing an integrated municipal action plan that addresses climate resilience, DRRM, land use, and urban development cohesively. Strengthening local governments technically and financially is crucial, as is the proactive role of provincial governments in facilitating local policymaking, fostering collaboration, and addressing regional climate issues. This approach ensures cohesive resource use and aligns local actions with broader regional and national strategies, creating a unified approach to address climate change and urban development challenges.

References

1. Adelekan I, Cartwright A, Chow W, Colenbrander S, Dawson R, Garschagen M, Haasnoot M, Hashizume M, Klaus I, Krishnaswamy J, Lemos MF, Ley D, McPhearson T, Pelling M, Kodira P P, Revi A, Sara L M, Simpson N P, Singh C, Solecki W, Thomas A, Trisos C (2022) What the latest science on impacts, adaptation and vulnerability means for cities and urban areas. Indian Institute for Human Settlements. https://doi.org/10.24943/SUPSV209.2022
2. Aghamohammadi N, Ramakreshnan L (2024) The impact of heat mitigation and adaptation technologies on urban health. In Aghamohammadi N, Santamouris M (eds) Mitigation and adaptation of urban overheating. Elsevier. https://doi.org/10.1016/B978-0-443-13502-6.00005-1
3. Han D, Zhang T, Qin Y, Tan Y, Liu J (2023) A comparative review on the mitigation strategies of urban heat island (UHI): A pathway for sustainable urban development. Climate Dev 15(5):379–403. https://doi.org/10.1080/17565529.2022.2092051
4. Rainey JL, Brody SD, Galloway GE, Highfield WE (2021) Assessment of the growing threat of urban flooding: A case study of a national survey. Urban Water J 18(5):375–381. https://doi.org/10.1080/1573062X.2021.1893356
5. Shahfahad TS, Rihan M, Hang HT, Bhaskaran S, Rahman A (2022) Modelling urban heat island (UHI) and thermal field variation and their relationship with land use indices over Delhi and Mumbai metro cities. Environ Dev Sustain 24:3762–3790. https://doi.org/10.1007/s10668-021-01587-7
6. Sohn W, Kim J-H, Li M-H, Brown RD, Jaber FH (2020) How does increasing impervious surfaces affect urban flooding in response to climate variability? Ecol Ind 118:106774. https://doi.org/10.1016/j.ecolind.2020.106774
7. Filho WL, Balogun A-L, Olayide OE, Azeiteiro UM, Ayal DY, Chavez Muñoz PD, Nagy GJ, Bynoe P, Oguge O, Toamukum NY, Saroar M, Li C (2019) Assessing the impacts of climate change in cities and their adaptive capacity: Towards transformative approaches to climate change adaptation and poverty reduction in urban areas in a set of developing countries. Sci Total Environ 692:1175–1190. https://doi.org/10.1016/j.scitotenv.2019.07.227
8. Mafuku S H, Musakwa W, Chirisa I (2021) Water, sanitation, and hygiene question of future cities of the developing world. In The Palgrave Encyclopedia of Urban and Regional Futures. Palgrave Macmillan, Cham. https://doi.org/10.1007/978-3-030-51812-7_71-1
9. Suhartini N, Jones P (2019) Urban governance and informal settlements. Springer Nature Switzerland AG. https://doi.org/10.1007/978-3-030-06094-7

10. UN-Habitat (2022) World cities report 2022: Envisaging the future of cities. United Nations Human Settlements Programme
11. Rangwala L, Chatterjee S, Agarwal A, Khanna B, Uri I, Shetty B, Palanichamy R B, Ramesh A (2024) Climate resilient cities: Assessing differential vulnerability to climate hazards in urban India. New Delhi: WRI India. https://doi.org/10.46830/wrirpt.22.00055
12. Bäckstrand K, Kuyper JW, Linnér BO, Lövbrand E (2017) Non-state actors in global climate governance: From Copenhagen to Paris and beyond. Environ Polit 26(4):561–579. https://doi.org/10.1080/09644016.2017.1327485
13. Biesbroek R, Lesnikowski A, Ford JD, Berrang-Ford L, Vink M (2018) Do administrative traditions matter for climate change adaptation policy? A comparative analysis of 32 high-income countries. Rev Policy Res 35(6):881–906. https://doi.org/10.1111/ropr.12309
14. Gordon DJ, Johnson CA (2018) City-networks, global climate governance, and the road to 1.5°C. Curr Opin Environ Sustain 30:35–41. https://doi.org/10.1016/j.cosust.2018.02.011
15. Heikkinen M, Karimo A, Klein J, Juhola S, Ylä-Anttila T (2020) Transnational municipal networks and climate change adaptation: A study of 377 cities. J Clean Prod 257:120474. https://doi.org/10.1016/j.jclepro.2020.120474
16. Lee T, Jung HY (2018) Mapping city-to-city networks for climate change action: Geographic bases, link modalities, functions, and activity. J Clean Prod 182:96–104. https://doi.org/10.1016/j.jclepro.2018.02.034
17. Woodruff SC (2018) City membership in climate change adaptation networks. Environ Sci Policy 84:60–68. https://doi.org/10.1016/j.envsci.2018.03.002
18. Eissa Y, Khalil HAEE (2022) Urban climate change governance within centralised governments: A case study of Giza, Egypt. Urban Forum 33:197–221. https://doi.org/10.1007/s12132-021-09441-9
19. Heidrich O, Reckien D, Olazabal M, Foley A, Salvia M, de Gregorio HS, Orru H, Flacke J, Geneletti D, Pietrapertosa F, Hamann JJ-P, Tiwary A, Feliu E, Dawson RJ (2016) National climate policies across Europe and their impacts on cities strategies. J Environ Manage 168:36–45. https://doi.org/10.1016/j.jenvman.2015.11.043
20. Pazhuhan M (2023) Institutionalizing urban climate governance in the global South? Evidence from Tehran urban management Iran. Clim Dev 15(9):793–807. https://doi.org/10.1080/17565529.2022.2161298
21. Vedeld T, Coly A, Ndour NM et al (2016) Climate adaptation at what scale? Multi-level governance, resilience, and coproduction in Saint Louis Senegal. Natural Hazards 82(Supplement 2):173–199. https://doi.org/10.1007/s11069-015-1875-7
22. Birchall SJ, Bonnett N, Kehler S (2023) The influence of governance structure on local resilience: Enabling and constraining factors for climate change adaptation in practice. Urban Climate 47:101348. https://doi.org/10.1016/j.uclim.2022.101348
23. Broto VC (2017) Urban governance and the politics of climate change. World Dev 93:1–15. https://doi.org/10.1016/j.worlddev.2016.12.031
24. Hughes S, Yordi, S, Besco L (2020) The role of pilot projects in urban climate change policy innovation. Policy Stud J 48(2):271–297. https://doi.org/10.1111/psj.12288
25. Hughes S (2017) The politics of urban climate change policy: Toward a research agenda. Urban Affairs Review 53(2):362–380. https://doi.org/10.1177/1078087416649756
26. Hughes S, Romero-Lankao P (2014) Science and institution building in urban climate-change policymaking. Environ Polit 23(6):1023–1042. https://doi.org/10.1080/09644016.2014.921459
27. Leck H, Roberts D (2015) What lies beneath: Understanding the invisible aspects of municipal climate change governance. Curr Opin Environ Sustain 13:61–67. https://doi.org/10.1016/j.cosust.2015.02.004
28. Lee T, Painter M (2015) Comprehensive local climate policy: The role of urban governance. Urban Climate 14(Part 4):566–577. https://doi.org/10.1016/j.uclim.2015.09.003
29. van der Heijden J (2019) Studying urban climate governance: Where to begin, what to look for, and how to make a meaningful contribution to scholarship and practice. Earth Syst GovAnce 1:100005. https://doi.org/10.1016/j.esg.2019.100005

30. FIARCC (2017) Unbundling/detailing of list of exclusive and concurrent powers of the federation, the state (province) and the local level provisioned in the schedule 5, 6, 7, 8, 9 of the constitution of Nepal. Federalism Implementation and Administration Restructuring Coordination Committee
31. Eckstein D, Künzel V, Schäfer L (2021) Global climate risk index 2021: Who suffers most from extreme weather events? Weather-related loss events in 2019 and 2000–2019. Germanwatch
32. GoN (2021) National Adaptation Plan (NAP) 2021–2050. Government of Nepal. https://doenv.gov.np/rules/NAP-2021-2050-1700841577.pdf
33. MoSTE (2014) Economic impact assessment of climate change in key sectors in Nepal. Ministry of Science, Technology and Environment, Government of Nepal. IDS Nepal, Global Climate Adaptation Partnership, and Practical Action
34. MoPE (2016) Nepal's strategic program for climate resilience:Consultative draft (20/11/10). Ministry of Population and Environment, Government of Nepal
35. MoFE (2019a) Climate change scenarios for Nepal for National Adaptation Plan. Ministry of Forest and Environment, Government of Nepal. Retrieved from https://lib.icimod.org/record/34554
36. MoFE (2021) Vulnerability and risk assessment and identifying adaptation options in rural and urban settlements. Ministry of Forests and Environment, Government of Nepal
37. World Bank (2022) Nepal country climate and development report. CCDR Series. World Bank
38. NSO (2024) Degree of urbanization (DEGURBA) in Nepal. National Statistics Office, Government of Nepal
39. Regmi B, Watkiss P, Gentle P, Bhandari D (2019) From risk to resilience: An integrated approach to harmonise climate change and disaster risk planning. Oxford Policy Management Limited
40. Carter JG (2018) Urban climate change adaptation: Exploring the implications of future land cover scenarios. Cities 77:73–80. https://doi.org/10.1016/j.cities.2018.01.014
41. Hurlimann A, Moosavi S, Browne GR (2021) Urban planning policy must do more to integrate climate change adaptation and mitigation actions. Land Use Policy 101:105188. https://doi.org/10.1016/j.landusepol.2020.105188
42. Acharya MR, Upadhyay P, Acharya A (2020) Assignment of functions across levels of government in NEPAL. The Asia Foundation, Kathmandu
43. MoUD (2016) Third United Nations Conference on Housing and Sustainable Urban Development (Habitat III): Nepal National Report. Ministry of Urban Development, Government of Nepal. https://moud.gov.np/storage/listies/July2019/HIII-Nepal-National-Report_Printed-version.pdf
44. MoUD (2017) National Urban Development Strategy 2017, Part A: Main document. Ministry of Urban Development, Government of Nepal
45. MoFE (2019b) Local Adaptation Plan of Action framework. Ministry of Forest and Environment, Government of Nepal (in Nepali)
46. CDSG Project (2020) A guidebook for preparation of local disaster and climate resilient plan. Deutsche Gesellschaft für Internationale Zusammenarbeit (GIZ) GmbH
47. Mercy Corps Nepal (2019) Key learnings and recommendations for creating local disaster and climate resilience plans (LDCRP): Experiences of Mercy Corps Nepal on local disaster risk management. Flood Resilience Alliance. Retrieved from https://dpnet.org.np/public/resource-detail/622
48. ICF (2023a) Butwal: Building a green, climate-resilient and livable city (2023–2030). SUDRIDH- Nepal Urban Resilience Project. Retrieved from https://sudridh-nurp.com/product/green-city-action-plan-2023-2030-butwal-full-report
49. ICF (2023b) Janakpurdham: Building a green, climate-resilient and livable city (2023–2030). SUDRIDH- Nepal Urban Resilience Project. Retrieved from https://sudridh-nurp.com/product/green-city-action-plan-2023-2030-janakpurdham-full-report
50. Subedi A, Khan R, Singh R (2023). Nepalgunj Heat Action Plan 2023. Red Cross Red Crescent Climate Centre

51. Robertson T (2021) Kathmandu's 'flash floods' are 4 decades in the making. Nepali Times (31 July). Retrieved from https://nepalitimes.com/here-now/kathmandus-flash-floods-are-4-decades-in-the-making
52. Uprety L P (2021) Land related policies and laws in Nepal: A discussion paper. Kathmandu, Nepal: Community Self-reliance Centre, International Land Coalition, and NES-Nepal.
53. Hada C, Shaw R (2022) Integrating disaster and climate change in risk-sensitive land use planning. In Shaw R (ed) Handbook on climate change and disasters. Edward Elgar Publishing. https://doi.org/10.4337/9781800371613
54. Bhandari D, Neupane S, Hayes P, Regmi B, Marker P (2020) Disaster risk reduction and management in Nepal: Delineation of roles and responsibilities. Oxford Policy Management Limited
55. Acharya KK, Zafarullah H (2020) Institutionalising federalism in Nepal: Operationalising obstacles, procrastinated progress. Public Adm Policy: Asia-Pac J 23(2):125–139. https://doi.org/10.1108/PAP-03-2020-0013

Chapter 13
Challenges of Implementing Urban Ecosystem-Based Adaptation to Climate Change in Developing Countries: A Systematic Review

Douglas Nyathi, Joram Ndlovu, and Samukele Hadebe

Abstract The contribution of urban areas to climate change is significant and accounts for approximately 70% of greenhouse gas emissions linked to energy consumption. Many people reside in urban areas, where the effects of global warming are starting to be felt. Urban areas possess the ability and duty to protect their citizens from the consequences of global warming while fostering resilient livelihoods and promoting the well-being of individuals, both mentally and physically. The research aims to explore the challenges associated with implementing EbA in urban settings in response to climate change. The opportunity to implement urban EbA practices to meet essential human needs and offer ecosystem services, such as recreation areas, stable food sources, cleaner air and water, and economic opportunities in urban green spaces, occurs when infrastructure is built, renovated, or updated. Nevertheless, there are several obstacles to adopting EbA, including a lack of government backing, inadequate institutional capacities, limited finances, a lack of decentralisation, and low levels of public participation. Although there is no one right way to solve complicated sustainability difficulties, it is critical to first determine the context and specific development challenge before considering the range of viable adaptation choices. Increasingly, multi-pronged approaches are being advocated. The chapter concludes that EbA is a more comprehensive adaptation strategy which should be included in planning, decision-making, and execution at all levels, from local to national. Helping communities develop urban EbA solutions to adapt and to make their neighbourhoods more liveable and climate resilient is vital for a sustainable future.

D. Nyathi (✉)
School of Public Management, Governance and Public Policy, University of Johannesburg, Gauteng, South Africa
e-mail: dnyathi@uj.ac.za

J. Ndlovu
School of Social Sciences, University of KwaZulu Natal, Howard College, Durban, South Africa

S. Hadebe
SARCHI Chair: African Diplomacy and Foreign Policy, Faculty of Humanities, University of Johannesburg and Department of Sociology, Rhodes University, Grahamstown, South Africa

S. Ghosh et al. (eds.), *Cities of Tomorrow: Urban Resilience and Climate Change Preparedness*, Urban Sustainability, https://doi.org/10.1007/978-981-97-9658-8_13

Keywords Nature-based solutions · Climate change adaptation · Urban planning · Ecosystem services · Ecosystem-based adaptation

13.1 Introduction

Since 2015 one of the top five global hazards with the potential to have an impact on the entire world is the inability to address and adapt to the fast-changing global climate [40, 50]. To concentrate existing efforts to adapt to climate change, countries and international organisations require a clear direction based on field evidence [34]. The significance of Ecosystem-based adaption (EbA) must be acknowledged as countries throughout the world work to improve the effectiveness of various adaption measures in response to climate change [40]. EbA is a comprehensive adaptation strategy that makes use of biodiversity and ecosystem services to assist humans in coping with the negative impacts of climate change [8, 40, 50]. EbA is a type of nature-based solution that aims to improve natural systems to increase biodiversity and help address societal concerns [39]. It is a well-established and widely acknowledged approach that effectively addresses the interrelated problems of poverty and climate change particularly in developing nations [22]. This approach is especially pertinent to areas where the economic and nutritional well-being of populations is greatly dependent on natural resources [34]. The United Nations Environment Assembly [47], the Intergovernmental Panel on Climate Change [20], and the Sendai Framework for Disaster Risk Reduction [44], all have supported techniques based on ecosystems for adaptation and initiatives for reducing disaster risks. An increasing number of nations and institutions are incorporating ecosystem-based adaptation (EbA) into their climate change development plans [22, 39, 50]. There are about 141 countries whose intended nationally determined contributions (INDCs) contain adaptation measures, and 49% of them refer to ecosystem-based approaches [34].

EbA receives a limited portion of adaptation funding in comparison to physical infrastructure solutions and is not consistently or widely implemented [36]. It is not mainstreamed sufficiently into policy processes both nationally and internationally [9, 34]. Several factors contribute to this situation, including the lack of clarity in financing for EbA, the discrepancy between immediate governance choices and the enduring effects of climate change, the intricate governance challenges stemming from the diverse and interconnected nature of EbA across sectors and scales, the challenges in addressing uncertainties related to climate and ecosystems, and the scarcity or disjointed nature of evidence regarding the success of EbA [31]. It is not usually clear or disclosed how much EbA activities cost, how difficult they are, or what causes the drawbacks. Comparably, little is known about the best means to put implementation strategies into practice, the extent of application that will maximise benefits, and the point at which ecosystems can no longer support adaptation to specific risks [15, 36]. Academic discussions frequently focus on the ecological, social, and economic impacts of EbA projects in comparison to alternative strategies. To assess the efficacy of EbA activities, these conversations highlight the significance of solid quantitative evidence or, at the absolute least, regularly gathered qualitative data [29, 40, 49].

13.2 Ecosystem-Based Adaptation: Conceptual Note

The concept of ecosystem services has seen a growth in adoption recently [18]. Ecosystem services refer to how natural ecosystems and their inhabitants support and enable human survival [11, 40]. The three primary components of the analytical framework include mitigating disaster risks, adapting to climate change, and delivering ecosystem services. They can be categorized into three primary groups: (1) offering services such as fuel, food, and fibre, (2) managing services such as local and global-scale water purification and climate regulation; and (3) sustaining services like water circulation and biodiversity [27]. An EbA strategy incorporates biodiversity and ecosystem services into a comprehensive strategy to assist individuals in adapting to the adverse impacts of climate change [34]. It comprises preserving, restoring, and managing ecosystems sustainably to assist communities and make them more resilient. Utilizing ecosystem services and functions is a crucial element of enhancing communities' and ecosystems' ability to adapt to the repercussions of climate change within this adaptation strategy [21, 36]. EbA encourages the preservation of biodiversity to improve ecosystem resilience and provide various advantages such as maintaining biodiversity, control over water flow, and defence against natural disasters [45]. Ecosystem-based adaptation (EbA) and the Convention on Biological Diversity (CBD) are closely related since they both place a high priority on biodiversity conservation and sustainable ecosystem management. The goal of the CBD is to ensure the equitable and fair sharing of benefits derived from genetic resources, advocate for the responsible utilization of natural resources and the preservation of biodiversity. EbA supports human adaptation to climate change by showcasing the pivotal role that biodiversity plays in enhancing ecosystem resilience and supporting the objectives of the CBD. It also promotes the use of Nature-Based Solutions for Adaptation (NBSA) and equitable conservation as well as sustainable resource utilisation. Lately, there has been a growing recognition of the significance of nature-centred approaches, like EbA, in addressing climate change adaptation efforts. EbA facilitates the repair and upkeep of natural habitats, which is critical for safeguarding ecosystem services and preserving biodiversity [34].

To strengthen coping strategies and recover from shocks brought on by climate change, urban EbA is essential to strengthen urban resilience, enhance urban biodiversity, offer ecosystem services, provide ecosystem services and involve communities [34]. Conversely, ecosystem services are thus how the ecosystem directly and indirectly contributes to human well-being and standard of living [2, 4, 36]. This includes the provision of food, and water, regulating climate, reducing green house gas concentrations in the atmosphere and cultural aspects such as reducing stress and anxiety. The view above is concurred by Dorkenoo [14] who asserts that wetlands play an essential role in reducing flood-related risks while forests have also favourable effects on humans such as reducing temperatures and sequestration of carbon dioxide to mitigate global warming. EbA emphasises the advantages that people receive from ecosystem services and biodiversity [25]. The strategy also seeks to determine how these advantages might be applied considering climate change. Although the idea of EbA is people-centred, it recognises that ecological integrity is vital to human resilience [14, 30, 50].

13.3 Urban Context: Risks and Vulnerability

Urban areas are densely populated areas with a concentration of resources and economic activity [30, 37]. Due to their high levels of economic activity, centralised infrastructure, and dense population, urban areas are especially susceptible to the impacts of climate change [36, 50]. Due to the increased frequency and intensity of extreme weather events, urban residents face heightened vulnerability to natural disasters which can cause fatalities, damage to infrastructure, and interruptions to vital services that include rising sea levels, storm surges, earthquakes, and floods [37]. Drought and excessive heat are further concerns associated with climate change [2, 17]. Living conditions that are too crowded, inadequate infrastructure and services, dangerous housing, poor nutrition, and ill health all increase one's exposure to risks [2]. A natural hazard or climate change can quickly become a disaster under these circumstances, with the loss of essential services, dwellings being damaged or destroyed, people losing their means of subsistence, starvation, illness, disability, and even death.

Low-lying infrastructure and coastal cities are at risk due to coastal erosion and rising sea levels [17]. Those in poverty are disproportionately impacted by natural disasters because of their urban living patterns and lack of reliable basic facilities [2]. During heatwaves, the urban heat island effect exacerbates heat-related diseases and mortality. Droughts and water scarcity can impact city sanitation and water supply systems. Poor air quality and air pollution hurt public health and exacerbate respiratory conditions [50]. Fundamental services from local governments are crucial to boosting the resilience of the urban poor. By integrating risk reduction into urban management, city leaders can increase resilience [2]. Modern urban designs that incorporate urban resilience (parks, green roofs, and urban forests are examples of urban green spaces that may be restored and preserved through EbA) and management strategies form the best strategy to address and sustain these efforts over time [4]. To make up for deficiencies in service delivery and the adaptation of basic infrastructure, local governments must make the most of both new and current resources [25]. In Latin America, there have been a lot of urban expansions over floodplains, up mountain slopes, and in other unsuitable locations for habitation, such as regions vulnerable to storms, flooding, sea surges, or other weather-related hazards [37]. These hazardous locations are typically inhabited by low-income people who are the most impacted by climate change [50].

Typically, the absence of infrastructure and services in these areas, together with their very vulnerable environmental conditions, has allowed low-income households to construct on them while keeping them undeveloped [2]. The poor typically lack official land tenure and are at risk of eviction in addition to environmental hazards. Without any other option, low-income groups live in cramped homes in densely populated neighbourhoods. However, plans for future road networks that would make it easier to establish utilities and regularise the area have frequently been included in the construction of informal communities [17]. In addition to heat waves, droughts, flooding, landslides, and overloaded water, drainage, and electrical supply systems, many people have become sensitive to climate change and the significance of encouraging urban climate resilience [25]. Consequently, severe weather poses a serious risk to the sizable urban population segment, yet these issues are frequently disregarded [17]. Those who live near hazardous locations and do not have the means or choices to lessen their susceptibility are especially vulnerable [25]. As a result of climate change, many urban residents especially the working class who do not have access to safe and adequate freshwater may have more difficulties. Climate change, thus adds another layer of stress to already vulnerable cities and populations, on top of the deficiencies in the coverage of water and sanitation, inadequate collection and treatment of solid and liquid waste, pollution, unemployment and poverty, low participation, and inadequate governance structures (including corruption) [4, 30].

13.4 Results and Analysis

13.4.1 The Case for Urban EbA to Climate Change

Essentially, EbA aims to leverage nature's strength to support ecosystems' resilience and improve humans' climate change adaptability ability [19, 43]. Several variables, such as the priorities of funders, the particular setting of the adaptation projects, and the availability of resources, might affect how much money is spent on ecosystem-based adaptation (EbA) as opposed to physical infrastructure for climate change adaptation. Local governments are essential in promoting comprehensive climate change adaptation since there are insufficient international responses to address its implications ([9, 40, 52]. Projects and efforts addressing climate change that seek to decrease greenhouse gas emissions enhance resilience to climate effects, and shift to a low-carbon and more sustainable economy are gaining support from both academics and practitioners [19, 35].

The strategies cover a wide range of topics, such as restoring mangroves to prevent storm surges, managing watersheds to prevent floods and droughts, managing rangelands to stop desertification, managing fish stocks, encouraging sustainable habits and preservation and restoration of ecosystems to guarantee food security [32]. Thus, the importance of healthy ecosystems for poor communities is becoming more recognised by the global legal frameworks on biodiversity and climate change as more vulnerable and impoverished people are impacted negatively by climate change [16, 48]. Although there are disagreements regarding its limits, the fact that ecosystems offer vital services for survival has drawn the interest of policymakers and communities globally, increasing the importance of environmental preservation on conservation agendas [38, 41]. Similar to how the ecosystem services framework can be extended, EbA can inspire action to combat climate change and recognise the contribution of nature to fight poverty, promote social adaptation and achieve sustainable Development Goals. For instance, in north-eastern Nigeria's Sahel savannah, the adverse impacts of upstream dams constructed for agricultural water supply during dry spells and restoration of Hadejia-Nguru wetlands have reduced the area's natural flooding cycle [3].

Even though the foundation of EbA's ecosystem services model acknowledges the interdependencies between humans and the natural world, it essentially takes an anthropocentric, utilitari an view of the connection between ecology and the well-being of humans [41, 42]. The ensuing debates about the importance of ecosystem services frameworks show the existing challenges in balancing anthropocentric and eco-centric perspectives on ecosystem management and biodiversity preservation in law, science, policy, and practice. When it comes to adaptation funding, EbA currently receives a lower share than hard infrastructure options. Although there are several positive experiences from throughout the globe, a strong theoretical argument, and an accepted multitude of additional advantages, there is no consistent application of EbA, nor is it being appropriately applied to policy processes both nationally and internationally [9].

13.4.2 Hurdles to Ecosystem-Based Adaptation Initiatives in Urban Areas

13.4.2.1 Insufficient Political Support

The political will is fundamental in driving the adoption of ecosystem-based adaptation initiatives. Without political support, it is difficult to implement effective ecosystem-based adaptation. Bhattarai et al. [5] posit that political support is often needed to secure funding, pass legislation and ensure compliance with regulations. Lack of political will poses serious challenges in the acquisition of necessary resources to implement adaptation measures. Hence, achieving long-term objectives in the implementation of ecosystem-based adaptation becomes a nightmare. According to Reid et al. [34], agencies are not mandated to work on EbA since it is not frequently given priority by national and local governments. When they detect conflicts, National governments frequently prioritise economic growth and industries that are less compatible with EbA, including mining or intensive agriculture. For instance, when it comes to their economic impact on Zimbabwe, the mining industry consistently receives greater recognition than other industries. However, the government typically provides so little support for EbA initiatives, like wetland conservation to improve adaptability to climate change, when ecosystems like wetlands are destroyed; no action is taken. Numerous studies have revealed that the government typically offers little assistance to devolve governance mechanisms and ensure that residents or indigenous peoples participate in decision-making [34].

13.4.2.2 Weak Institutions of Government and Poor Governance

One of the main concerns mentioned is weak governance, which makes implementation challenging due to poor local government leadership and institutions [33]. For instance, some countries have no legal authority for government representatives to enforce fishing laws and regulations. Higher government levels may also lack the capacity to include EbA in national adaptation policies and planning procedures [34]. Lax grazing law enforcement in Kenya, lack of government backing for mining in Peru, and illegal extraction of natural resources in Bangladesh have been entirely facilitated through corruption and all the other factors that have hampered the implementation of the EbA [34, 46]. Similarly, in Zimbabwe, the Mines and Minerals Act tends to be given priority against EbA initiatives. Weak traditional leadership and community groups may not have the technical know-how to execute EbA. Research carried out in Bangladesh reveals that the main issues with the local government's implementation of the incentive-based Hilsa conservation strategy are a lack of technical expertise, manpower, financial resources, and logistical assistance [33].

13.4.2.3 Lack of Decentralization

The failure to decentralise and lack of bottom-up procedure shave paused a major challenge to ecosystem-based adaptation. Local authority and programme-implementing partners in Bangladesh seem to have limited authority in formulating adaptation strategies at the community level [33]. For fisher community associations to effectively manage and conserve fisheries, programme implementing partners have advised that they should be included in the process and given the power to influence policy direction. According to Reid et al. [33], evolution is rare in Bangladesh and local voices now face significant obstacles in their attempts to influence policy and planning at higher levels. In response, Dorkenoo [14] contends that more decentralised governance structures could help in addressing some issues impeding the program's advancement, including bribery of local law enforcement officials, illegal fishing, and compliance with fishery regulations. Devolution of authority can also save transaction and administrative costs, raise regulatory awareness, and strengthen institutional frameworks. Reid et al. [33] provide more evidence to support the claim that the top-down management strategy used to build sanctuaries appears to ignore community involvement. This strategy has to be modified because fishermen's participation is essential to sanctuaries' success. Therefore, the top-down approach to policy and decision-making has affected the successful implementation of EbA to climate change. Stakeholders like the general public are often excluded in decision-making to climate change adaptation yet they possess rich indigenous knowledge that can be harnessed to steer ecosystem-based adaptation [38]. The general population does not have the authority to take the necessary actions to guarantee that the ecosystem-based adaptation approach is implemented effectively [2].

To effectively execute ecosystem-based adaptation solutions, it is necessary to support stakeholder ideas [5, 38]. Similarly, in Bangladesh, it was difficult to implement EbA due to a lack of support for the bottom-up strategy and decentralisation. However, in Cambodia, the government emphasises giving local levels of decision-making authority which presents a chance to improve community involvement and establish a system of local rights crucial to their lives allowing them to access all resources and services inside protected areas. Nonetheless, continued encroachment into protected areas and the destruction of natural assets, such as through illegal timber harvesting, are caused by the lax application of the law and the contradiction between official laws and customary standards.

13.4.2.4 Insufficient Funding and Collaboration

Overall, US$ 30 billion, or 12% of all public climate money worldwide, was allocated to adaptation in 2018. An additional US$12 billion was allocated for initiatives that would benefit both mitigation and adaptation. Development finance institutions provide the majority of adaptation-specific funding, or roughly 23 billion per year, or 79% of the total [7, 26]. Despite recent increases, funding is still limited, especially for EbA. Drawing from the latest OECD statistics, the WRI study demonstrates that ODA backing for NBSA increased from USD 2.1 to 4.1 billion (inclusive of lower and higher estimate ranges) in 2012 and from USD 3.8 to 8.7 billion in 2018. However, this only represented 1.5% of public climate finance allocations. Generally, there is 0.6% of all climate flows, and 9% of public adaptation financing globally. The UNEP Adaptation Gap 2020 report indicates that initiatives that incorporated components relevant to adaptation-related NbS or collectively secured 13% of funding from four climate funds, namely: the Green Climate Fund (GCF), Global Environment Facility (GEF), Adaptation Fund (AF), and Germany's International Climate Initiative (IKI) [26]. In comparison to other adaptation initiatives, particularly engineering-based infrastructure projects for adaptation, EbA has reportedly had difficulty attracting investment [10]. EbA usually involves several sectors and calls for cooperation between various governmental levels. However, governments are often sector-specific, and political rivalry or instability can make cooperation difficult [39]. Zimbabwe is an excellent example, where stakeholders frequently view citizen engagement as inadequate in the areas where decisions are made and local government agencies or technical services frequently operate independently from one another.

Similar to this, governance is frequently divided and compartmentalised at the provincial or regional government levels [34]. Nationally, there is frequently insufficient cooperation amongst the agencies in charge of disaster assistance, prevention, and climate change. There are several regions where transboundary collaboration has to be improved. For instance, in Zimbabwe, the Matobo Hills are a part of the Limpopo River basin, which greatly enhances the region's water security. National and international policy processes do not sufficiently mainstream ecosystem-based adaptation, nor is it being applied broadly or consistently which is another serious challenge. Currently, ecosystem-based adaptation receives a smaller share of adaptation money than approaches involving hard infrastructure [26, 33]. EbA is a multipurpose initiative that calls for many cross-sectoral interventions, ideally through longer-term initiatives. Due to the necessity of involving numerous parties, transaction costs may be substantial. In finance arrangements where targeted, short-term initiatives are frequently preferred, such variables may serve as disincentives.

13.4.2.5 Knowledge Gaps and Inadequate Knowledge Sharing

Knowledge gaps and inadequate sharing of knowledge are some of the challenges facing the advancement of EbA. There has not been much awareness among the public about the advantages of environmental preservation and EbA [34]. Numerous reports have indicated that a more robust scientific evidence base is required for economic cost–benefit analysis [34], especially for quantitative socioeconomic assessments. Such gaps in information might make it challenging to monitor, evaluate, and get solid impact data. However, it can be difficult to fully quantify the entire spectrum of social and economic advantages resulting from EbA, and these benefits are frequently underestimated [39]. The body of studies compiling the EbA evidence foundation is lacking. This has inhibited academics and practitioners from addressing specific problems about EbA and raised doubts about its efficacy compared to other strategies [28]. Further investigation that is guided by those required to implement EbA is required. A notable deficiency in knowledge pertains to a more comprehensive comprehension of ecosystems capacity to maintain adaption benefits in light of climate change [29]. Biological knowledge and accurate data on ecosystem hazards are frequently neglected, especially when it comes to delivering the environmental services required by humans (such as fuelwood in rural South Africa). EbA is not possible if the very ecosystems that underlie this approach to adaptation are at risk from climate change or high levels of utilisation [23, 28].

13.4.2.6 Social Equity, Local Knowledge and Participation

Social issues can significantly impact the implementation of successful EbA initiatives. These issues include social equity, local knowledge and participation, and social norms and values [25]. To overcome these obstacles, tactics that encourage diversity, involving local communities, and incorporating traditional knowledge into decision-making procedures are needed [50]. Social equality is a social problem that can impact how EbA programmes are implemented. When resources and power are not distributed fairly, marginalised communities and vulnerable groups like women and children may have less access to resources and be disproportionately affected by the outcomes of climate change and adaptation measures. Nevertheless, the best way of achieving EbA is through participatory approaches, engagement with local communities, and the establishment of inclusive governance structures. This will go a long way in ensuring that the vulnerable and marginalised groups' needs and viewpoints are included in the decision-making processes [25].

Participation and local knowledge are significant social aspects that affect how EbA activities are carried out. Local populations are deeply aware of the changes taking place in their surroundings and retain invaluable traditional knowledge about their ecosystems. Local knowledge also goes hand in glove with social norms and values as they significantly shape individual and collective behaviours related to environmental conservation and adaptation. Cultural values, beliefs, and norms can influence the acceptance and adoption of EbA measures. Increasing awareness of the significance of EbA and its conformity to local cultural values is crucial for overcoming obstacles to social norms and values [43]. This can be done through community education programs, cultural events, and the involvement of local leaders and influencers [6] in promoting EbA initiatives. For instance, integrating traditional ecological knowledge and local engagement has been essential to the implementation of EbA programmes in the Solomon Islands, northeast of Australia. Local communities have been actively involved in mangrove restoration projects, utilizing their traditional knowledge to guide decision-making processes and ensure initiatives' sustainability [1, 5, 26].

13.4.3 Insufficient Legal and Policy Backing for EbA

While numerous countries have established policies and strategies to address climate change and reduce natural disasters, EbA is frequently not well integrated into these frameworks, so governmental support is seldom and frequently minimal [23, 34, 46]. For instance, while numerous countries have established policies and strategies to address climate change, there is acknowledgement and support for EbA over and above the plans for mitigating climate change in the fishing industry [46]. Other places may also have poor policy and legal backing, which could undermine EbA. These domains include the usage and extraction of water as well as the funding of ecosystem services initiatives [34, 46]. Government subsidies for mining, deforestation, and intensive agriculture can encourage these activities, thereby compromising EbA [39]. Stakeholders in several nations believed that government actions undermined local agencies, such as when there is restricted devolution and decentralised governance or neglected to uphold user rights on common land [34]. Furthermore, policies may be top-down and inappropriate for the situation in a certain area.

Ecosystems and the services they offer may not receive the essential protection from a weak or badly written legal framework, leaving them more vulnerable to the negative effects of climate change [24, 29, 51]. Strategies for effective climate change adaptation must be a part of the country's climate action policy. A cross-sectoral legal framework must be created to ensure that various industries, including forestry, energy, water, and agriculture, work together to put EbA plans into action to lessen the adverse effects of climate change [5, 26, 29]. Both Cape Town and Durban need a strong, well-designed regulatory and legislative framework that is backed by sufficient funding and long-term planning to provide for efficient governance and strategic direction in the coastal management sectors. Furthermore, there is a lack of information on jurisdictional clarity and relevant institutional arrangements between municipal, provincial and national governments. While Cambodia's NAPA identifies several EbA-focused projects, the absence of integration and deficiencies in the larger legal and legislative framework on climate change hinder the advancement of successful EbA [24]. To be successful, EbA must promote the significance of ecosystems in lowering climate risks, and consult and coordinate a variety of government agencies and other stakeholders [5]. For example, the approach to the EbA Project to Maintain Water Security in Critical Water Catchment areas of Mongolia was financed by the Adaptation Fund and the UNDP. The project promoted the establishment of river basin councils, which improved district activity coordination at the basin scale, and strengthened public servants' capacity to promote ecosystem-based adaption options [23].

13.4.3.1 High Levels of Poverty and Poor Infrastructure

Numerous obstacles brought about by poverty can prevent EbA projects from being implemented successfully and sustainably [40]. Social vulnerability, such as small social networks, a lack of institutional support, and marginalisation, is frequently linked to poverty [22]. These can make it challenging for impoverished communities to mobilize resources, advocate for their needs and effectively engage in decision-making processes related to ecosystem-based adaptation initiatives. Due to high unemployment rates and poverty, impoverished people rely significantly on ecosystem services for their livelihoods, including agriculture, fishing, and the gathering of non-timber forest products [22]. Their income source may be directly impacted by climate change effects like biodiversity loss or coastal erosion, which would increase poverty and make it more difficult for them to fund ecosystem-based adaptation projects [5, 7]. For example, limited transportation networks made it difficult to reach marketplaces, high rates of illiteracy and unemployment, intermittent cell phone coverage, and insufficient water supplies were all considered to reduce adaptive ability [39, 43]. Some believed that being heavily indebted made it harder to follow regulations about the exploitation of natural resources [34]. Lack of funding for environmental benefit assessments (EbA) at the municipal, regional, and national levels might impede the implementation of EbA even in the presence of plans and policies that prioritise environmental justice [39]. Although some outside funders fund EbA, they might not do so through official government channels or bodies [5, 34]. This can undermine nationally determined adaptation priorities.

13.4.3.2 Increasing the Scale of Ecosystem-Based Urban Climate Change Adaptation

The adoption of EbA can be aided by the national, provincial, and regional governments prioritising climate change and EbA. Numerous nations such as Chile's Ministry of Environment's Climate Change office have established national-level organisations specifically focused on addressing climate change [22, 34]. For instance, the government of Nepal has increased funding to address climate change and ordered that 80% of the funding for adaptation be sent to local levels [23, 34, 43, 46]. Additionally, efforts have been made by other nations to allocate greater financial resources to combating climate change. Protection of the environment, reducing the risk of disasters, providing services, creating jobs, and reducing poverty are usually the domain of local government entities. They routinely work together across departments to guarantee the cross-sector cooperation necessary for EbA [34]. However, local government bylaws and institutions frequently support the implementation of EbA. Effective support for EbA implementation can be provided by local governments when they possess this degree of capability [22].

In several countries such as Nepal, Kenya and South Africa, the national and subnational climate change policies have evolved and they have the potential to support EbA [46]. For instance, Nepal developed a National Framework for Local Adaptation Plans for action, Kenya's legal and constitutional framework permits county and local government organisations to allocate funds and make adaptation plans; and the legislative framework and environmental governance laws of South Africa promote EbA. Some non-climate change-related policies were crucial in enabling EbA, especially those that favoured decentralisation. Other noteworthy ones are those that manage and conserve forests and watersheds, acknowledge the land rights of traditional communities, and protect indigenous knowledge systems in Peru [34, 46]. Examples of how knowledge creation and sharing help the establishment and implementation of EbA in China and Peru, meetings between and amongst farmers, participatory plant breeding, and the significance of exchange trips in China, Burkina Faso, Senegal, and Uganda [34, 46]. Combining local and scientific knowledge is also important for facilitating EbA implementation [22, 34, 46, 51]. For instance, the initiative in China frequently conducted on-site research in association with academic institutions like the Yunnan Academy of Agricultural Science and the Guangxi Maize Research Institute [34, 46].

13.5 Discussion and Implications

Therefore, carefully considered adaptation strategies have the potential to improve ecosystem resilience. Enhancing ecosystem services management can boost both adaptation and mitigation efforts, particularly when a favourable correlation exists between biodiversity levels, hydrological services, and carbon spatial distributions [24]. For instance, mangrove regeneration and conservation work together to store significant amounts of carbon while safeguarding coastal habitats [13]. Within the battle against climate change, initiatives like REDD+ (Reducing Emissions from Deforestation and Forest Degradation) are instrumental in placing a high priority on the preservation and sustainable management of forests. Forests act as carbon sinks, capturing and retaining substantial quantities of atmospheric carbon dioxide. However, by raising greenhouse gas emissions, deforestation and forest degradation worsen climate change (World Bank 2010). By sheltering people from hydrological changes and offering more locally regulated ecosystem services, local livelihoods and resilience can be improved. In addition to reducing human pressure on forests, REDD+ projects can enhance climate change adaptability, enhance connection among forest zones and safeguard biodiversity hotspots [24] Considering that climate change is frequently within the purview of weak environmental ministries, getting the required cross-government support for EbA is challenging [23]. Since environmental laws are frequently sparse, EbA lacks the legal support necessary to produce the desired results. Furthermore, even in instances when EbA is bolstered by diverse policies and initiatives, such as in South Africa [12] and Peru [34, 46], often, human resources are not adequately available for its implementation. One notable skill shortage is in EbA monitoring and evaluation; another typical issue is government employee turnover [34, 51].

The power to make decisions at higher levels of government remains concentrated in central government in the absence of decentralisation. This may lead to a lack of ownership and involvement in the conception, creation, and execution of ecosystem-based adaptation projects. Local communities and stakeholders may feel excluded from key decisions, leading to a lack of understanding, buy-in and commitment to the initiatives. By redefining public works initiatives in South Africa in light of the EbA, it is possible to make the programmes' current components comply with the requirements of the EbA, strengthen their resilience to climate change uncertainty, and possibly open up new funding avenues. One possible solution to this would be to simplify the science behind EbA so that policymakers could grasp it better and emphasise EbA more in the national curriculum and higher education [34]. EbA efforts can be made more effective and sustainable by utilising local expertise, norms, values and decision-making processes. It is possible to promote ownership and guarantee that adaptation measures are locally relevant and context-appropriate by involving local communities in the planning, implementation, and monitoring processes [6, 51]. Policies and strategies regarding disaster management and climate change do exist, but they are fragmented, with the latter focusing mostly on emergency relief after a disaster and showing little integration with national or international climate change policies [24, 43, 51]. The possibility of EbA in facilitating climate change adaptation is significantly limited by poverty and inadequate infrastructure [34, 39, 40]. Encouraging and bolstering regional planning procedures and organisations is also critical to enabling successful EbA implementation [34]. In some cases, this means creating entirely new groups, such as climate change planning committees, established collective governance bodies, and local risk reduction committees.

13.6 Conclusion

In conclusion, the EbA method has helped communities that are more susceptible to the harmful effects of climate change, which has been praised for improving resilience, increasing adaptive capacity, and lowering vulnerability. Ecosystems and biodiversity are essential to life as a whole, and as such, they are essential elements in building cities that are resilient enough to endure the negative consequences of climate change. Scientists dealing with climate change and policy officials have supported the implementation of EbA approaches ever since they gained traction in recent years. Consequently, the strategy offers ecological advantages since it encourages biodiversity ecosystem management and sustainable conservation. Nevertheless, the above-mentioned social, economic, institutional, and political considerations have made it difficult to implement ecosystem-based adaptation strategies effectively. In addition to providing long-term support for human adaptation, sustainably restored and preserved ecosystems can act as a first line of defence against the immediate effects of global warming. The threat posed by climate change on natural ecosystems is becoming more widely acknowledged globally. As more is learned about climate change, particularly in underdeveloped countries, ecosystems are becoming increasingly important in shaping policy. Recent years have shown the significance of using natural remedies, including EbA, in the fight against climate change adaptation. Sometimes action is undermined because intentions are rarely translated into solid, well-informed, and quantifiable targets. Even though signatories to the Paris Agreement update their Nationally Determined Contributions on an annual basis it is important to remember that adopting natural solutions is essential in achieving the global targets for mitigating climate change and sustainable development.

13.7 Recommendations

Therefore, in determining the goals that can benefit people and the ecosystems they depend on, officials must collaborate with scientists. Therefore, EbA should be integrated into a more comprehensive adaptation strategy by involving a variety of stakeholders in decision-making, planning, and implementation at all levels, as well as by searching for ways to build on and work with existing resources.

- Promote ownership and guarantee that adaptation measures are locally relevant and context-appropriate by involving local communities in the planning, implementation, and monitoring processes. Shelter people from hydrological changes and offer more locally regulated ecosystem services. This may involve participatory planning, community workshops, and engaging with residents to ensure that projects align with their needs such as improving funding mechanisms, enhancing governance structures, building technical capacity, and fostering community participation.
- Redefine public works initiatives to make the programmes' current components comply with the requirements of the EbA. Conduct training programs and workshops for local authorities, stakeholders, and community members to enhance their comprehension of ecosystem-based adaptation approaches as part of capacity building and knowledge sharing. This includes sharing best practices, case studies, and technical expertise.
- Encourage and bolstering regional planning procedures is critical to enabling successful EbA implementation by integrating urban ecosystem-based adaptation strategies and policies into current urban planning and development policies. This integration and alignment ensure that adaptation efforts are in line with broader sustainability objectives and are coherent with national climate change policies.
- Invest in the establishment of green infrastructure initiatives like green roofs, urban forests, and permeable surfaces to bolster ecosystem resilience and climate adaptation. Green Infrastructure Development endeavours to enhance urban aesthetics and biodiversity.
- Establish comprehensive monitoring and evaluation frameworks to gauge the effectiveness of ecosystem-based adaptation measures. Regular monitoring aids in tracking progress, pinpointing obstacles, and adjusting strategies to evolving climate conditions and urban landscapes.

References

1. Adger WN, Barnett J, Ellemor H (2009) Unique and valued places at risk. In: Schneider SH, Rosencranz A, Mastrandrea M (eds) Climate change science and policy. Island, Washington DC
2. Ahmed SA, Diffenbaugh NS, Hertel TW (2009) 'Climate volatility deepens poverty vulnerability in developing countries. Environ Res Lett 4(3):034004. https://doi.org/10.1088/1748-9326/4/3/034004
3. Andrade C, Santos JA, Pinto JG and Corte-Real J (2011) Large-scale atmospheric dynamics of the wet winter 2009–2010 and its impact on hydrology in Portugal. Clim Res 46:29–41. https://doi.org/10.3354/cr00945
4. Avashia V, Garg A (2020) Evaluation of classification techniques for land use change mapping of Indian cities. J Indian Soc Remote Sens 48:877–908. https://doi.org/10.1007/s12524-020-01122-7
5. Bhattarai S, Regmi BR, Pant B, Uprety DR, Maraseni T (2021) Sustaining ecosystem-based adaptation: the lessons from policy and practices in Nepal. Land Use Policy 104:105391

6. Berkes F (2009) Evolution of co-management: Role of knowledge generation, bridging organizations and social learning. J Environ Manage 90(5):1692–1702
7. Buscher B, Dressler W, Fletcher R (2014) Environmental conservation in the neoliberal age. University of Arizona Press, Tucson. Available at https://doi.org/10.2307/j.ctt183pdh
8. CBD: Convention on Biological Diversity (2009) Connecting biodiversity and climate change mitigation and adaptation. Report of the Second Ad Hoc Technical Expert Group on Biodiversity and Climate Change. CBD Technical Series No. 41. Montreal, Canada
9. Chong J (2014) Ecosystem-based approaches to climate change adaptation: progress and challenges. Int Environ Agreements: Polit Law Econ 14(4):391–405
10. Cohen-Shacham E, Andrade A, Dalton J, Dudley N, Jones M, Kumar C, Walters G (2019) Core principles for successfully implementing and upscaling nature-based solutions. Environ Sci Policy 98:20–29
11. Daily GC (1997) Nature's services—societal dependence on natural ecosystems. Island Press, Washington, D.C., USA
12. De Villiers A (2013) Ecosystem-based adaptation to climate change in Namaqualand, South Africa: cost-effectiveness of rangeland rehabilitation for erosion control. Conservation South Africa, Cape Town
13. Donato DC, Kauffman JB, Murdiyarso D (2011) Mangroves are among the most carbon-rich forests in the tropics. Nat Geosci 4:293–297
14. Dorkenoo K (2015) Bridging gaps: investigating private sector financing of ecosystem-based adaptation to climate change. A closer look at the Greater Mekong Sub region. IIIEE Master thesis
15. Doswald N, Munroe R, Roe D, Giuliani A, Castelli I, Stephens J, Möller I, Spencer T, Vira B, Reid H (2014) Effectiveness of ecosystem-based approaches for adaptation: review of the evidence-base. Clim Dev 6(2):185–201. https://doi.org/10.1080/17565529.2013.867247
16. Dudley N, Mansourian S, Stolton, Suksawan S (2011) Do protected areas contribute to poverty reduction? Biodiversity 11:5–7
17. Fekete A, Subramanian S (2023) Multi-risks attributed to climate change and urbanization in East Africa: a bibliometric analysis of a science gap. Afr Geogr Rev 1–23. https://doi.org/10.1080/19376812.2023.2256696
18. Hubacek K, Kronenberg J (2013) Synthesizing different perspectives on the value of urban ecosystem services. Landsc Urban Plan 109(1):1–6. https://doi.org/10.1016/j.landurbplan.2012.10.010
19. IPCC (2014) Climate change 2014: Mitigation of climate change. Contribution of working group III to the fifth assessment report of the intergovernmental panel on climate change. Cambridge University Press, Cambridge, United Kingdom and New York, NY, USA
20. IPCC (2018) Global warming of 1.5 °C. An IPCC special report on the impacts of global warming of 1.5 °C above pre-industrial levels and related global greenhouse gas emission pathways, in the context of strengthening the global response to the threat of climate change, sustainable development, and efforts to eradicate poverty. In: Masson-Delmotte V, Zhai P, Pörtner H, Roberts D, Skea J, Shukla P, Pirani A, Moufouma-Okia W, Péan C, Pidcock R, Connors S, Matthews J, Chen Y, Zhou X, Gomis M, Lonnoy E, Maycock T, Tignor M, Waterfield T (eds) World Meteorological Organization, Geneva
21. IUCN and UNEP (2010) Ecosystem-based adaptation: a natural response to climate change. Retrieved from https://www.iucn.org/sites/dev/files/import/downloads/eba_natural_response_to_cc.pdf
22. Kyriakopoulos GL, Sebos I (2023) Enhancing climate neutrality and resilience through coordinated climate action: review of the synergies between mitigation and adaptation actions. Climate 11(5):105. https://doi.org/10.3390/cli11050105
23. Ladekjær Gravesen M, Funder M (2021) Nature-based solutions to development and climate change challenges: understanding ecosystem-based adaptation approaches (No. 2021: 09). DIIS Working Paper
24. Locatelli B (2015) Ecosystem services and climate change. In: Potschin M, Haines-Young R, Fish R, Kerry Turner R (eds) Routledge handbook of ecosystem services. Routledge, pp 481–490. ISBN: 978-1-13-802508-0. https://www.routledge.com/products/9781138025080

25. Marrengane N, Croese S (2021) Reframing the urban challenge in Africa: knowledge co-production from the south. Taylor & Francis. https://doi.org/10.4324/9781003008385
26. Magnan A, Dale TW (2020) Adaptation gap. Environ Sci Policy 89:357–364
27. Millennium Ecosystem Assessment (2005) Ecosystems and human well-being: synthesis. Island Press, Washington, DC
28. Munang R, Thiaw I, Alverson K, Mumba M, Liu J, Rivington M (2013) Climate change and ecosystem-based adaptation: a new pragmatic approach to buffering climate change impacts. Curr Opin Environ Sustain 5(1):67–71
29. Nalau J, Becken S, Mackey B (2018) Ecosystem-based adaptation: a review of the constraints. Environ Sci Policy 89:357–364
30. Nguyen TT, Pittock J, Nguyen BH (2017) Integration of ecosystem-based adaptation to climate change policies in Viet Nam. Clim Change 142(1–2):97–111
31. Ojea E (2015) Challenges for mainstreaming ecosystem-based adaptation into the international climate agenda. Curr Opin Environ Sustain 14:41–48
32. Parry ML, Canziani OF, Palutikof JP, van der Linden PJ and Hanson CE (2007) Climate change 2007: Impacts, adaptation and vulnerability, contribution of working group II to the fourth assessment report of IPCC, Cambridge University Press, Cambridge
33. Reid H, Scorgie S, Halcyone M, and Bourne A (2018) Ecosystem-based approaches to adaptation: strengthening the evidence and informing policy. Research results from the Climate Resilient Livestock production on communal lands project, South Africa. Available from: http://pubs.iied.org/17622IIED
34. Reid H, Hou Jones X, Porras I, Hicks C, Wicander S, Seddon N, Kapos V, Rizvi A R, Roe D (2019) Is ecosystem-based adaptation effective? perceptions and lessons learned from 13 project sites. IIED Research Report. IIED, London
35. Roberts D (2008) Post-conflict state building and state legitimacy: from negative to positive peace? Dev Change 39(4):537–555. https://doi.org/10.1111/j.1467-7660.2008.00495.x
36. Sircar A, Thomas A, Serdeczny O, Jattansingh S (2024) A review of loss and damage in the Caribbean (1994 to 2024). Clim Anal
37. Satyal P, Corbera E, Dawson N, Dhungana H, Maskey G (2020) Justice-related impacts and social differentiation dynamics in Nepal's REDD+ projects. For Policy Econ 117:102203
38. Sebesvari Z, Rodrigues S, Renaud F (2017) Mainstreaming ecosystem-based climate change adaptation into integrated water resources management in the Mekong region. Reg Environ Change 17:1907–1920
39. Seddon N, Daniels E, Davis R, Harris R, Hou-Jones X, Huq S, Kapos V, Mace G, Rizvi A, Reid H, Roe D, Wicander S (2019) Global recognition of the importance of nature-based solutions to climate change impacts. In press, Nature Climate Change
40. Sherpa TO (2024) Integration of urban ecosystem-based adaptation in Nepal: a policy. https://doi.org/10.1371/journal.pone.0297786
41. Skroch M, López-Hoffman L (2010) Saving nature under the big tent of ecosystem services: A response to Adams and Redford. Conserv Biol 24(1):325–327. http://www.jstor.org/stable/40419659
42. Spash C, Stagl S, Getzner M (2005) Exploring alternatives for environmental valuation. In: Spash C, Stagl S (eds) Alternatives for environmental valuation, London, Routledge, pp 1–27
43. Stergios-Aristoteles M, Dan VB, Sotirios A (2023) Sustainability and climate resilience metrics and trade-offs in transport infrastructure asset recovery. Transp Res Part D: Transp Environ 121:103800. https://doi.org/10.1016/j.trd.2023.103800
44. UN (2015) Sendai framework for disaster risk reduction: 2015–2030. Switzerland, Geneva
45. UNDP (2009) Human development report 2009. Overcoming barriers: human mobility and development. United Nations Development Programme, New York, NY
46. UNDP (2015) Making the case for ecosystem-based adaptation: the global mountain EbA programme in Nepal, Peru, and Uganda. New York
47. UNEA (2014) Proceedings of the United Nations Environment Assembly of the United Nations Environment Programme at its first session, Nairobi, 23–27 June 2014. UN, Nairobi

48. UNFCCC (2011) Durban climate change conference. November/December 2011. http://unfccc.int/meetings/durban_nov_2011/meeting/6245.php
49. UNFCCC (2017) Adaptation planning, implementation and evaluation addressing ecosystems and areas such as water resources. Synthesis report by the secretariat for the Subsidiary Body for Scientific and Technological Advice, 46th session, Bonn, 8–18 May 2017
50. UNFCCC (2022) Sharm El-Sheikh implementation plan. Available online: https://unfccc.int/sites/default/files/resource/cop27_auv_2_cover%20decision.pdf. Accessed on 18 April 2024
51. Vasseur L (2021) How ecosystem-based adaptation to climate change can help coastal communities through a participatory approach. Sustainability 13(4):234
52. Wu J (2014) The antecedents of corporate social and environmental irresponsibility. Corp Soc Responsib Environ Manage 21:286–300. https://doi.org/10.1002/csr.1335

Part V
Technological Innovations for Urban Resilience

Chapter 14
Developing a Climate Resilience Framework for Cities with an Emphasis on Innovative Technology Applications

Reza Mirzaei and Elnaz Lakizadeh

Abstract In the twenty-first century, climate change is one of the most challenging and threatening global issues. Therefore, it is necessary to pursue a multilateral approach to address both the inevitable causes and effects. Today, the future of cities' resilience against climate change lies in the innovative application of technologies, making them a sustainable and promising solution. Therefore, the current research takes a practical approach by gathering information from articles and documents to present the framework of climate resilience in cities. The focus is on the application of innovative technologies through the examination of concepts, dimensions, indicators, and experiences. Experiences are a fundamental pillar in research. Studies of Denmark, the United States of America, Singapore, Japan, and the Netherlands demonstrate that innovative technology has been utilized in the fields of water management, energy, beaches, buildings, and green systems. The application process also takes place in four steps: planning, data collection and processing, validation, and policy. The output of these steps includes identification, technological needs assessment, information base, environment with potential for technology application, capacity building for technology application, technology application mechanisms, and all actors. Innovative technologies can be considered a driving force in enhancing climate resilience. The climate resilience framework includes the goals of mechanisms (institutional framework and resilient innovative technology framework), processes, and outputs.

Keywords Climate change · Climate resilience · Innovative technology · Institutional framework · Climate resilience framework

R. Mirzaei (✉) · E. Lakizadeh
Faculty of Architecture and Urban Planning, Tehran University of Art, Tehran, Iran
e-mail: Eng.RezaMirzaeii@gmail.com

S. Ghosh et al. (eds.), *Cities of Tomorrow: Urban Resilience and Climate Change Preparedness*, Urban Sustainability, https://doi.org/10.1007/978-981-97-9658-8_14

14.1 Introduction

Industrialization and the resulting economic hardships, with the slogan of improving human societies' quality of life, were the main factors that initiated the world's exposure to climate change [1]. Climate change, environmental problems, and the disturbances caused by them, along with the increasing growth of urbanization and uncontrolled expansion of cities due to population increase, have become significant topics of discussion in urban studies circles. These issues also play a crucial role in influencing people's decisions [25]. The intersection of climate change and urbanization manifests in various ways, such as droughts, storms, floods, heat islands, air pollution, extreme heat, and cold, all of which have adverse effects on citizens. Therefore, the evidence of climate change has made it imperative to consider solutions to address the factors influencing climate change and, on the other hand, the adverse effects of climate change [12].

To address climate change, the concept of climate resilience has been at the forefront for planners and policymakers, becoming the central focus of future city policies. This approach aims to mitigate the impacts of climate change or adapt to the changing climate, ensuring that future generations can also experience a favorable climate quality similar to the present time [20]. Environmental systems consist of people who use or manage them, organizations, institutions, various elements, laws, and frameworks in use. Each element in this system reacts to risks based on its nature and the laws it operates under. This interaction is linked to the overall system's resilience and has an impact on it. Therefore, it can be said that the climate resilience of an environmental system as a whole depends on a set of capacities, institutional characteristics, factors, and organizations that influence the system during times of crisis [14]. Therefore, it can be said that climate change is a controversial concept, making it susceptible to engaging in discussions that encompass a wide array of new approaches, elements, actors, stakeholders, influential factors, and diverse measures in the realms of planning, decision-making, and policy-making for resilient and sustainable climate change management.

Cities are acknowledged as hubs that embody the dual nature of both harm and innovation in the crucial discourse on climate change. This places cities at the forefront of innovation in addressing environmental crises and challenges [4].

Innovation in the form of advanced technologies highlights the creative utilization of tools and applications to mitigate environmental damages and crises, offering a significant opportunity for efficient organizations and institutions. Therefore, technological innovations are pursued to solve problems related to specific issues such as the environment, and to exert control over nature in a way that may be considered inhumane through the application of innovations that cater to the needs of human societies in novel ways [8]. Perhaps, despite its many cumulative effects in various economic, social, and political dimensions, climate change can be considered as one of the sectors where technology has been underutilized. The importance of utilizing innovative technologies in this sector lies in the fact that these technologies are the outcome of the most recent knowledge related to climate change and are closely associated with the top scientific centers worldwide. Therefore, the significant improvement in the field of innovative technologies, which encompasses human resources, information, material, and organizational resources, in the context of climate change, leads to the emergence of new processes [9].

Despite the increasing number of crises and hazards targeting city residents worldwide, and the specific measures taken to prevent and address climate change in urban areas, there is a lack of research focusing on the use of innovative technologies to enhance climate resilience in cities. The scarcity of studies focusing on the utilization of innovative technologies in climate planning, decision-making, and policy-making is a subject of debate. This is primarily attributed to the neglect of management structures towards practical, procedural, and framework-based strategies for enhancing climate resilience. These studies typically concentrate solely on the negative and traumatic impacts of climate change on daily life and urban areas. They often delve into theoretical concepts more aligned with adaptation rather than resilience. The key distinction lies in their approaches: adaptation emphasizes absorption, avoidance, or anticipation, whereas resilience is centered on anticipation, coping, recovery, and progression. The studies in this field have been very limited, focusing solely on examining the climate resilience of cities based on existing urban management practices. These studies have primarily addressed prevention strategies related to physical protection and traditional programs. There is a lack of theoretical frameworks and conceptual tools for the identification of technological needs, assessment of information bases, environments with potential for technology application, capacity building for technology application, technology application mechanisms, and all actors. This limitation hinders the empirical understanding of various institutionalized modes of governance in terms of institutional characteristics, agents, and organizations, and how they influence the capacity to utilize technology in different circumstances.

Finally, the absence of an effective and efficient framework for applying innovative technologies to enhance the city's climate resilience is the missing link in contemporary urban management. Considering the importance and necessity of the innovative climate resilience framework of the city and addressing the existing gaps, this research meets the aforementioned need.

This study aims to establish an effective and efficient framework for applying innovative technologies to enhance the climate resilience of cities. This will be achieved through a practical approach supported by citations from relevant articles and documents. The text aims to answer a fundamental question: What is the most effective framework for enhancing the innovative climate resilience of cities? To answer this question and achieve the mentioned goal, the concepts, dimensions, and components of the city's climate resilience and innovative technologies are identified. This study aligns with the overarching theme of the book, focusing on the cities of the future: urban resilience and readiness for climate change.

14.2 Background Knowledge

14.2.1 Definitions and Concepts

In the following section, we will examine the definitions, concepts, and evolution of approaches to climate resilience.

14.2.1.1 City's Climatic Resilience

Climate changes occur either as a result of gradual events such as the slow and continuous rise in sea levels and droughts, or sudden shocks like heatwaves and tropical storms. These changes impact the health of the environmental system and other related systems, as well as the response capacity of each. Climate resilience systems consist of various infrastructures, institutions, actors, and activities aimed at predicting, reacting, confronting, recovering, adapting, and moving forward regarding climate change. These systems strengthen and enhance the ability of management in all stages of crisis absorption, adaptation, and transformation of shocks into success. In this way, the capacities of learning, adaptation, and transformation are maintained [24].

Climate resilience is a crucial factor in decision-making, resource allocation, and multilateral governance concerning climate change. It encompasses the quality of project design and outcomes, offering the opportunity to adopt a contextual and tailored approach to enhance weather and climate resilience. This involves addressing the need to adapt to different time scales while maintaining a proactive stance in the face of uncertainties related to future climate conditions. Therefore, in climate resilience, by identifying the gaps in policy and planning, as well as the problems, challenges, and information surrounding it, the necessary inputs such as financial resources and human resources are determined. Subsequently, appropriate activities are carried out. Finally, the expected outputs, results, and effects are achieved. This set constitutes a robust system that emerges from the convergence of adaptation and resilience, enabling it to withstand shocks or factors related to climate change [3].

Climate resilience involves absorbing disturbances caused by exposure to damage, which helps maintain the fundamental structure of environmental systems and their functionality. This process also enhances the capacity of these systems to fully or partially recover.

14.2.1.2 Innovative Technologies

Innovative technologies are among the contextual approaches mentioned in climate resilience. Innovative technologies are tools that consider the diverse impacts of climate change on economic, social, and environmental dimensions. They aim to mitigate the adverse effects and enhance environmental conditions. Considering the negative effects that human activities have on the climate, countries worldwide should prioritize enhancing innovation and adopting innovative technologies to address climate change. Therefore, investing in innovation to address this environmental concern is inevitable. Therefore, innovative technologies provide strong support for climate resilience, playing an effective role in transitioning from climate change and its adverse effects to sustainability [2].

Societies are influenced by their economic, social, and environmental factors, which vary across different national, regional, temporal, and spatial contexts. Innovative technologies offer operational solutions to adapt and mitigate threats and impacts. Innovative technologies, with their moderating role and managerial nature, have been proposed to eliminate harm to humans and the environment. In the intersection of climate resilience and innovative technologies, it is evident that the goal of integrating innovation in meteorology and climate science is to achieve precise, specialized, and ongoing forecasting of climate change. Innovative technologies in the field of climate are classified into three categories: measuring tools, tools for collecting data and transmitting forecasts, and tools for advanced imaging of the atmosphere and sending it to the ground, which leads to predictable and long-term policies [17].

14.2.2 Evolution of Approaches to Climate Change

Technology is one of the main pillars in the field of implementing measures to adapt to climate change. The start of innovative measures in the field of climate can be traced back to the 1990s, and between 1995 and 2015, the application of technological innovations intensified. Countries such as the United States, Western Europe, China, South Korea, and Japan are leaders in this field and are categorized as having strong technology to address climate change and mitigate its effects. In recent years, advancements at the intersection of climate and technology have been made in the fields of coastal, river, and marine environments, drought management, storm prediction, water resource management, infrastructure development, agriculture, and healthcare (see Table 14.1).

As depicted in the table above, climate change policies have become more comprehensive over time. Innovative technologies are now being utilized in the modern era to align measures for climate change prevention with modernization.

14.3 Methodology

This study employs a qualitative-applied approach to gather data from documents, with two specific goals in mind. The first goal is to review the conceptual literature on urban climate resilience and innovative technologies, as well as examine the experiences of various countries. In the second step, the aim is to identify weaknesses

Table. 14.1 Evolution of approaches to climate change

A variety of approaches	The reason for its origin	Goals	Potential impacts on climate resilience with an emphasis on risk mitigation technology
Sustainable development	Overcoming economic growth and excessive productivity of existing resources are essential to address the failure of previous agreements in maintaining current conditions for future generations	Creating economic, social, and environmental sustainability can act as a catalyst for improving the performance of executive bodies and motivating policy makers	*Negative effects*: turning sustainable development into theoretical issues and a lack of application in action *Positive effects*: By focusing on climate change and addressing the associated challenges, it paves the way for guiding the actions of different actors and stakeholders in the realm of environmental sustainability
Climate resilience	Vulnerability, hazards, and risks are the fundamental pillars and the starting point for building climate resilience and the necessity to implement it	Strengthening and stabilizing the link between vulnerability, risk, and climate resilience involves complementary and alternative approaches to risk assessment and management. It requires adopting a forward-looking perspective, fostering private mobilization, and utilizing public finance	*Negative effects*: may include creating a negative impression on policy makers by focusing solely on environmental protection without considering other dimensions *Positive effects*: Creating transformative and adaptive capacity compared to risk management, offering more assurance in the face of future uncertainty, reducing costly retrofits, disclosing climate risks, and implementing cost-effective approaches through innovation

(continued)

Table. 14.1 (continued)

A variety of approaches	The reason for its origin	Goals	Potential impacts on climate resilience with an emphasis on risk mitigation technology
Nature-based solutions	Complementing the approach of sustainable development and resilience is the need to strengthen resilience to ensure that future generations can enjoy the current climate quality	Enhancing resilience, enhancing the role of nature in climate change solutions, aligning resilience with sustainability to mitigate climate change, and implementing measures to address urban challenges and climate change	*Negative effects*: include the requirement of a lengthy period of time, the necessity of collaboration among various actors and stakeholders with diverse approaches, and dependence solely on social acceptance *Positive effects*: include improving resilience, developing personal norms and values, supporting the role of nature in adapting to climate change, affordability, and considering all economic, social, and environmental benefits

(continued)

Table. 14.1 (continued)

A variety of approaches	The reason for its origin	Goals	Potential impacts on climate resilience with an emphasis on risk mitigation technology
Market-based solutions	Limiting global warming involves implementing measures related to the carbon market, complementing and strengthening environmental resilience and sustainability, and aligning with nature-based solutions	Development of operational tools and creation of climate and blue carbon initiatives for the field of fisheries	*Negative effects*: Special attention focused solely on one aspect of the environment *Positive effects*: preventing harmful actions and activities and their destructive effects in economic, social, and biological dimensions, preventing the loss of marine biodiversity, and safeguarding ecosystem services and food supply
Innovative technology	The necessity of using new models and technologies to address the complex climate issues and challenges of urban areas, the importance of keeping up with modernization, exploring new environmentally-friendly approaches, enhancing the integration of information and communication technology for the efficient use of natural resources, and the imperative of employing smart technologies as part of environmental ethics	The advancement of technological solutions, changes in political, economic, and social structures, the development of smart network infrastructure integrated with environmental concepts, environmental modernization, and the collaboration of experts from various fields to create smart environmental infrastructure	*Negative effects*: The prevalence of non-democratic thinking that relies solely on technological solutions to address the complex issues associated with climate change in urban areas. This approach has limitations in offering solutions, poses risks and uncertainties in implementation, and raises concerns about potential outcomes and supply security *Positive effects*: Stimulating climate resilience, aligning technology and innovation with climate goals, creating sustainable cities, fostering alignment among producers, providers, planners, decision-makers, and policymakers, and offering sustainable services and goods

Source Authors adapted from: Montero [15], Subiyanto et al. [22], Quitzow and Rohde [19], Sari et al. [20], Prellezo et al. [18]

and needs by reviewing conceptual literature and existing public discourse, and then present them in the form of an innovative climate resilience framework.

14.4 Climate Resilience with a Focus on Innovative Technologies

The first section argues that the need for resilience in the world's cities stems from the fact that, now more than at any other time in history, climate change threatens cities and countries in the form of shocks. The ability of these regions to resist, recover, and move forward following these events depends on their climate resilience. On the other hand, climate resilience depends on a framework of diversity, technology, and organization. The second part discusses the innovative technologies that various countries have incorporated into their strategies for climate resilience. The third part argues that the integrated thinking of climate resilience, along with the application of innovative technologies, in the planning and policy-making of countries has not been institutionalized as it should be. The final section argues that developing a climate resilience framework can ensure a focus on incorporating innovative technologies into climate resilience policies within urban and national structures.

14.4.1 Innovative Technology: Driver of Climate Resilience

The existing literature in the field of theoretical and practical concepts of climate resilience depicts a complex and evolving context marked by emerging challenges. These concepts, which play a significant role in the development of climate resilience policies and the utilization of innovative technologies to promote them, are accompanied by a wide range of criticisms. This is because the primary foundation of the existing concepts is rooted in reverting to past lifestyles. These concepts are not compatible with the changing conditions as they have a traditional and short-term view.

The approach of the present study should be recognized and valued as a hybrid approach. This approach exerts a positive influence on planning, decision-making, policy formulation, and, most importantly, on enhancing climate resilience and promoting the adoption of innovative technologies on a broad scale through the pursuit of a climate resilience framework. Finally, it can be said that the criteria for climate resilience in cities and countries can be defined by the dimensions and components of innovative technologies. In the following, an integrated view of climate resilience and its driving force, i.e., innovative technologies, is presented. It states that the concept of climate resilience comprises mechanisms, goals, and enablers, with technology being one of the most crucial enablers. Table 14.2 examines the indicators and components of climate resilience and its driver, i.e., innovative technologies (see Table 14.2).

As shown in the table above, the dimensions and components associated with climate resilience often have a managerial nature. To utilize and implement them effectively, there is a requirement for infrastructures equipped with innovative technologies. In the following section, we will discuss global experiences related to the role of innovative technologies in climate resilience.

14.4.2 Rotterdam, Netherlands

Rotterdam has implemented comprehensive policies to enhance the adaptation and resilience of its infrastructure and governance structures to climate change. These policies address various dimensions, including governance arrangements, infrastructure, waterways, and climate-related challenges such as floods, rivers, and rainfall. The Netherlands is implementing the Rotterdam Climate Stabilization Program (see Table 14.3).

14.4.3 New York, USA

New York has implemented various policies to enhance adaptation and resilience to climate change. Two key programs, PlaNYC and OneNYC, focus on sustainability and resilience, addressing climate challenges like extreme cold, heat waves, coastal storms, energy, electricity, and air quality. The main goal is to protect cities, neighborhoods, economies, and services from the impacts of climate change. The Climate Innovation Center in New York is dedicated to combating climate change by fostering a green economy ecosystem to promote climate-focused strategies. This effort is typically characterized by dimensions like governance arrangements, involvement of diverse elements, and the shift towards a green economy (see Table 14.4).

14.4.4 Copenhagen, Denmark

Copenhagen has implemented various policies to promote adaptation to climate change, including governance arrangements, infrastructure improvements, and addressing climate-related issues such as energy and precipitation. Copenhagen is known as one of the most digitally advanced cities in the world, which has also embraced digitalization in the realm of climate. Copenhagen has advanced the Copenhagen Climate Plan and the CPH 2025 Climate Plan, focusing on sustainability (see Table 14.5).

Table. 14.2 Dimensions and components of climate resilience and innovative technologies

Dimensions and components	Innovative technologies	Climate resilience
Climate management (beaches, rivers, seas, lagoons, etc.)	Artificial reefs, beach nourishment, breakwaters, reef stabilization, coastal zone monitoring, dykes, dams, flood prevention, jetties, permanent flood barriers, coral reef protection/repair, flood discharge pump, seawalls, storm-proof vessels, temporary flood barriers, uninterruptible power supply, aquaculture technology	Watershed protection, coastal structures, and groundwater recharge
Infrastructure management	Computerized flood control, evacuation/floodplain containment systems, real-time meteorological measurements, resilient generators, flexible IT infrastructure, sediment management, storm shelters/undergrounds, climate databases/geographic modeling; weather monitoring systems, early warning systems	Systems infrastructure, technology, data, and information, resilience of settlements, capacity to utilize information, online platforms and tools, technical and organizational capacity, and capital and financial resources for resilient infrastructure
Management of facilities and equipment	Very weather-resistant electrical equipment for floating houses	Versatility, modularity, safety equipment

(continued)

Table. 14.2 (continued)

Dimensions and components	Innovative technologies	Climate resilience
Management of natural resources and the natural environment	Rebuilding or creating hills, implementing greenhouse technologies, selecting plant species compatible with the environment, and enhancing ventilation technology	The protection of forests and green areas is crucial for agriculture and food security
Administrative-institutional	Disaster preparedness plans, productivity (number of storm casualties), storm damage rate, availability of technology, storm-exposed population, and storm-exposed assets	Responsiveness (improving and strengthening supervision, facilitating access to early warning information), being resourceful (understanding capacities, paying attention to different parts of planning, paying attention to different income groups), learning (training in school and city programs); The right to access information, new organizations, limiting development plans in areas at risk, participation of different elements in the resilient planning process, improving warning services, climate coordination offices, promoting embodied communities, long-term and integrated planning; Private sector participation, monitoring and evaluation, financial management, policy cycle, multi-level governance; Information and awareness, efficiency and transparency, forward process, adaptive capacity, flexibility, integration, fairness, human resources, expression and valuation of uncertainties, public finance, adaptive management

Source Authors' studies

14.4.5 Singapore

Singapore has implemented policies to enhance adaptation and resilience to climate change, focusing on aspects like governance structures, infrastructure, and climate-related matters such as energy. Singapore has adopted the Singapore 2030 Green Plan and the SolarNova program (see Table 14.6).

Table. 14.3 Application of innovative technology to improve climate resilience in Rotterdam

The name of the technology	Field of application	Application	Equipment
Monitoring system	Flood	Water level control, providing early warning services	Sensor, camera
Intelligent networks	Energy	Optimizing energy consumption and enhancing resilience against power outages	ICT, IoT
Green systems	Green roof and wall	Improving air quality, reducing the effects of heat islands, and absorbing rainwater	Plant species suitable for the climate, infrastructure, and solar panels
A floating building	Floating offices	Optimal energy efficiency, utilization of water for heating and cooling in offices, sustainable and low-consumption buildings, promotion of climate-oriented approaches in urban construction, advocacy for environmentally friendly construction practices, mitigation of the ecological footprint of buildings, and reduction of greenhouse gases	Anchors, connections, smart air conditioners, solar cells

Source Gemeente Rotterdam [10], ICLEI [11]

14.4.6 Tokyo, Japan

Tokyo has implemented policies to enhance adaptation and flexibility in response to climate change. These policies often encompass governance structures, infrastructure development, and climate-related concerns like energy. In Tokyo, there is a campaign called the HTT campaign, which focuses on saving, production, and re-saving (See Table 14.7).

Examining the experiences of using innovative technologies to enhance climate resilience revealed that most of the experiences are in coastal management, water resources management, energy management, green infrastructure, and buildings. The process of applying innovative technologies in climate resilience can be considered to include four stages: planning, data collection and processing, validation, and making decisions and policies. As the review of indicators and experiences has shown, activists are one of the main pillars in enhancing climate resilience and addressing climate change. They play a crucial role in all stages from planning to policy-making, involving the government, private sector, and citizens. In the first stage, which is planning, the participation of the other two pillars, namely the private sector and civil societies, is required. The legal support from the public sector is essential for organizing a group of activists. At the round table, there will be a discussion about climate challenges and their effects. The output of the planning stage leads

Table. 14.4 Application of innovative technology to improve climate resilience in New York

The name of the technology	Field of application	Application	Equipment
FloodNet	Flood	Flood prediction, investigation, and control of rainfall and water levels	A network of sensors
Green systems	Green roofs, green infrastructure, green areas, and green building	Reducing the effects of heat islands	Plant species should be selected based on the climate, infrastructure, and panels, as well as the irrigation system and intelligent monitoring
Resilient infrastructure	Extreme climatic events	Preparation and installation of critical infrastructure for resilience and resistance to climatic events	All necessary infrastructure
Electric vehicle charging station	Transportation and charging of electric vehicles	Facilitating and institutionalizing the green economy	Public sites and lands, smart equipment, sensors, cameras, batteries
Clean path NY	Clean energy	Enhancing the city's electricity grid, decreasing suspended particle emissions, offsetting carbon emissions, linking communities to renewable energy sources such as wind and solar power, delivering clean electricity services, decreasing reliance on fossil fuels, and enhancing public–private collaboration	Underground transmission lines, operational and maintenance facilities, turbines, meteorological towers, and substations

Source City of New York [6, 7]

to the identification of challenges and the extraction of requirements. In the data collection and processing stage, which is the second phase of the process mentioned, a database is created to prioritize the identified challenges and share these priorities among a group of stakeholders. The output of this stage leads to the identification and needs assessment of the environment and the necessary capacities. In the next step, the validation of the results from the previous two steps is conducted. Finally, policies are formulated and implemented by a group of actors, and the mechanisms for application are determined for implementation. By following this process, it ensures absorption capacity, change capacity, and transformation capacity in various economic, social, physical, environmental, and organizational dimensions.

Table. 14.5 Application of innovative technology to improve climate resilience in Copenhagen

The name of the technology	Field of application	Application	Equipment
Smart management	Water/flood	Monitoring and controlling water resources, optimizing resource utilization, implementing flood prevention measures, and tracking the city's water consumption using diverse data sources	A sound leak sensor network, smart water meter, smart valves, smart pumps, and modeling software are essential components
Smart building	Municipal offices	Enhancing the quality of municipal buildings by incorporating green roofs and promoting urban design suitable for the climate	Sensor network

Source City guide to Copenhagen [5]

Table. 14.6 Application of innovative technology to promote climate resilience in Singapore

The name of the technology	Field of application	Application	Equipment
Innovative beaches management	Beaches	Beaches protection against sea level rise and storms, and providing early warning services	Seawalls, breakwaters, drainage systems, and early warning systems
Green building	Building	Paying attention to reducing water consumption and saving in buildings, utilizing low-consumption technologies, and incorporating solar photovoltaic systems in buildings due to limited space	Plant species should be selected based on the climate, infrastructure, and panels, as well as the irrigation system and intelligent monitoring

Source Singapore Green Plan 2030 [21], National Parks Board [16]

14.5 Discussion and Implications

Climate Resilience Framework with an Emphasis on the Application of Innovative Technologies

The framework of climate resilience, with a focus on innovative urban technologies, presents a conceptual and multi-layered map that outlines all the essential elements for implementing innovative technology to enhance and fortify cities' climate resilience against climate change. In this framework, the negative effects of using innovative technologies are significantly mitigated. The elements within the climate resilience framework are interconnected, continuous, and mutually influential. The existing framework is summarized in four sections. The first part outlines the goals aimed at applying innovative technologies to address climate change and

Table. 14.7 Application of innovative technology to improve climate resilience in Tokyo

The name of the technology	Field of application	Application	Equipment
Green systems	Green roof, cool pavement, forest	Improving air quality, reducing the effects of heat islands, and absorbing rainwater	Plant species suitable for the climate, infrastructure, and solar panels
Crisis management system	Disaster preparedness	Providing early warning and evacuation planning services	Emergency response infrastructure
Energy management	Energy	Calculation and reporting of greenhouse gas emissions, as well as promoting and improving energy efficiency in factories and buildings	Solar panels, electronic charging systems, storage batteries

Source Ministry of the Environment (環境省) [13], Tokyo Metropolitan Environment Bureau (東京都環境局) [23]

enhance climate resilience. Optimizing energy efficiency, enhancing air quality, minimizing environmental impact, reducing heat island effects, monitoring and controlling resources, and providing early warning services are some of the most crucial goals being pursued. The second part discusses the constructive mechanisms of this framework, which comprises two components: the institutional framework and the resilient innovative technology framework. The institutional framework includes (institutional characteristics, characteristics of goods and services, characteristics of activists, and characteristics of people) and the framework of resilient innovative technology includes (management-institutional dimensions, infrastructures, facilities, and areas of technology application). The following section illustrates the application process and its output of the process (see Fig. 14.1). In the following, the mechanism section and its subsections will be examined.

- **Institutional Framework**

<u>*Risk Management*</u>: Managing any system, such as a city, always involves theoretical and practical complications, and its success depends on several factors. One of the important pillars and foundations in the successful management of a city is risk management. Risk management and city resilience are closely related. In this way, risk management can be considered a prerequisite for the city's climate resilience. In risk management, the various types of risks that pose a threat to a system are identified, analyzed, prioritized, evaluated, controlled, and monitored. The various types of these risks have different origins, which when combined, are known to pose a serious threat to the city, impacting safety. The important point is that risk management is a continuous process that does not stop; rather, it works continuously to identify, evaluate, confront, and prevent threats and damages. Attention to

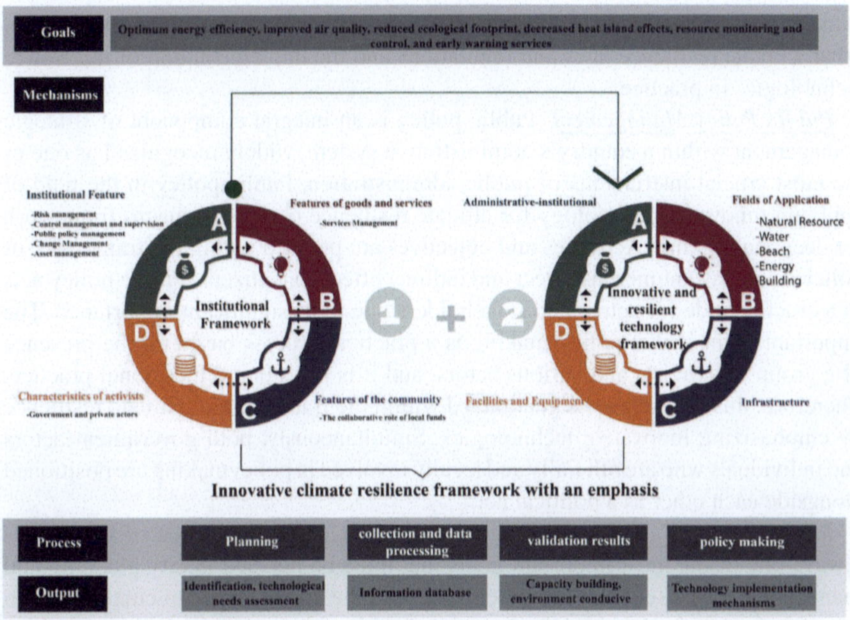

Fig. 14.1 limate resilience framework with a focus on the utilization of innovative technologies (Illustration by the Authors)

this aspect of the city's climate resilience is ensured and is consistently maintained through the option of receiving feedback. Risk management is an interactive process that involves strategic risk management, financial risk management, risk management of urban climate resilience projects (including the application of innovative technologies), operational risk management, and risk management of laws and regulations. Strategic risk management is one of the most crucial types of risk management and should be prioritized as it directly aligns with the strategic goals of effectively implementing innovative technology to enhance climate resilience. In financial risk management, threats to the liquidity of organizations responsible for innovative technology and climate resilience are identified, and efforts are made to control them. One of the key pillars in ensuring a city's climate resilience is that these responsible institutions have the financial capacity to manage crises and address damages. In the risk management of urban climate resilience projects, with a focus on innovative technologies, risks and threats that could lead to project failure are identified, analyzed, assessed, and mitigated. Operational risk management involves enhancing organizational processes and key factors in enhancing the city's climate resilience. This type of risk management involves identifying and analyzing work processes, technical issues, system malfunctions, and daily activities within the organization, and monitoring defects in each area to make necessary adjustments. Operational risk management is a crucial factor in assessing the climate resilience of a city, ensuring effective measures are taken to enhance resilience. Finally, there is the risk management of laws and regulations, which is crucial in ensuring compliance with urban climate resilience laws and regulations and the implementation of innovative technologies in practice.

Public Policy Management: Public policy is an integral component of strategic management within a country's administrative system, widely recognized as one of the most crucial instruments of public administration. Public policy in the field of applying innovative technology for climate resilience provides a macro framework for decision-making. Activities and objectives are pursued within the framework of policies and have numerous direct and indirect effects on citizens. Public policy acts as a crucial guide for actors and stakeholders due to its significant importance. The important point is that policy-making is a practical process based on the presence of a group of activists and various actors, and it is not a one-dimensional practice. Therefore, this importance is guaranteed within the framework of climate resilience by emphasizing innovative technologies. Simultaneously, both government actors and individuals who are officially and legally involved in policymaking are positioned alongside each other as a political act.

Management of change and transformation: Climate resilience of the city is always one of the most important goals that must be pursued in city planning and management because it plays an essential role in the quality of life of citizens. Given the complexity of urban climate issues, it is essential to change attitudes towards the application of innovative technologies. The city's management system requires changes and transformations in this area to develop innovative measures and initiatives. These changes aim to enhance performance, ensure climate resilience, and improve the city's capacity to recover while progressing forward. Change management often focuses on human processes and dimensions, which leads to the design of new work processes. The platform ensures the application of innovative technology, which is evident through organizational change management and human resources practices within it.

Management, Control, and Supervision: The discussion of control and supervision is a fundamental topic that underpins the management structure in weather-related contexts. In the elaboration of this concept, it can be stated that continuous control and monitoring aim to ensure that all actions are executed in alignment with policies and prescribed procedures. In the control department, it is also possible to review and modify actions. When the output of the actions does not comply with the policies, corrective actions must be taken. This stage involves providing feedback in the climate resilience process, which includes revising and updating policies.

Asset management: Asset management can be defined as a comprehensive process that includes a wide range of properties and rights that benefit from them. The intersection of asset management, climate resilience, and innovative technology encompasses a wide range of benefits, values, and investments.

Service features: The management of services related to urban climate resilience includes various aspects such as economic, socio-cultural, and physical infrastructure, access to urban facilities and equipment, as well as health and environmental considerations. It is important to pay attention to a set of factors in various urban contexts, including marginal areas, informal settlements, dilapidated areas, historical districts, modern areas, and new cities. To deliver effective services, it is crucial to provide the necessary services based on needs assessments and the consensus of the unit.

Characteristics of actors: Actors and activists play a crucial role in shaping policies, interests, values, and actions related to the use of innovative technologies to enhance climate resilience based on their perspectives.

Characteristics of society: Civil societies and citizens are at the heart of the city's climate resilience. Their involvement in all stages of planning, data collection, processing, validation of results, and policy-making for the implementation of innovative technology to enhance climate resilience is crucial.

- **Innovative and resilient technology framework**

Areas of Application of Technology: Surveys indicate that the most commonly considered areas include beaches, rivers, seas, buildings, green infrastructure, and energy.

Infrastructure, Facilities, and Equipment: The essential infrastructure and facilities in this field encompass sensors, cameras, solar panels, electronic charging systems, storage batteries, seawalls, breakwaters, drainage systems, intelligent early warning systems, sound leakage sensor networks, smart water meter, smart valves, smart pumps, modeling software, computer control, and database.

Managerial-institutional: The managerial-institutional dimension overlaps significantly with the institutional framework, particularly within the sub-sectors of actors and society. This pillar is responsible for making decisions, formulating policies, and overseeing their implementation.

14.6 Concluding Remarks

In recent years, the intensification of environmental instabilities and impacts on the environment have negatively affected cities and countries. The approaches aimed at climate resilience have made the city the central focus of policymaking, especially in utilizing innovative technologies to mitigate the effects of climate change. It is important to pay attention to the effects of climate change so that future generations can enjoy the same quality of life as the current generation. For this reason, it is important to pay attention to climate resilience and its drivers. Innovative technologies are defined as a stimulus for climate resilience. They provide strong support for climate resilience and guarantee it due to their impact on coastal management, water resources management, energy management, building management, and green management.

Innovative technologies in the field of climate change are a controversial concept that makes them susceptible to participation in discourses defined by the framework of climate resilience, with an emphasis on innovative technologies. This framework is one of the crucial pillars that ensure the city's climate resilience. In the present study, the framework of climate resilience encompasses four levels of goals (optimal energy efficiency, improvement of air quality, reduction of environmental impact, reduction of heat island effects, monitoring and control of resources, and early warning services), Mechanisms (institutional framework (comprising institutional characteristics, characteristics of goods and services, characteristics of activists, and characteristics of society) and an innovative and resilient technology framework (encompassing fields of technology application, infrastructure and services, facilities, and managerial-institutional aspects). The process (planning, data collection, and processing, validation of results, and policy-making) and output (identification, technology needs assessment, information base, an environment conducive to technology use, capacity building for technology use, technology implementation mechanisms, and involvement of all stakeholders) was expanded and showed that these dimensions are interconnected and are influenced by an efficient institutional framework of management components necessary for the application of innovative technologies. This causes cities and countries to become climate-resilient regions that can maintain their ability to resist crisis and damage. The presented framework facilitates the shift from traditional crisis and risk management to modern crisis management by leveraging innovative technologies. It envisions climate-resilient urban areas with multi-level governance, functional diversity of technology tailored to specific environmental needs, and technical-organizational capacities.

Acknowledgements In this section, we would like to express our gratitude to Professor Majumdar and all the editors for providing us with this opportunity.

References

1. Ali M, Awe EO, Mohammed SS, Isah KO (2024) Industrialization, FDI inflow and climate change in Africa: a scenario analysis. Chin J Urban Environ Stud 2450003
2. Apostu SA, Nichita EM, Manea CL, Irimescu AM, Vulpoi M (2023) Exploring the influence of innovation and technology on climate change. Energies 16(17):6408
3. Bank AII (2019) A framework and principles for climate resilience metrics in financing operations
4. Bellinson R, Chu E (2019) Learning pathways and the governance of innovations in urban climate change resilience and adaptation. J Environ Plan Policy Manage 21(1):76–89
5. City guide to Copenhagen (2024). Copenhagen. https://www.copenhagen.com/copenhagen-climate-plan/
6. City of New York (2024a). https://www.nyc.gov/site/planning/plans/one-nyc/one-nyc.page/
7. City of New York (2024b) https://www.nyc.gov/site/planning/plans/plaNYC/plaNYC.page/
8. Coccia M (2021) Technological innovation. Innovations 11:I12
9. Diaconu M (2011) Technological innovation: concept, process, typology and implications in the economy. Theor Appl Econ 18(10)
10. Gemeente Rotterdam (2024). https://www.rotterdam.nl/english/climate-proof/
11. ICLEI (2024). ICLEI. https://www.iclei.org/en/case-studies/rotterdam-climate-proof.html/
12. Li C, Yang H, Yao Q, An N, Meng H (2024) Governing urban climate resilience (UCR): systems, agents, and Institutions in Shanghai, China. Sustainability 16(7):2648
13. Ministry of the Environment (環境省) (2024). https://www.env.go.jp/earth/ondanka/adaptation/adaptation_plan/tokyo_e.pdf/

14. Moench M (2018) Experiences applying the climate resilience framework: Linking theory with practice. In: Climate change adaptation and development. Routledge, pp 17–34
15. Montero S (2020) Leveraging Bogotá: Sustainable development, global philanthropy and the rise of urban solutionism. Urban Stud 57(11):2263–2281
16. National Parks Board (2024) National parks board. https://www.nparks.gov.sg/gardens-parks-and-nature/singaporegreenplan/
17. Nwankwo W, Ukhurebor KE, Aigbe UO (2020) Climate change and innovation technology: a review. Renew Energy 20:23
18. Prellezo R, Da-Rocha JM, Palomares ML, Sumaila UR, Villasante S (2023) Building climate resilience, social sustainability and equity in global fisheries. npj Ocean Sustain 2(1):10
19. Quitzow L, Rohde F (2022) Imagining the smart city through smart grids? urban energy futures between technological experimentation and the imagined low-carbon city. Urban Stud 59(2):341–359
20. Sari R, Soytas U, Kanoglu-Ozkan DG, Sivrikaya A (2023) Improving the climate resilience of European cities via socially acceptable nature-based solutions. npj Urban Sustain 3(1):9
21. Singapore Green Plan 2030 (2024). https://www.greenplan.gov.sg/
22. Subiyanto A, Boer R, Aldrian E, Kinseng R (2020) Climate resilience: concepts, theory and methods of measuring. Environ Asia 13(1)
23. Tokyo Metropolitan Environment Bureau (東京都環境局) (2024) https://www.kankyo.metro.tokyo.lg.jp/en/climate/climate_change_strategy.html/
24. World Health Organization (2022) Measuring the climate resilience of health systems. World Health Organization
25. Yang J (2018) Climate change and domestic migration in China. Chin J Urban Environ Stud 6(03):1850020

Chapter 15
Resilience Planning: Metamorphosis of the Zorrotzaurre Peninsula, Bilbao (Spain)

Olivera Radovanović and Svetlana Perović

Abstract This chapter refers to the treatment of urban planning as a mechanism in the context of the resilience of cities to floods, the most common natural disaster caused by climate change. The focus of the research is the Zorrotzaurre peninsula in Bilbao, a city undergoing constant urban transformation and a significant industrial and tourist city in Spain. It is a peninsula that has been identified as a degraded industrial area, with a significant degree of contamination and which is subject to flooding. Relying on the research done within the master's thesis, this chapter will present a proposal for concrete urban regeneration—metamorphosis—of the peninsula mentioned above, to create a more flood-resistant place. In addition, through the urban regeneration of the Zorrotzaurre peninsula, the goal is to point out the importance of urban planning and urban architectural design in mitigating climate change. Several goals were set in the research: to understand the necessity of having a clearly defined action policy, clearly defined measures, adaptation, and mitigation strategies; to reduce the impact of climate change on the environment, that is, to increase the degree of resilience in architecture, infrastructure, and urban space, and, in the end, to contribute to increasing the degree of development of awareness of the importance of everything previously mentioned, and then to the definition of practical solutions, applicable not only in the area of Bilbao but also wider, bearing in mind, all specific territorial, social and economic aspects. Several methods were used in the work: after a theoretical review of the place that is the subject of research, the identification of specifics followed, and then practical guidelines were given for the urban regeneration—metamorphosis—of the same, all to create a more resilient place to live.

Keywords Climate change · Urban regeneration · Resilience planning · Sustainability · Zorrotzaurre (Bilbao)

O. Radovanović (✉) · S. Perović
Faculty of Architecture, University of Montenegro, Džordža Vašingtona bb, 81 000 Podgorica, Montenegro
e-mail: olivera.r@ucg.ac.me

S. Perović
e-mail: svetlana.perovic@ucg.ac.me

S. Ghosh et al. (eds.), *Cities of Tomorrow: Urban Resilience and Climate Change Preparedness*, Urban Sustainability, https://doi.org/10.1007/978-981-97-9658-8_15

15.1 Introduction

As a global process, urbanization manifests itself through several phenomena, such as changes in the density of the human population, technological development, etc., being characterized by constant growth. In addition, the urban environment faces increasing uncertainty caused by climate change, with an evolution from the traditional process of planning and designing for the foreseeable future, to planning and designing for the unpredictable [11]. The inability to predict changes or future threats leads to the development of strategies based on resilience, flexibility, and adaptability, and these will be the focus of this paper.

The comfort and pleasantness of living in the urban environment are directly conditioned by climate change. The effects of climate change are visible in every part of the planet, but they are significantly intensified precisely by the actions of man, his carelessness, and disrespect for the true nature of the place [6, 35].

It is essential to respond to the challenges that climate change and the resulting rise in water levels bring with it. Until now, different approaches have been studied, and different tools and different strategies for "climate-friendly" urban planning and urban architectural design have been proposed. Nevertheless, the fact that even the experts who deal with climate change do not have fully agreed opinions regarding the exact causes and consequences of the same is worrying.

Socio-ecological resilience, as a concept that primarily originates from ecology, as Cathy [44] claims, becomes a potential new conceptual ground for urban planning and interdisciplinary discourse on the management and planning of cities as related to socio-ecological systems [1, 9]. These systems are further conceptualized as complex adaptive systems, which is consistent with emerging interests in planning for the impacts of non-linear ecosystem dynamics [41, 44].

Although so far there is not much interdisciplinary research across the aforementioned areas, in general, the use of resilience is rapidly expanding across various fields of urban studies including the following: climate change mitigation and adaptation [28], disaster planning, management, and recovery [15, 24] energy and environmental security [18, 27], land use change on local flooding [3], urban water management [12]; integral land use and traffic planning [10], urban design [41]. The degree to which socio-ecological resilience affects each of these areas varies from study to study. Those who take it explicitly, as a planning approach, are significant because, among other things, they explore the potential of resilience in terms of achieving a successful integration between ecology, urban planning, and urban design [34].

Given that the peninsula identified as a degraded industrial area was taken as the subject of the research, at the very beginning, it is interesting to mention a few architects who promoted a new approach, a new attitude toward these so-called "urban voids".

Eduard [5] states that the urban void is a key element in any discussion of the contemporary city, and proposes that architecture and urban planning be based on metamorphosis, on the recycling of materials, forms, and uses. He believes that urban emptiness is a problem of the modern city and that it represents a black spot in the urban fabric [5].

Ignasi de Solà Morales [8] uses the term desert (French: "Terrain Vague") to define the complex nature of the urban void. At a time when few recognized the value and potential of urban empty and abandoned spaces, Solà Morales wrote about his fascination with them. He was against violent transformation, emphasizing that architecture should not become an aggressive instrument of power, but should preserve the unique nature of those strange and alternative spaces by introducing new elements and preserving the existing ones [8].

In connection with urban voids and their urban regeneration, Rem Koolhaas points out: "Where there is nothing, everything is possible. Where there is architecture, nothing (else) is possible" ([20], p. 199).

15.2 Impacts of Climate Change on Urban Sustainability

Climate change is attributed to activities that change the composition of the atmosphere, which are recorded over a longer time interval. The term "climate change" is used to describe a climate change that occurs as a result of natural and/or human factors [25]. In the 4th Report of the Intergovernmental Panel on Climate Change [IPCC], a group of scientists and experts on climate change claims: "The climate is changing and these climate changes are mainly caused by human activity" ([16], p. 950).

15.2.1 Global Trends and Projections

The effects of climate change are evident and are the subject of numerous discussions around the world. It is believed that in the coming decades, they will become more and more intense and frequent.

Namely, from the beginning of thermometer measurements, the temperatures recorded in the period between 1980 and 2010 turned out to be the highest in the last 1400 years, including the last decade as the warmest ever [17, 35]. Then, the amount of precipitation increased significantly, both in terms of frequency and intensity [17, 35]. The frequency of extreme events has also increased, the number of cold days and cold nights has significantly decreased, while the sea surface temperature has significantly increased [17, 35]. The cryosphere, that is, the parts of the Earth's surface where water is continuously represented in its solid state, for at least one part of the year, is constantly shrinking in response to global warming [17, 35]. Due to global warming, the number of glaciers, as well as the surface of permanent ice and snow, is decreasing [17, 35]. When it comes to global sea level, it rose by 0.19 m in the last century [17, 35]. Carbon dioxide emissions, which were mainly caused by fossil fuel burning and deforestation, have increased by 40% [17, 35].

Increasingly intense greenhouse gas emissions, caused by human activities, significantly affect the increased warming of the atmosphere [17, 35]. Therefore, as a critical issue, the question arises of how to diminish the impact of man on the climate, and how to adapt to the changes that have already occurred or are emerging, bearing in mind that the mentioned processes affect globally: the planet, but also all living things in the world, including man. Planet Earth and its ecosystem have become very vulnerable, with man as its part. It is important to mention that, due to climate change, flora and fauna are dying out [17, 35]. They also cause significant problems in the functioning of human society, primarily when it comes to the production and availability of food and water and people's health [17, 35]. Climate change could cause conditions and numerous social problems such as mass migrations, and major conflicts over the little remaining arable land, water resources, etc.

15.2.2 Urban Floods as a Global Phenomenon

The challenge that many places around the world are facing today, which has a growing character and is becoming more and more serious, is flooding. Floods cause significant damage, among which the biggest damages are certainly lost human lives [14, 19, 29, 42]. These are the most common natural disasters, the frequency of which has particularly increased in the last twenty years [19, 29, 42]. All of the above points to the importance of better understanding, and then better and more efficient management of existing and future flood risks.

Despite the higher frequency of floods in the last twenty years, the number of lost lives is decreasing. This points to more successful, efficient flood risk management. However, although this fact is encouraging, the number is still high [19, 42]. Analyzing the impact of floods around the world, it is clear that urban areas—cities, are particularly affected by them. The current and predicted levels of impact point to the urgency of defining adequate measures, i.e. strategies for flood risk management in urban areas.

We distinguish between direct and indirect impacts of floods. The direct impacts of floods include the impact on human lives and the economy. Unlike indirect impacts, they are easier to identify and quantify. Indirect impacts include various diseases, reduced opportunities for nutrition and education, loss of livelihood, etc. [14, 19, 42].

As mentioned earlier, the urban environment is facing increasing urbanization. It manifests itself, among other things, through a change in the density of the human population, thus influencing increased flood risk. In fact, the urban population represents and will continue to represent a larger share of the world's population. Therefore, due to the size of the exposed population, the focus of discussions on defining adequate measures and strategies for flood risk management is precisely the floods that affect urban areas.

What also affects increasing the degree of flood risk is the urban expansion of non-plane character. Namely, precisely to adapt to the increase in population, cities grow in those areas that are prone to floods [19, 42]. "Slums", i.e. dense, informal settlements that are characterized by poor living conditions (lack of water for drinks, pollution, overcrowding, poverty, unemployment, crime), are most often at the highest flood risk [19].

The intensification of the consequences of floods is also influenced by the increasing percentage of impervious surfaces, overcrowding, congestion, higher housing density, inadequate infrastructure, i.e. equipment, but also the lack of activities aimed at flood risk management [19].

Finally, what is particularly important to mention about this research and what directly causes the appearance of floods and influences the intensification of their consequences are changes in meteorological patterns, i.e. climate change [19, 45]. What does the impact of climate change represent? For example, changes in precipitation patterns cause frequent phenomena of river floods; flood damage in removable areas, due to the growth of sea level, is increasing; the increased frequency of the storm results in a higher frequency of march waves while changing the frequency and duration of dry periods leads to groundwater pumping and land subsidence [19, 42].

It is clear that the management of flood risk is today and in the future of special importance. Each city that is exposed to the mentioned risk requires a different approach. But, in general, it should strive to achieve a balance between approaches that reduce the impacts of floods and better management and far-sighted approaches that involve new infrastructure or reshaping an urban environment to mitigate floods [19, 42, 45].

15.2.3 Understanding the Causes and Risks of Floods

The success of managing current and future flood risks depends, above all, on understanding floods and the dangers they can cause. In fact, this understanding is of essential importance for defining adequate measures, that is, strategies that could mitigate, or even completely prevent, the damage caused by certain types of floods.

What does it mean to understand floods? This means, first of all, knowing what are the causes of floods, what types of floods occur, and how, depending on the type, floods manifest. Important inputs include: knowing which areas are most exposed to the risk of floods, how many inhabitants they have, how much area they occupy, what the degree of vulnerability of those areas is, whether they have the property of resilience, and what they are doing to reduce the risk to which they are exposed. Understanding all of the above is crucial to understanding the urgency of action [14, 19, 32, 45].

In addition to everything previously mentioned, information on the probability of occurrence of a flood event plays an important role in the successful management of its risk. However, very often, excessive reliance on this information is not entirely appropriate. And why is that so? The probability of the occurrence of a flood event is most often calculated based on historical data, and the question is how adequate these data are for making such conclusions. In the case of inadequate data, communication of flood risk in the form of probabilities calculated based on the same and their application in defining decisions related to flood risk management can lead to misconceptions [19, 21]. Therefore, flood hazard and risk maps have been identified as a significant aid in defining the previously mentioned decisions.

Hazard maps are tools that help in disaster risk assessment. Considering flood risk assessment and management, they are significant because they show how a certain flood event can develop and what kind of damage it can cause. They also provide information on the number of potentially endangered residents, that is, on the potentially endangered area and its degree of vulnerability [2, 19, 30].

To save lives and property, people have used another tool since ancient times—forecasting flood risk based on past patterns. In fact, they believed that the future could be predicted from the past. However, as Jha [19] points out, such predictions can only contribute to the damage caused by a flood event because they are based on stationarity, and today the areas that are most exposed to the risk of floods—cities, i.e. urban areas are dominantly characterized by non-stationarity. It is, in fact, a consequence of urbanization, i.e. rapid growth and development of cities, but also increasingly frequent changes in weather patterns due to climate change [20, 21].

Therefore, it is clear that approaches to managing current and future flood risks must be more flexible, and more adaptable, that is, they must deal with the greater uncertainty of the occurrence of a flood event [20].

15.3 Bilbao's Vulnerability to the Climate Change Impacts

Bilbao, in the heart of the Basque Country, is the largest city and the largest port on the long Spanish coast of the Atlantic Ocean, bordered on three sides by steep slopes, which lead to the Pyrenees in the east, and to the sea in the north [36]. Climate change and the resulting rise in water levels have affected many areas in the Bilbao area and the Basque region in general. River and pluvial floods have been recognized as the main climate hazards, and the main impacts of climate change in the Bilbao area are divided into direct and indirect [36, 38].

When it comes to direct impacts, we distinguish impacts on local climate conditions, impacts on the built environment, and impacts on living conditions and biodiversity. Among the impacts on local climate conditions, it is significant to mention the rise in temperature and the increased frequency of storms and floods. Furthermore, the increase in temperature and the greater frequency of extreme weather conditions directly cause discomfort and changes in the physical environment, affecting living conditions and biodiversity. The pressure on ecosystems and habitats is increasing, and there are also changes in the water regime. When it comes to the built environment and the climate change impacts on it, it is important to mention, first of all, structural problems, i.e. the climate change impact on buildings, on the stability of bridges and other infrastructure that is threatened by the last flow of river flow and waste accumulation. There are interruptions and failures in the power system, the number of landslides is increasing, and a lack of water, directly affecting the city's greenery [38].

In the Bilbao area, as a result of climate change, there are also changes in how land is used. Some activities are relocated, and transportation is interrupted. Households are getting into problems due to the increase in costs, and there is an increasing demand for water for drinking and irrigation [38].

It is clear how important it is for Bilbao to improve its action policy and how important urban planning, that is, urban-architectural design, is in the process of reducing all the previously mentioned negative climate change impacts.

15.4 Assessing Vulnerabilities and Planning for Resilience: Metamorphosis of the Zorrotzaurre Peninsula

The Zorrotzaurre peninsula, with an area of about 838,781 m^2, was formed during the 1950s and the 1960s, when the Deusto Canal was built, to facilitate navigation and provide access for larger ships [37, 46]. Due to its low topography, surrounded by water, the peninsula has always been exposed to a high risk of floods, which, as already said, represent the main identified climatic hazards in the Bilbao area. In the 1970s, experiencing social and industrial decline, it became an impoverished, abandoned area with contaminated soil and only 500 inhabitants [37, 46].

The urban regeneration of this area began in 1995, when the General Urban Plan of Bilbao was repurposed from, so to speak, "industrial" to "residential". A significant step in the aforementioned urban regeneration process was the opening of the Deusto Canal in 2018. It was also the beginning of the metamorphosis of the Zorrotzaurre peninsula and the first step taken to reduce the flood risk to which this area is exposed [37, 46].

15.4.1 Abandoned Place as an Impulse

Once this area used to be essential to the industrial city of Bilbao, today the aforementioned Zorrotzaurre peninsula is the epitome of the loss of identity, loneliness, and distance from reality. Nowadays, it is the environment that oppresses and depresses its inhabitants. However, thanks to its privileged position and the opportunities it provides, the Zorrotzaurre peninsula has always been, and is still today, a unique enclave. It is a paradigm of the connection between, on the one hand, Bilbao and its metropolitan area, and on the other hand, an articulating element between the right and left banks of the Nervion River [37, 46].

Despite the danger it was exposed to due to floods, the Zorrotzaurre peninsula corresponded to the interests and needs of every era. Within the last two decades, it has been left out of the dynamics aimed at urban regeneration, that is, the transformation of the city of Bilbao. Today it is more a space of the past and future, than of the present [37, 46].

The image of this unique urban space and perhaps the last remnant of Bilbao's industrial past today is made up of empty lots, ruins, and mostly abandoned buildings which, in some cases, are of great architectural importance. The absence of public spaces and facilities, a small percentage of green areas, poor infrastructural equipment, and isolation from the surrounding urban fabric, are among the other aspects that make the Zorrotzaurre peninsula a place that should be avoided, rather than visited. Hard concrete edges dating back to the period of development of shipping and industry weakened the connection between the peninsula and the river. All of the above, together with some other socio-economic and geophysical factors, affect the sustainability and resilience of this area [37].

It is clear that the peninsula, i.e. today the island of Zorrotzaurre, due to the weakened credit rating in terms of construction and loss of functions, negatively affects its narrower and wider environment, not only in the urban and economic sense, but also in the aesthetic, psychological, social and ecological sense. It is one of those brownfield areas that are the products of social and global processes that caused a break in the continuities of urban images. Such surfaces, i.e. spaces that dominantly depicted unified and uniform spatial components throughout the industrial period, are potent to become identity carriers, key symbols, places of social processes, and visual dominants through regeneration [33]. The island of Zorrotzaurre has additional potential because it sublimates memories of former life and traces of past times [26].

15.4.2 Costura Urbana

Areas such as the island of Zorrotzaurre represent interruptions in the continuity of the built fabric, insufficiently recognizable by identity, and programmatically attractive by disposition. As such, they require a new approach to planning, the goal of which is the creation of a layered but integral, harmonious, humane living environment with authentic values and opportunities [26].

After analyses of the current situation, and finding characteristic fragments of space for urban regeneration, it was concluded that the island of Zorrotzaurre needs to be enriched and refined in terms of content, functionality, design, structure, aesthetics, spirituality, psychology, technology and other aspects to meet the diverse needs of the existing and new space users. It was necessary to create a layered and integral, harmonious, humane living environment, with authentic values and opportunities. It was necessary to create a project that would be based on the originality and specificity of the design expression, the unity of diversity, multifunctionality, and layering of space, the integrity of the natural and the built, as well as the interaction between parts and the whole [26].

Throughout the design process, flow was thought of as an unavoidable factor in the high-quality functioning of space, as a balance between established and overemphasized, as a factor that offers security and durability, as a consequence of a well-designed and functionally harmonious urban environment. Attention was paid to the location, the environment in which it is located, preservation, and maximum integration [26].

Due to its very specific position within the city of Bilbao, the island of Zorrotzaurre was seen as an urban seam—*Costura Urbana* (Fig. 15.1), that is, as an opportunity to establish cohesion between all parts of the city [26]. From the very beginning, the idea was to form a public, open space that gives freedom and encourages community and inclusion; the idea was to form a layered, yet integral, harmonious, both still dynamic and humane living environment of the highest values and possibilities.

Enhancing the entire space with greenery, connecting it with the Nervión River, as the greatest natural potential of the location in question, but also with other contact zones, the island acquires a new dimension and the function of a popular public space where each new step brings a new experience, a new encounter while enriching entire feeling. Man moves—everything moves, everything changes. Man is the basic starting point and benchmark, being particularly important and included in every part of architecture as well as the formation of space [26].

Using the potential of the island of Zorrotzourre, a space whose indeterminacy gives different possibilities of its use, the aim was to form a public space that will be

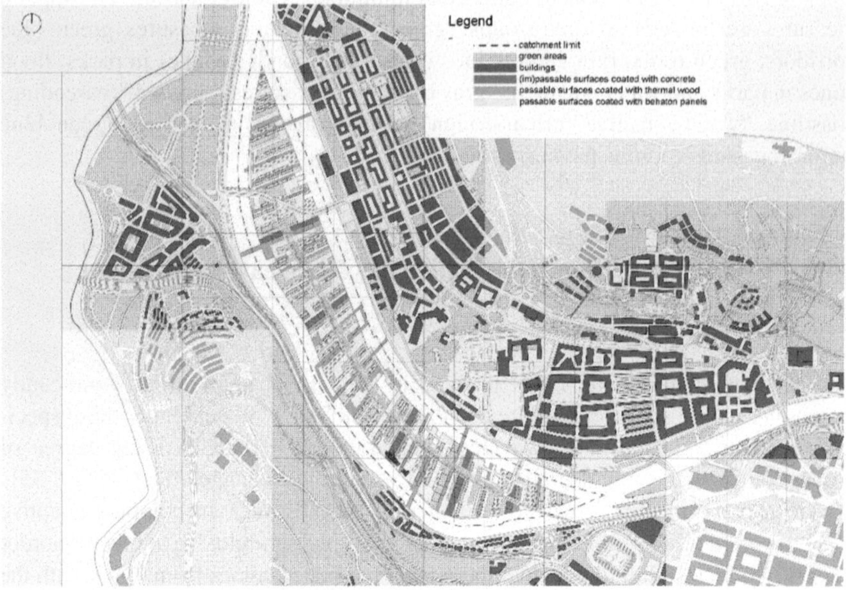

Fig. 15.1 Master plan: urban regeneration—metamorphosis—of the Zorrotzaurre peninsula—*Costura Urbana* (*Source* Mirković [26])

deeply correlated with the context, but also that will establish changes in the environment and serve as an example for new standards that push the limits of the existing ones. As it is clear that global processes, changeability, dynamism, and complexity of the world we live in require authentic, innovative, and daring urban-architectural solutions, the result of such solutions is a new identity, a new energy. This is exactly the case with this proposed solution. Zorrotzaurre Island has been given a completely new identity: universality, authenticity, and interweaving of natural and artificial. Less physical structure and human energy have made Zorrotzaurre Island a space that not only revives the post-industrial landscape of the city of Bilbao but also represents the emblem of the urban regeneration process [26].

Within the framework of the project in question, seven basic zones are defined, interwoven, and connected: zone for recreational activities; zone for sports activities; zone for ecological educational activities; zone for contemplative activities; "blue" zone (interaction with water); "green" zone (interaction with greenery) and zone for social activities. The "green" zone occupies the largest percentage, 50% of the total area (≈300,000 m^2). Furthermore, the zone for contemplative activities is 15% (≈90,000 m^2); the zone for social activities is 12.5% (≈75,000 m^2); the "blue" zone is 10% (≈60,000 m^2); the zone for environmental and educational activities is 5% (≈30,000 m^2); zone for recreational activities also 5% (≈30,000 m^2) and zone for sports activities 2.5% (≈15,000 m^2) [26].

In addition to the previously mentioned "division" of the island into zones depending on the purpose, it is also divided into zones depending on the measures taken to reduce climate change impact and minimize its consequences. The applied measures are divided into three groups: "green–blue dynamic" measures (green–blue corridors; green roofs; rain garden; bioswales; large ponds or lakes in parks; flood zones in parks and water squares); "gray dynamic" measures ("graded" (cascading) coastline, "water-sensitive" urban design and land raising) as well as "green–blue static" measures (porous paving) [26].

15.5 Living with Floods—Guidelines to Create a Resilient Place for Life

The impacts, that is, the consequences of climate change differ significantly depending on the specificity of the context to which they relate. Under these specificities, among other things, we mean vulnerability. It represents the degree of susceptibility, that is, the degree of sensitivity to climate change effects [7, 17, 35].

An increase in the degree of resilience to one disorder, in complex adaptive systems, often affects an increase in the degree of vulnerability to another disorder [23, 39]. In fact, it should be remembered that resilience comes from living with the disorder [13, 23, 43].

Building resilience in cities characterized by the presence of the river is the process of adapting. Instead of opposing the river and its nature, cities, like Bilbao, should learn to live with it. Periodic floods should serve as a lesson to the city, to learn from them to be more ready and more resilient to the impacts of those more extreme floods [23, 40]. As [23] states "It is a paradigm shift from *resistant* to *resilient* cities with the management agenda redirected from *safety against floods* to *safety at floods*" (p. 9).

In countries such as Bangladesh, Cambodia, and Egypt, a built environment and lifestyle adapt to river dynamics [4, 22, 23]. These communities use previously mentioned periodic floods, i.e. those less dangerous floods to increase productivity in individual activities [24]. This is the essence of planning cities like Bilbao, all to increase the degree of resilience.

The perception that cities, as urban areas, and floods, cannot coexist, has been rooted for a long time. This is precisely what inhibits the ability of cities to be flexible, adaptable, and able to cope with the greater uncertainty of the occurrence of a flood event.

By changing the perception, that is, by changing the way of thinking and approach to planning, cities would live with floods. Through this chapter, the aim was to point out just that. Buildings can be elevated, floating, or protected in such a way that the lower floors will be used to store water. The coast can be designed in steps or cascades, which, in addition to forming a flexible, dynamic border, would ensure contact between the city and the water, man and the water, regardless of the level of the same. Large retention lakes, bioswales, rain gardens, and green–blue corridors can become an integral part of the urban landscape. In general, when planning, that is, designing, one should strive for dynamism, instead of stability and permanence. As Liao states [23]: "The resilient city is always a work in progress" (p. 10). Therefore, projects of an adaptable, removable, and changeable character are the future, especially in cities that face flooding.

Future research should show that the previously mentioned claim is correct—it should be focused on long-term monitoring of the impact of the transformations carried out on the Zorrotzzaure peninsula or on comparative studies with other cities where transformations of the same or similar nature have been carried out, to identify their (in)efficiency and possibilities of application in other, similar projects.

15.6 Discussion

Through the process of urban regeneration (metamorphosis) of the Zorrotzaurre peninsula, one of the coastal areas in the city of Bilbao, which are most affected by floods caused by climate change, the aim was to contribute to increasing the level of awareness development of how everything previously mentioned is important and then defining practical solutions, applicable not only in the Bilbao area but also wider, keeping in mind, of course, all specific territorial, social and economic aspects. The modern development of both the city of Bilbao itself and the peninsula (today's island of Zorrotzaurre) faces the ubiquitous global challenge of sustainability, which is greatly influenced by the current process of climate change. Bilbao did not fully use the potential of the spatial disposition, i.e. the position of the aforementioned peninsula, even though it is precisely in the areas as they are that the potential for the renewal of the urban landscape lies in projects of a temporary/occasional character that could also respond to the acute problems of climate change.

Namely, the once-dominant industrial character of the city of Bilbao, after a series of crises and economic difficulties, experienced a collapse. However, its traces remained in the form of abandoned spaces, as well as the consequences that such spaces left on the city viewed in the light of climate change. Most of the mentioned spaces have gone through the process of urban regeneration and today they complete the image of Bilbao as a post-industrial metropolis. The area of the Zorrotzaurre peninsula is not one of them.

Bearing in mind the position it occupies, and the importance that the Zorrotzaurre peninsula has had for Bilbao throughout history, this paper aimed to point out the necessity of approaching the incorporation process into the development of Bilbao as a modern urban city, implementing action strategies in the process of adapting to climate change and its mitigation.

To round off the process of metamorphosis, the Zorrotzaurre peninsula was seen as an urban seam, longitudinal and transverse, representing an opportunity to establish cohesion between all parts of the city. Radically transforming the landscape and, more importantly, symbolizing the post-industrial, innovative, and competitive Bilbao, the area of this peninsula has become an emblem of the process of urban regeneration, i.e. revitalization of the city. The infrastructure has been upgraded, and the entire area has been refined with green areas and is connected to the Nervión River, as the greatest natural potential of the location in question. In the end, a derelict, devastated area was regenerated and as such represents an exemplary example of an urban megaproject.

15.7 Conclusions

By posing the climate change problem and (re)placing the topic in the context of one of its causes (industry and its abandoned spaces), the aim is, first of all, to change the consciousness of the individual and to think more soberly about the modern city and the actions of each individual in it, especially in the process of climate change. Although it is usually considered that climate change is something that happens on a global level and that greenhouse gases are its main cause, it is a man who accelerates and additionally intensifies it with his activity. Therefore, to achieve and mitigate the ultimate goal, a change in collective awareness about climate change is necessary. In combination with successful and sustainable planning and design, the existence of an effective action policy, effective adaptation, and mitigation strategies, can provide optimal results.

Successful and sustainable planning implies planning that will be based on thinking about resilience. This is indicated by the problems faced by cities around the world, which arise as a result of climate change. To create resilient cities, the practice of planning must be radically changed—a balance must be established between different actors, especially planners; key priorities must be clearly defined and beliefs and value systems changed to create resilient cities for the future.

References

1. Adger WN (2000) Social and ecological resilience: are they related? Prog Hum Geogr 24(3):347–364
2. Antony R, Rahiman KA, Vishnudas S (2021) Flood hazard assessment and flood inundation mapping—a review. Curr Trends Civ Eng 209–218
3. Avashia V, Garg A (2020) Implications of land use transitions and climate change on local flooding in urban areas: An assessment of 42 Indian cities. Land use policy 95:104571
4. Berkes F (2007) Understanding uncertainty and reducing vulnerability: lessons from resilience thinking. Nat Hazards 41(2):283–295
5. Bru E (1989) El vacíourbano = The urban void. Quaderns d'arquitectura i urbanisme, Núm 183:50–53
6. Cacciamani C (2013) È l'uomo a cambiare il clima, la scienza conferma. Ecoscienza, 5, ARPA Emilia Romagna
7. Cutter SL (2012) Vulnerability to environmental hazards. Routledge, pp 99–110
8. De Sola Morales I (2013) Terrain vague. Routledge, pp 24–30
9. Deppisch S, Hasibovic S (2013) Social-ecological resilience thinking as a bridging concept in transdisciplinary research on climate-change adaptation. Nat Hazards 67:117–127
10. Eraydin A (2012) "Resilience thinking" for planning. Resilience thinking in urban planning. Springer, Netherlands, Dordrecht, pp 17–37
11. Ernstson H, Van Der Leeuw SE, Redman CL, Meffert DJ, Davis G, Alfsen C, Elmqvist T (2010) Urban transitions: on urban resilience and human-dominated ecosystems. Ambio 39:531–545
12. Feilberg M, Mark O (2016) Integrated urban water management: Improve efficient water management and climate change resilience in cities. Sustain Water Manage Urban Environ 1–32
13. Gunderson LH (2000) Ecological resilience—in theory and application. Annu Rev Ecol Syst 31:425–439
14. Hammond MJ, Chen AS, Djordjevic S, Butler D, Mark O (2015) Urban flood impact assessment: a state-of-the-art review. Urban Water J 12(1):14–29
15. Huq ME, Sarker MNI, Prasad R, Kormoker T, Hossain MA, Rahman MM, Al Dughairi AA (2021) Resilience for disaster management: opportunities and challenges. Clim vulnerability Resil Glob South: Hum Adapt Sustain Futures 425–442
16. IPCC (2007) Climate change 2007: the physical science basis
17. IPCC (2013) Climate change 2013: the physical science basis
18. Jasiūnas J, Lund PD, Mikkola J (2021) Energy system resilience—a review. Renew Sustain Energy Rev 150:111476
19. Jha AK, Bloch R, Lamond J (2012) Cities and flooding: a guide to integrated urban flood risk management for the 21st century. World Bank Publications
20. Koolhaas R, Mau B (1995) S, m, l, xl. The Monacelli Press, LLC
21. Kuspilić N, Gilja G, Ocvirk E (2015) Calculation of flood event exceedance probability and related flood risk. In: Proceedings of the 6th Croatian water conference with international participation, Opatija, Croatia, pp 20–23
22. Laituri MJ (2000) Cultural perspectives of flooding. In: Wohl EE (ed) Inland flood hazards: human, riparian, and aquatic communities. Cambridge University Press, Cambridge, UK, pp 451–468
23. Liao KH (2012) A theory on urban resilience to floods—a basis for alternative planning practices. Ecol Soc 17(4)

24. Mayer B (2019) A review of the literature on community resilience and disaster recovery. Curr Env Health Rep 6:167–173
25. Mihajlović P (2016) Uticaj klimatskih promjena na arhitekturu gradova. In: 4th international conference contemporary achivements in civil engineering. Subotica, Serbia, pp 947–956
26. Mirković O (2021) Retreat, defend, or attack?—metamorphosis of the Zorrotzaurre peninsula, Bilbao, Spain. Master Thesis
27. Molyneaux L, Brown C, Wagner L, Foster J (2016) Measuring resilience in energy systems: Insights from a range of disciplines. Renew Sustain Energy Rev 59:1068–1079
28. Nelson DR (2011) Adaptation and resilience: responding to a changing climate. Wiley Interdiscip Rev Clim Change 2(1):113–120
29. Nkwunonwo UC, Whitworth M, Baily B (2020) A review of the current status of flood modeling for urban flood risk management in developing countries. Sci Afr 7:e00269
30. Nones M (2017) Flood hazard maps in the European context. Water Int 42(3):324–332
31. Norris FH, Stevens SP, Pfefferbaum B, Wyche KF, Pfefferbaum RL (2008) Community resilience as a metaphor, theory, set of capacities, and strategy for disaster readiness. Am J Community Psychol 41:127–150
32. Parker DJ (2014) Floods. Routledge
33. Perović S, Kurtović Folić N (2012) Brownfield regeneration–imperative for sustainable urban development. Građevinar 64(05):373–383
34. Pickett ST, Cadenasso ML, McGrath B (eds) (2013) Resilience in ecology and urban design: Linking theory and practice for sustainable cities, vol 3. Springer Science & Business Media
35. Pinto F (2014) Urban planning and climate change: adaptation and mitigation strategies. TeMA J Land Use Mobil Environ
36. Power A (2016) Bilbao city story
37. Quintana San Vicente A (2016) De la Ribera de Deusto a Zorrotzaurre. Un siglo de planeamiento urbanístico en la vega de Deusto. Doctoral Thesis
38. RAMSES (2013) Science for cities in transition. Project RAMSES—Reconciling Adaptation, Mitigation and Sustainable Development for Cities
39. Roberge M (2002) Human modification of the geomorphically unstable Salt River in metropolitan Phoenix. Prof Geogr 54(2):175–189
40. Schielen RMJ, Roovers G (2008) Adaptation as a way for flood management. In: Proceedings of the 4th international symposium on flood defence: managing flood risk, reliability and vulnerability. Toronto, Ontario, Canada
41. Simić, I. (2016). *The Possibilities for improvement of urban form resilience to climate change.* Univerzitet u Beogradu;
42. Tingsanchali, T. (2012). *Urban flood disaster management*. Procedia Engineering, 32, 25–37
43. Walker B, Holling CS, Carpenter SR, Kinzig A (2004) Resilience, adaptability, and transformability in social-ecological systems. Ecol Soc 9(2):5
44. Wilkinson C (2012) Social-ecological resilience: Insights and issues for planning theory. Plan Theory 11(2):148–169
45. World Health Organization [WHO] (2002) Floods: climate change and adaptation strategies for human health. In: Report on a WHO meeting. WHO, London
46. Ziarrusta IV, Gaztañaga AL (2016) Zorrozaurre: from a port and industrial Peninsula to a 'dream' islet for the new tertiary-residential Bilbao. Horiz Soc Urban, Spat Transf. Kobie Cult Anthropol 20:53–70

Chapter 16
Inspired Applications of Artificial Intelligence in Disaster-Resilient Regional Construction for Climate Risk

Yuxuan Liao, Wanying Liao, Dongzi Wu, Chenhe Zhang, Jiajun Xu, and Sushobhan Majumdar

Abstract Artificial Intelligence (AI) technologies are crucial in advancing smart cities and sustainable development practices. Conventional approaches to disaster-resilient construction face significant challenges, such as limited access to real-time data, resulting in inefficiencies, inadequate treatment methodologies, and deficiencies in effective decision-making processes. Innovative methods are therefore necessary to address these challenges effectively. Integrating big data and AI stands out as a leading technological frontier in disaster-resilient construction, offering solutions to traditional response management dilemmas and providing actionable insights for climate risk prevention and response strategies. Intelligent disaster-resilient construction leverages sensor-equipped Internet of Things (IoT) devices, integrated data streams, big data analytics, and machine learning to enhance resilience in disaster response systems. These intelligent technologies enable real-time monitoring, spatial simulation, optimized decision-making, and digital system enhancements, improving responsiveness and decision-making efficiency. This chapter reviews practices and

Y. Liao
School of Information and Control Engineering, China University of Mining and Technology, Xuzhou 221116, Jiangsu, China

W. Liao
Faculty of Arts and Social Sciences, Universiti Malaya, 50603 Kuala Lumpur, Malaysia

D. Wu
New Frontiers International Planning & Design (Hubei) Co., Ltd., Wuhan 430023, Hubei, China

C. Zhang
School of Humanities and Communication, Wuhan Polytechnic University, Wuhan 430048, Hubei, China

J. Xu (✉)
Institute for Advanced Studies, Universiti Malaya, 50603 Kuala Lumpur, Malaysia
e-mail: jiajunxu2000@gmail.com

S. Majumdar
Pandit Raghunath Murmu Smriti Mahavidyalaya, Bankura University, Bankura 722150, India

S. Ghosh et al. (eds.), *Cities of Tomorrow: Urban Resilience and Climate Change Preparedness*, Urban Sustainability, https://doi.org/10.1007/978-981-97-9658-8_16

case studies in AI applications for intelligent disaster-resilient construction, highlighting critical perspectives on the framework, recommendations, challenges, and opportunities for effective implementation. The aim is to foster innovation in spatial intelligence technology systems, providing the direction towards a sustainable future.

Keywords Artificial Intelligence (AI) · Disaster-resilient construction · Climate risk · Experience cases · Sustainable development

16.1 Introduction

In recent years, the World Meteorological Organization (WMO) and climate experts have continuously warned countries about the increasingly apparent climate disasters that could cause incalculable damage to life and the ecological environment on Earth [10]. Historical evidence suggests that extreme heat, rampant floods, and cold snaps have swept across nations worldwide [10]. As complex systems, town areas face significant disaster issues due to various disaster prevention facilities' exposure, concentration, and vulnerability [21]. Some scholars argue that once climate damage reaches the critical limit of disaster prevention facilities, solely relying on and considering the redundancy of engineering measures in traditional disaster resistance will significantly increase the vulnerability of regions when facing extreme climate events [21]. Furthermore, various towns' structural layouts and climate disaster risk prevention systems lead to differing climate risks these crucial nodes face under extreme conditions. Therefore, when dealing with the highly diverse, complex, and uncertain disaster prevention systems in town areas, traditional methods may overlook key parameters, ignore real-time changes in disasters, and need help to make timely and effective decisions, causing deficiencies in climate disaster resistance.

Artificial Intelligence (AI), as a significant branch of computer science, typically refers to the intelligent technologies exhibited by computer programs or controlled machines created by humans [22]. With recent advancements in science and technology, AI has gradually evolved, developed, and refined into key branches such as machine learning, natural language processing, image recognition and processing, and intelligent robotics. Compared to traditional mathematical or data models, AI offers numerous absolute advantages, including real-time responsiveness, fast decision-making speed, and the ability to fit nonlinear patterns [20]. After extensive data training, AI can provide appropriate and valuable outputs based on user needs, sometimes even without a clear understanding of the underlying mechanisms [4]. Moreover, by analyzing sensor data and integrating image recognition and processing technologies, AI can achieve segmentation and classification of specific elements. Compared to manual sorting, AI excels in handling cumbersome and large datasets [16]. Therefore, AI holds potential advantages in applications dealing with highly complex issues such as climate prediction and disaster prevention.

Currently, AI, leveraging its capabilities in big data analysis, deep learning, intelligent prediction, and emerging technologies such as drones, has achieved significant accomplishments in fields like healthcare, manufacturing, transportation, and construction. In the realm of smart cities, integrating AI with traditional disaster resistance infrastructure presents strong application prospects. Specifically, AI can enhance response speed during emergency disaster management and improve the efficiency of post-disaster communication, rescue, and recovery. This, in turn, facilitates the rational allocation of disaster relief resources, reduces casualties, and mitigates the damage and losses caused by disasters. Therefore, this chapter systematically analyzes and discusses AI's technical methods and practical application cases in disaster resistance construction. The aim is to explore insights and future innovations in applying AI technology to address climate risks and optimize disaster resistance in town areas.

16.2 Literature Review

With the rapid development of AI, its application scope continues to expand across various industries and fields. In disaster management, AI technology has proven particularly outstanding, becoming a crucial tool for responding to emergencies and minimizing losses. In many countries, AI has already been applied significantly in disaster management. Specifically, through techniques such as data analysis, predictive models, image recognition, and natural language processing, AI provides support and solutions at all stages of disaster prevention, monitoring, response, and recovery [19].

16.2.1 AI in Pre-disaster Warning Applications

By integrating AI technology with aerial and remote sensing information, the generated high-resolution maps have improved accuracy by approximately 1,000 times compared to traditional maps. These maps cover a wide range of information, including natural terrain feature classification, land use monitoring, groundwater exploration, environmental disaster assessment, and town area planning [14].

On May 27, 2018, Ellicott City, Maryland, United States, experienced a once-in-a-millennium rainstorm. Days before the event, the U.S. Department of Homeland Security had effectively used AI technology combined with spatial remote sensing information to issue warnings. In August 2018, the U.S. government and Harvard University researchers developed an AI model to predict aftershock locations up to a year after a major earthquake. This model was trained on data from 199 major earthquakes and approximately 130,000 aftershocks over the past decades. The results showed that the AI model was more accurate than the Coulomb stress change baseline method commonly used in aftershock research [3].

Researchers at Brigham Young University in the United States improved the global flood early warning system by incorporating information from small rivers and tributaries that were previously unrecorded into the AI model. They also created an online application to display flood warnings for different rivers at various times. This provided local governments with unprecedented detailed and nearly real-time information, which helped protect regions from floods [1].

Some scholars have pointed out that India accounts for 20% of the global deaths caused by flood disasters. To address this issue, Google collaborated with the Indian government to create an AI flood prediction model and improve India's flood alerts through Google Public Alerts. Since the first flood warning in September 2018, this model has significantly improved its accuracy in predicting disaster locations, timing, and severity. The AI model can also accurately predict daily water flow in basins without monitoring stations and provide longer warning times to enhance regional flood response capabilities [9].

Furthermore, SenSat Technology used drones to collect over 600 million data points and utilized AI algorithms to reconstruct rivers digitally and potential flood zones [1]. This resulted in a map of a 12 km stretch of the River Aire floodplain, which assisted planning and emergency management departments in making more rational decisions and more efficiently using limited public funds to protect vulnerable areas and mitigate flood damage [1]. These AI technologies provide innovative approaches for pre-disaster warnings, demonstrating that integrating AI into climate disaster prevention can improve prediction accuracy and reduce the impact of disasters.

16.2.2 AI in Disaster Rescue Applications

On August 8, 2017, a 7.0 magnitude earthquake struck Zhaigou County, China, resulting in 2,318 aftershocks and causing widespread communication outages in the region. Generally, within the crucial 72 h golden rescue period, rapidly locating survivors and establishing effective communication links are essential. However, the earthquake-triggered floods and landslides caused road damage and continuous rockfalls, rendering traditional emergency communication vehicles completely inaccessible. In this urgent situation, China Mobile innovatively deployed the AirStar drone-based aerial base station for the first time, restoring mobile communication within 50 h and effectively covering 100 square kilometers. This instantaneously communicated with 1,200 local users, ensuring basic communication needs for disaster victims and rescue teams [18].

Similarly, in Luding County, China, an earthquake of magnitude 6.8; the disaster resulted in communication and road disruptions across multiple areas in 2022, creating several "isolated" regions. The Chinese government's emergency management department swiftly deployed the Large high-altitude full-network communication drone and the Wing Loong UAV. These drones provided real-time disaster information, ensuring that the emergency command center could make rapid and accurate decisions, thereby enhancing rescue efficiency [13]. These drones played a crucial role in disaster response, offering innovative technological solutions for rescue operations and demonstrating that integrating AI in climate disaster prevention can significantly improve rescue efficiency.

16.2.3 AI in Post-disaster Governance Applications

In November 2018, the Paradise wildfire in California claimed 86 lives and burned a total area of 1.89 million acres. SilviaTerra, a San Francisco-located company, used AI for forest mapping to provide planners with data resources and help reduce the risk of future fires. The company utilized satellite data, aerial imagery, and laser scanning technology to measure vegetation growth and analyze different fire risk levels around the Paradise area. AI identified and marked the highest-risk areas, providing detailed information for government authorities to take further interventions, such as setting fire barriers or cutting down trees to block the fire spread [1].

In September 2019, Severe Typhoon Faxai (International Code: 1915) hit Japan, causing widespread power outages in Chiba City [2]. The Chiba government used AI technology to collect disaster-related information from social media platforms like Twitter. Specifically, they searched for keywords such as "power outage," "flooding," and "geological disaster" to analyze and classify the affected locations and situations based on the information posted by users. The goal was to use this information to guide disaster relief and post-disaster reconstruction efforts [2].

In January 2024, following an earthquake in Japan's Noto Peninsula, Japan Airlines conducted post-disaster damage assessments. They used drones for detailed aerial photography and spatial information collection, facilitating emergency response and assisting disaster recovery departments in establishing temporary housing [5]. In Wajima City, Japan, drones inspected over 400 bridges to quickly assess structural damage, demonstrating the efficiency of drone applications in disaster response and infrastructure maintenance [5].

In summary, based on the literature review of multiple case studies, AI has significant application prospects in pre-disaster prevention, disaster response, and post-disaster management. AI applications make it possible to achieve intelligent disaster resistance from multiple levels and perspectives, including prevention, emergency response, decision-making, and assessment. These technologies actively support the scientific governance of climate disasters in town areas. However, the current AI-based intelligent disaster resistance applications mostly assist management in preventing, responding to, and managing climate disasters, showcasing a certain degree of autonomy but still heavily relying on human intervention for specific operations.

16.3 Results and Discussion

16.3.1 Integrating AI into Multi-Dimensional Disaster-Resistant Construction

Climate disasters are characterized by diversity, uncertainty, and dynamism, making disaster management and disaster resistance construction highly challenging. Establishing an AI-based disaster prevention and control framework will help enhance multidimensional process collaboration in disaster resistance. This framework can aid in identifying overlooked disaster hazards in town areas, provide intuitive and scientific disaster relief decisions, and offer references for practical disaster resistance construction. Specifically, this intelligent disaster prevention and control model framework can be built from the following four dimensions (Fig. 16.1).

16.3.1.1 Regional Geosciences Dimension

Geographic environment and location are the crucial factors influencing climate and natural disasters. Even if two town areas have identical construction styles, the types of natural disasters they may encounter and their disaster resilience capabilities can vary significantly due to their different locations. AI's extensive databases can store and analyze local climate and disaster records from the past few decades, thereby identifying the specific climate characteristics and potential natural disasters for a town area.

By utilizing AI's image recognition and segmentation technologies, town areas can be delineated, and maps with precise locations of buildings, roads, and critical infrastructure can be generated. Combining these maps with climate characteristics and employing geospatial analysis through Geographic Information Systems (GIS),

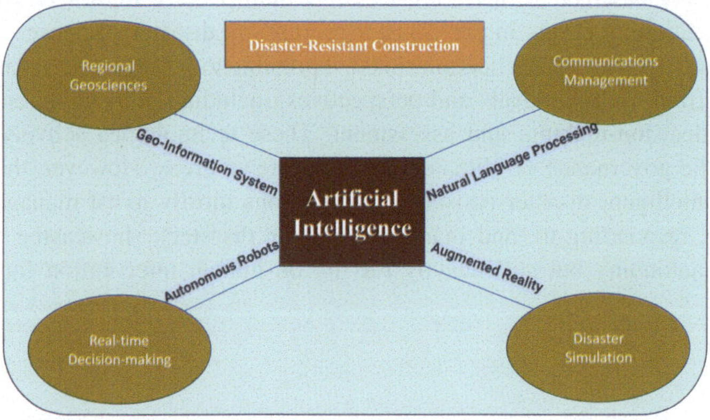

Fig. 16.1 Intelligent disaster prevention and control model framework

AI can assess disaster risks and help formulate disaster prevention and mitigation policies. This approach effectively enhances disaster management at all stages and provides guidance for planning disaster resistance construction in town areas.

16.3.1.2 Communications Management Dimension

In the era of the Internet of Things (IoT), interconnected devices enable AI to collect and analyze diverse data sources such as sensors, satellite imagery, and social media. This capability allows AI to continuously monitor parameters like water levels, air temperature, seismic activity, and the health status of infrastructure in town areas, issuing timely warnings for sudden town disasters. Through Natural Language Processing (NLP), AI promptly sends information to disaster management departments and alerts all residents via their mobile phones, computers, and other communication devices, urging and assisting them to evacuate promptly. Moreover, AI manages a large influx of emergency calls during disasters, categorizing them based on urgency and type of need and directing them to appropriate emergency responders. Automated response systems also provide solutions to common queries from callers, thereby alleviating the burden on human customer service personnel.

16.3.1.3 Real-Time Decision-Making Dimension

In highly dynamic and complex disaster scenarios, AI plays a crucial role in assisting individuals to make accurate and scientifically informed decisions. Equipped with a degree of autonomy, rescue robots, though still reliant on human operation, possess inherent capabilities to navigate extremely hazardous environments for human rescue personnel, such as extreme heat and dense smoke during events like fires or earthquakes. In such instances, rescue personnel may face life-threatening situations due to smoke inhalation or aftershocks, whereas rescue robots are designed with flexibility to adapt to diverse and harsh conditions. Concurrently, the use of drones for reconnaissance enables rapid coverage of extensive disaster areas, facilitating efficient search and rescue operations and enhancing the success rate of locating survivors. AI contributes significantly to disaster management by optimizing routes for rescue vehicles and effectively allocating personnel and resources, thereby providing disaster managers and the public with efficient rescue solutions. This technology not only demonstrates its decision-support capabilities in disaster prevention and emergency response but also plays a proactive role in post-disaster reconstruction. As AI technology advances, ongoing improvements and optimizations of related systems are essential to better cope with unpredictable disaster scenarios.

16.3.1.4 Disaster Simulation Dimension

Simulating disasters in advance through disaster drills is a common method for disaster preparedness. However, in reality, these simulations lack the unpredictability and chaos of actual disasters, as participants are forewarned and aware of the exercise. Moreover, conducting large-scale drills consumes significant human resources and financial resources. AI offers a more convenient and scientifically rigorous approach to disaster drills by providing simulated scenarios that are economically feasible and effective.

AI facilitates the creation of virtual disaster zones in mapped areas and simulates the movement of populations during town and regional disasters. This approach allows for economical, efficient, and convenient disaster simulations, offering visual insights to town governance departments. Furthermore, integrating mature Augmented Reality (AR) and Virtual Reality (VR) technologies enables residents to experience realistic disaster scenarios, enhancing their understanding of disaster preparedness and self-rescue techniques. This immersive experience aims to reduce casualties and economic losses during actual disaster events.

16.3.2 Opportunities and Challenges of Introducing AI into Disaster-Resistant Construction

The long-term scientific principles of climate change exhibit significant complexity and uncertainty. Current climate change predictions and models also need to be revised. The methods for establishing relationships between climate indicators and simulation results are inherently uncertain, leading to uncertainties in forecasting global climate trends [12]. AI cannot achieve perfect predictions when facing the intrinsically chaotic nature of complex and dynamic climate change. There is a need for further improvements in AI climate prediction accuracy, stability, diversity of forecasting products, and refinement of requirements [8]. AI algorithms may lack transparency, as black-box models lack interpretability and cannot predict potential outputs. The feasibility and practicality of AI networks depend heavily on the quantity and quality of training data, requiring substantial and costly computational resources [6].

The development of AI may bring about societal disorder, conflicts, and inequalities. Stephen William Hawking warned that the comprehensive development of AI could ultimately threaten humanity. Despite its advantages in risk management, unconscious machines may pose greater risks than the human brain, potentially surpassing human control and leading to more considerable societal risks. Furthermore, AI could exacerbate social divisions, cultural fragmentation, unemployment, privacy violations, family structure disruption, gender discrimination, concentration of power, and misuse of technology. Thus, while AI plays a role in addressing societal risks, it also has the potential to create new social challenges.

Undeniably, AI offers new possibilities for addressing climate security risks, particularly in managing climate disaster risks, vulnerabilities, and exposure-related risks. Its applications not only enhance the effectiveness of disaster early warning systems but also improve the efficiency of long-term climate disaster modeling [6]. In the realm of intelligent town development, AI technology has become an indispensable component. AI can establish virtual reality simulation models to simulate town operational scenarios and development scenarios under different decision-making contexts, providing insights for town planning and management decisions. Additionally, leveraging AI and deep learning technologies enables autonomous processing of massive real-time data for automated decision-making and intelligent control of the physical world, facilitating autonomous learning and innovative development of town areas [17].

Looking ahead, AI will continue to evolve into strong AI and super AI. Entrusting significant governance technological issues to computers helps overcome the limitations of traditional governance technologies and mindsets. In the fields of climate risk governance and intelligent disaster response, AI demonstrates promising prospects for the future.

16.4 Conclusion

The intensification of global warming has significantly increased the frequency of natural disasters and risks, posing severe challenges to human society [11, 15]. The instability of climate change represents one of the major global threats, encompassing destructive impacts from fluctuations in climate elements such as temperature, rainfall patterns, sea levels, and atmospheric CO_2 concentrations [7, 15]. As different countries and regions increasingly adopt AI to govern climate change and enhance town disaster resilience, societal acceptance of AI is poised to grow. AI technologies, particularly in natural language processing, have advanced rapidly, exemplified by innovations like OpenAI's ChatGPT, which have garnered significant social attention and media coverage. Continuous updates and optimizations of AI algorithms in climate change and town disaster resilience domains, coupled with reporting on more prosperous disaster prevention case studies, will further underscore AI's potential in intelligent disaster management.

This chapter explores AI technology's feasibility, theoretical frameworks, and practical cases in addressing climate change and integrating traditional disaster resilience methods to forge new innovative town developments. Positioned at the intersection of AI and disaster governance research, this study offers both theoretical innovation and practical value, suggesting inspirational directions for town development and disaster governance; looking ahead, with advancements in environmental sensing and AI technology, intelligent disaster management is poised for broader research and application, promising to enhance global resilience to natural disasters and climate challenges.

References

1. Beall A (2019) How to bring wildfires under control. BBC. https://www.bbc.com/future/article/20190226-how-to-bring-wildfires-back-under-control. Accessed 6 July 2024
2. China News (2020) Japan's Chiba government plans to use AI to collect and analyze disaster information for post-disaster reconstruction. Sina News. https://news.sina.cn/2020-02-14/detail-iimxxstf1453268.d.html. Accessed 6 July 2024
3. DeVries PMR, Viégas F, Wattenberg M, Meade BJ (2018) Deep learning of aftershock patterns following large earthquakes. Nature 560:632–634. https://doi.org/10.1038/s41586-018-0438-y
4. Guo Y (2017) On deep teaching: origin, foundation and concept. Educ Res Exp 3:1–11 [In Chinese] // 郭元祥 (2017) 论深度教学:源起、基础与理念, 教育研究与实验, 3:1–11
5. Japan Airlines (2024) Japan airlines and KDDI SmartDrone partner to support post-earthquake damage assessment in Noto Peninsula. https://press.jal.co.jp/en/release/202403/007972.html. Accessed 6 July 2024
6. Kim K, Boulanin V (2023) Artificial intelligence for climate security: possibilities and challenges. Stockholm International Peace Research Institute (SIPRI). https://doi.org/10.55163/QDSE8934
7. Kogo BK, Kumar L, Koech R (2021) Climate change and variability in Kenya: a review of impacts on agriculture and food security. Environ Dev Sustain 23:23–43. https://doi.org/10.1007/s10668-020-00589-1
8. Liu G, Tang W (2023) A preliminary study on the application of artificial intelligence in climate prediction. iChina 10:29–35 [In Chinese] // 刘冠州&唐伟 (2023) 人工智能在气候预测领域应用初探, 中国信息化, 10:29–35
9. Nearing G, Cohen D, Dube V, Gauch M, Gilon O, Harrigan S, Hassidim A, Klotz D, Kratzert F, Metzger A, Nevo S, Pappenberger F, Prudhomme C, Shalev G, Shenzis S, Tekalign TY, Weitzner D, Matias Y (2024) Global prediction of extreme floods in ungauged watersheds. Nature 627:559–563. https://doi.org/10.1038/s41586-024-07145-1
10. Pan B (2022) Global climate conditions continue to deteriorate in 2021. Ecol Econ 38(2):1–4 [In Chinese] // 潘博煌 (2022) 2021 年全球气候状况继续恶化, 生态经济 38(2):1–4
11. Qu H, Li J, Wang W, Xin W, Zhou C, He Z (2022) New insight into the coupled grain-disaster-economy system based on a multilayer network: an empirical study in China. IISPRS Int J Geo-Inf 11(1):59. https://doi.org/10.3390/ijgi11010059
12. Shen Y, Wang G (2013) Key findings and assessment results of IPCC WGI fifth assessment report. J Glaciol Geocryol 35(5):1068–1076. https://doi.org/10.7522/j.issn.1000-0240.2013.0120
13. Sichuan Provincial Department of Emergency Management (2022) Live coverage of the rescue and relief efforts after the 6.8-magnitude earthquake in Luding, Sichuan: every second counts, these high-tech technologies save more lives. Ministry of Emergency Management of the People's Republic of China. https://www.mem.gov.cn/xw/gdyj/202209/t20220911_422182.shtml accessed: 6 July 2024
14. Singh V, Kumar G, Sabherwal D (2015) A comparative study of computational intelligence based techniques in the field of remote sensing image classification. In: 2015 2nd international conference on computing for sustainable global development (INDIACom). IEEE, pp 1402–1408. Retrieved from https://ieeexplore.ieee.org/document/7100479/
15. Wang H, Xu J, Lim NHHL, Liao W, Fong CS (2024) The impact of climate change on China's central region grain production: evidence from spatiotemporal pattern evolution. AIMS Geosci 10(3):460–483. https://doi.org/10.3934/geosci.2024024
16. Xu J (2023) Risk assessment of town waterlogging disaster based on complex network. Master's Thesis, Inner Mongolia University of Science & Technology. https://doi.org/10.27724/d.cnki.gnmgk.2023.000136
17. Zhang X, Li S, Zhou Q, Sun Y (2021) The rationale and innovative thinking of building Digital Twin City. Sci Surv Map 46(3):147–152. https://doi.org/10.16251/j.cnki.1009-2307.2021.03.022

18. Zhou L, Long Z (2017) The innovation of early warning of disaster in the age of big data: take the center of information issuing of early warning of disaster of Yangjiang City for example. Wuhan Univ J (Soc Sci) 70(3):121–132. https://doi.org/10.14086/j.cnki.wujss.2017.03.011
19. Zhou L, Gu Y (2021) Innovation of social risk governance in the era of artificial intelligence. J Hohai Univ (Philos Soc Sci) 23(3):38–45. https://doi.org/10.3876/j.issn.1671-4970.2021.03.006
20. Zou F, Jin L Dong J (2017) Review of convolutional neural network. Chin J Comput 40(6):1229–1251 [In Chinese] // 周飞燕, 金林鹏&董军 (2017) 卷积神经网络研究综述, 计算机学报, 40(6):1229–1251
21. Zou J, Duan C, Chen B, Wang H (2024) Town waterlogging resilience assessment based on quantification of economic losses under climate change. Water Resour Prot 1–17. Retrieved from http://kns.cnki.net/kcms/detail/32.1356.tv.20240704.1702.021.html
22. Zou L, Zhang X (2012) Application and development of the artificial intelligence. Netinfo Secur 2:11–13. https://doi.org/10.3969/j.issn.1671-1122.2012.02.003

SPRINGER NATURE

GPSR Compliance

The European Union's (EU) General Product Safety Regulation (GPSR) is a set of rules that requires consumer products to be safe and our obligations to ensure this.

If you have any concerns about our products, you can contact us on ProductSafety@springernature.com

In case Publisher is established outside the EU, the EU authorized representative is:

Springer Nature Customer Service Center GmbH
Europaplatz 3
69115 Heidelberg, Germany

The manufacturer's authorised representative in the EU is Springer Nature Customer Service Centre GmbH, Europaplatz 3, 69115 Heidelberg, Germany. If you have any concerns regarding our products, please contact ProductSafety@springernature.com

Printed and bound by CPI Group (UK) Ltd, Croydon, CR0 4YY
01/10/2026
02225859-0008